METHODS IN MOLECULAR BIOLOGY

Series Editor
**John M. Walker
School of Life and Medical Sciences
University of Hertfordshire
Hatfield, Hertfordshire, AL10 9AB, UK**

For further volumes:
http://www.springer.com/series/7651

Hypoxia

Methods and Protocols

Edited by

L. Eric Huang

Department of Neurosurgery, Clinical Neurosciences Center, University of Utah,
Salt Lake City, UT, USA

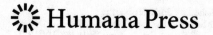

Editor
L. Eric Huang
Department of Neurosurgery
Clinical Neurosciences Center
University of Utah
Salt Lake City, UT, USA

ISSN 1064-3745   ISSN 1940-6029   (electronic)
Methods in Molecular Biology
ISBN 978-1-4939-8538-8   ISBN 978-1-4939-7665-2   (eBook)
https://doi.org/10.1007/978-1-4939-7665-2

Printed on acid-free paper

This Humana Press imprint is published by Springer Nature
The registered company is Springer Science+Business Media, LLC
The registered company address is: 233 Spring Street, New York, NY 10013, U.S.A.

Preface

Oxygen is an essential molecule that serves as an enzyme substrate in bioenergetics and metabolic reactions. Adaptation to oxygen deprivation—hypoxia—is a critical process in development, physiology, and pathophysiology. This volume offers a comprehensive coverage of step-by-step lab protocols that are widely used and those that are developed more recently for studying hypoxic responses in physiology and diseases. All the protocols contain extensive detail intended to help competent scientists in the field of biomedical sciences carry out unfamiliar molecular biology techniques successfully by simply following the detailed procedures. Practical notes are explained in greater detail for additional help. Methods included range from the application of hypoxic conditions to the techniques of investigating hydroxylase activities with various approaches including CRISPR/Cas9, substrate hydroxylation with mass spectrometry, transcriptional activation of hypoxia-response genes with a dynamic luciferase assay, chromatin—immunoprecipitation, and RNA sequencing, metabolic changes in vitro and in vivo, sensing of hypoxia signal in animal, placental and erythrocyte development, hypoxic responses in *C. elegans* and *D. rerio*, and in tumor tissues, manipulation of hypoxic responses with RCAS/TVA tumor models, cancer-stem cell enrichment and quantification, and tumor growth with various imaging modalities. This volume provides a valuable set of tools that can be utilized to study hypoxia and beyond.

Salt Lake City, UT, USA   *L. Eric Huang*

Contents

Contributors

JAIME ABREGO • *The Eppley Institute for Research in Cancer and Allied Diseases, University of Nebraska Medical Center, Omaha, NE, USA*

TILL ACKER • *Institute of Neuropathology, University of Giessen, Giessen, Germany*

IGNACIO ARIAS-MAYENCO • *Instituto de Biomedicina de Sevilla, Hospital Universitario Virgen del Rocío/CSIC/Universidad de Sevilla, Seville, Spain; Departamento de Fisiología Médica y Biofísica, Facultad de Medicina, Universidad de Sevilla, Seville, Spain; Centro de Investigación Biomédica, en Red sobre Enfermedades, Neurodegenerativas, Madrid, Spain*

KULDEEP S. ATTRI • *The Eppley Institute for Research in Cancer and Allied Diseases, University of Nebraska Medical Center, Omaha, NE, USA*

NURAY BÖGÜRCÜ • *Institute of Neuropathology, University of Giessen, Giessen, Germany*

DANIELLE L. BROOKS • *Department of Pathology and Laboratory Medicine, Center for Cancer Research, University of Tennessee Health Science Center, Memphis, TN, USA; Women's Malignancies Branch, Center for Cancer Research, National Cancer Institute, Bethesda, MD, USA*

LANCE D. BURRELL • *Center for Quantitative Cancer Imaging, Huntsman Cancer Institute, University of Utah, Salt Lake City, UT, USA*

CANDELA CABALLERO • *Instituto de Biomedicina de Sevilla, Hospital Universitario Virgen del Rocío/CSIC/Universidad de Sevilla, Seville, Spain; Facultad de Medicina, Departamento de Fisiología Médica y Biofísica, Universidad de Sevilla, Seville, Spain; Centro de Investigación Biomédica en Red sobre Enfermedades Neurodegenerativas, Madrid, Spain; Division of Pulmonary and Critical Care Medicine, Department of Medicine, Johns Hopkins University School of Medicine, Baltimore, MD, USA*

MIGUEL A.S. CAVADAS • *Systems Biology Ireland, University College Dublin, Dublin, Ireland; Conway Institute of Biomolecular and Biomedical Research, School of Medicine, University College Dublin, Dublin, Ireland; Instituto Gulbenkian de Ciencia, Oeiras, Portugal*

DAMAYANTI CHAKRABORTY • *Departments of Pathology & Laboratory Medicine and Pediatrics, Institute for Reproductive Health and Regenerative Medicine, University of Kansas Medical Center, Kansas City, KS, USA; Massachusetts General Hospital, Cancer Center, Charlestown, MA, USA*

ALEX CHEONG • *Systems Biology Ireland, University College Dublin, Dublin, Ireland; Conway Institute of Biomolecular and Biomedical Research, School of Medicine, University College Dublin, Dublin, Ireland; Life and Health Sciences, Aston University, Birmingham, UK*

ANEESHA DASGUPTA • *Department of Biochemistry and Molecular Biology, University of Nebraska Medical Center, Omaha, NE, USA*

CUNMING DUAN • *Department of Molecular, Cellular, and Developmental Biology, University of Michigan, Ann Arbor, MI, USA*

JOACHIM FANDREY • *Institute of Physiology, University of Duisburg-Essen, Essen, Germany*

LIN GAO • *Instituto de Biomedicina de Sevilla, Hospital Universitario Virgen del Rocío/ CSIC/Universidad de Sevilla, Seville, Spain; Departamento de Fisiología Médica y Biofísica, Facultad de Medicina, Universidad de Sevilla, Seville, Spain; Centro de Investigación Biomédica en Red sobre Enfermedades Neurodegenerativas, Madrid, Spain*

BOYAN K. GARVALOV • *Institute of Neuropathology, University of Giessen, Giessen, Germany*

DAVID L. GILLESPIE • *Department of Neurosurgery, University of Utah, Salt Lake City, UT, USA*

GENNIFER GOODE • *The Eppley Institute for Research in Cancer and Allied Diseases, University of Nebraska Medical Center, Omaha, NE, USA*

VENUGOPAL GUNDA • *The Eppley Institute for Research in Cancer and Allied Diseases, University of Nebraska Medical Center, Omaha, NE, USA*

KEVIN P. HORN • *Center for Quantitative Cancer Imaging, Huntsman Cancer Institute, University of Utah, Salt Lake City, UT, USA*

L. ERIC HUANG • *Department of Neurosurgery, Clinical Neurosciences Center, University of Utah, Salt Lake City, UT, USA; Department of Oncological Sciences, University of Utah, Salt Lake City, UT, USA*

RANDY L. JENSEN • *Department of Neurosurgery, University of Utah, Salt Lake City, UT, USA; Department of Neurosurgery, Clinical Neurosciences Center, University of Utah, Salt Lake City, UT, USA*

DAE-EUN JEONG • *Department of Life Sciences, Pohang University of Science and Technology, Pohang, Gyeongbuk, South Korea*

HIROYASU KAMEI • *Faculty of Natural System, Noto Marine Laboratory, Institute of Science and Engineering, Kanazawa University, Noto, Ishikawa, Japan*

MICHAEL KARSY • *Department of Neurosurgery, University of Utah, Salt Lake City, UT, USA*

RYAN J. KING • *The Eppley Institute for Research in Cancer and Allied Diseases, University of Nebraska Medical Center, Omaha, NE, USA*

PEPPI KOIVUNEN • *Oulu Center for Cell-Matrix Research, Biocenter Oulu, and Faculty of Biochemistry and Molecular Medicine, University of Oulu, Oulu, Finland*

SUSHIL KUMAR • *The Eppley Institute for Research in Cancer and Allied Diseases, University of Nebraska Medical Center, Omaha, NE, USA*

KYOUNG EUN LEE • *Abramson Family Cancer Research Institute, University of Pennsylvania, Philadelphia, PA, USA; Perelman School of Medicine at the University of Pennsylvania, Philadelphia, PA, USA*

SEUNG-JAE V. LEE • *Department of Life Sciences, School of Interdisciplinary Bioscience and Bioengineering, Pohang University of Science and Technology, Pohang, Gyeongbuk, South Korea; School of Interdisciplinary Bioscience and Bioengineering, Pohang University of Science and Technology, Pohang, Gyeongbuk, South Korea*

YUJIN LEE • *Department of Life Sciences, Pohang University of Science and Technology, Pohang, Gyeongbuk, South Korea*

FUMING LI • *Abramson Family Cancer Research Institute, University of Pennsylvania, Philadelphia, PA, USA; Perelman School of Medicine at the University of Pennsylvania, Philadelphia, PA, USA*

JOSÉ LÓPEZ-BARNEO • *Instituto de Biomedicina de Sevilla, Hospital Universitario Virgen del Rocío/CSIC/Universidad de Sevilla, Seville, Spain; Departamento de Fisiología Médica y Biofísica, Facultad de Medicina, Universidad de Sevilla, Seville, Spain; Centro de Investigación Biomédica en Red sobre Enfermedades Neurodegenerativas, Madrid, Spain*

SÉAN B. LYNE • *Department of Neurosurgery, Clinical Neurosciences Center, University of Utah, Salt Lake City, UT, USA*

KAMIYA MEHLA • *The Eppley Institute for Research in Cancer and Allied Diseases, University of Nebraska Medical Center, Omaha, NE, USA*

SCOTT E. MULDER • *The Eppley Institute for Research in Cancer and Allied Diseases, University of Nebraska Medical Center, Omaha, NE, USA*

ANA MARÍA MUÑOZ-CABELLO • *Instituto de Biomedicina de Sevilla, Hospital Universitario Virgen del Rocío/CSIC/Universidad de Sevilla, Seville, Spain; Facultad de Medicina, Departamento de Fisiología Médica y Biofísica, Universidad de Sevilla, Seville, Spain; Centro de Investigación Biomédica, en Red sobre Enfermedades, Neurodegenerativas, Madrid, Spain*

DIVYA MURTHY • *The Eppley Institute for Research in Cancer and Allied Diseases, University of Nebraska Medical Center, Omaha, NE, USA*

JOHANNA MYLLYHARJU • *Oulu Center for Cell-Matrix Research, Biocenter Oulu, and Faculty of Biochemistry and Molecular Medicine, University of Oulu, Oulu, Finland*

JAYASRI NANDURI • *Institute for Integrative Physiology, Center for Systems Biology of Oxygen Sensing, Biological Sciences Division, University of Chicago, Chicago, IL, USA*

PATRICIA ORTEGA-SÁENZ • *Instituto de Biomedicina de Sevilla, Hospital Universitario Virgen del Rocío/CSIC/Universidad de Sevilla, Seville, Spain; Facultad de Medicina, Departamento de Fisiología Médica y Biofísica, Universidad de Sevilla, Seville, Spain; Centro de Investigación Biomédica en Red sobre Enfermedades Neurodegenerativas, Madrid, Spain*

YING-JIE PENG • *Institute for Integrative Physiology, Center for Systems Biology of Oxygen Sensing, Biological Sciences Division, University of Chicago, Chicago, IL, USA*

CHRISTINA PICKEL • *Institute of Physiology, University of Zurich, Zurich, Switzerland*

NANDURI R. PRABHAKAR • *Institute for Integrative Physiology, Center for Systems Biology of Oxygen Sensing, Biological Sciences Division, University of Chicago, Chicago, IL, USA*

JOSEF T. PRCHAL • *Division of Hematology, Department of Internal Medicine, University of Utah School of Medicine, Salt Lake City, UT, USA*

KATRIN PROST-FINGERLE • *Institute of Physiology, University of Duisburg-Essen, Essen, Germany*

JAVIER RODRIGUEZ • *Edinburgh Cancer Research Centre, IGMM, University of Edinburgh, Edinburgh, UK*

DEBANGSHU SAMANTA • *McKusick-Nathans Institute of Genetic Medicine, Johns Hopkins University School of Medicine, Baltimore, MD, USA; Johns Hopkins Institute for Cell Engineering, Johns Hopkins University School of Medicine, Baltimore, MD, USA*

CARSTEN C. SCHOLZ • *Institute of Physiology, University of Zurich, Zurich, Switzerland*

VERA SCHÜTZHOLD • *Institute of Physiology, University of Duisburg-Essen, Essen, Germany*

REGAN L. SCOTT • *Departments of Pathology & Laboratory Medicine and Pediatrics, Institute for Reproductive Health and Regenerative Medicine, University of Kansas Medical Center, Kansas City, KS, USA*

TIFFANY N. SEAGROVES • *Department of Pathology and Laboratory Medicine, Center for Cancer Research, University of Tennessee Health Science Center, Memphis, TN, USA*

SASCHA SEIDEL • *Institute of Neuropathology, University of Giessen, Giessen, Germany; Institute of Cell Biology and Neuroscience and Buchmann Institute for Molecular Life Sciences, Goethe University, Frankfurt, Germany*

GREGG L. SEMENZA • *McKusick-Nathans Institute of Genetic Medicine, Johns Hopkins University School of Medicine, Baltimore, MD, USA; Johns Hopkins Institute for Cell Engineering, Johns Hopkins University School of Medicine, Baltimore, MD, USA; Departments of Pediatrics, Medicine, Oncology, Radiation Oncology, and Biological Chemistry, Johns Hopkins University School of Medicine, Baltimore, MD, USA*

SURENDRA K. SHUKLA • *The Eppley Institute for Research in Cancer and Allied Diseases, University of Nebraska Medical Center, Omaha, NE, USA*

M. CELESTE SIMON • *Abramson Family Cancer Research Institute, University of Pennsylvania, Philadelphia, PA, USA; Perelman School of Medicine at the University of Pennsylvania, Philadelphia, PA, USA*

PANKAJ K. SINGH • *The Eppley Institute for Research in Cancer and Allied Diseases, University of Nebraska Medical Center, Omaha, NE, USA; Department of Pathology and Microbiology, University of Nebraska Medical Center, Omaha, NE, USA; Department of Genetics, Cell Biology and Anatomy, University of Nebraska Medical Center, Omaha, NE, USA; Department of Biochemistry and Molecular Biology, University of Nebraska Medical Center, Omaha, NE, USA*

MICHAEL J. SOARES • *Departments of Pathology & Laboratory Medicine and Pediatrics, Institute for Reproductive Health and Regenerative Medicine, University of Kansas Medical Center, Kansas City, KS, USA; Fetal Health Research, Children's Research Institute, Children's Mercy, Kansas City, MO, USA*

JIHYUN SONG • *Division of Hematology, Department of Internal Medicine, University of Utah School of Medicine, Salt Lake City, UT, USA*

CORMAC T. TAYLOR • *Systems Biology Ireland, University College Dublin, Dublin, Ireland; Conway Institute of Biomolecular and Biomedical Research, School of Medicine, University College Dublin, Dublin, Ireland*

PATRICIA D.B. TIBURCIO • *Department of Neurosurgery, Clinical Neurosciences Center, University of Utah, Salt Lake City, UT, USA*

HORTENSIA TORRES-TORRELO • *Instituto de Biomedicina de Sevilla, Hospital Universitario Virgen del Rocío/CSIC/Universidad de Sevilla, Seville, Spain; Facultad de Medicina, Departamento de Fisiología Médica y Biofísica, Universidad de Sevilla, Seville, Spain; Centro de Investigación Biomédica, en Red sobre Enfermedades, Neurodegenerativas, Madrid, Spain*

ENZA VERNUCCI • *The Eppley Institute for Research in Cancer and Allied Diseases, University of Nebraska Medical Center, Omaha, NE, USA*

ALEX VON KRIEGSHEIM • *Edinburgh Cancer Research Centre, IGMM, University of Edinburgh, Edinburgh, UK*

YI XIN • *Department of Molecular, Cellular, and Developmental Biology, University of Michigan, Ann Arbor, MI, USA*

JEFFERY T. YAP • *Department of Radiology and Imaging Sciences, Center for Quantitative Cancer Imaging, Huntsman Cancer Institute, University of Utah, Salt Lake City, UT, USA*

Chapter 1

Genetic Knockdown and Pharmacologic Inhibition of Hypoxia-Inducible Factor (HIF) Hydroxylases

Christina Pickel, Cormac T. Taylor, and Carsten C. Scholz

Abstract

Reduced oxygen supply that does not satisfy tissue and cellular demand (hypoxia) regularly occurs both in health and disease. Hence, the capacity for cellular oxygen sensing is of vital importance for each cell to be able to alter its energy metabolism and promote adaptation to hypoxia. The hypoxia-inducible factor (HIF) prolyl hydroxylases 1–3 (PHD1–3) and the asparagine hydroxylase factor-inhibiting HIF (FIH) are the primary cellular oxygen sensors, which confer cellular oxygen-dependent sensitivity upon HIF as well as other hypoxia-sensitive pathways, such as nuclear factor κB (NF-κB). Studying these enzymes allows us to understand the oxygen-dependent regulation of cellular processes and has led to the development of several putative novel therapeutics, which are currently in clinical trials for the treatment of anemia associated with kidney disease. Pharmacologic inhibition and genetic knockdown are commonly established techniques in protein biochemistry and are used to investigate the activity and function of proteins. Here, we describe specific protocols for the knockdown and inhibition of the HIF prolyl hydroxylases 1–3 (PHD1–3) and the asparagine hydroxylase factor-inhibiting HIF (FIH) using RNA interference (RNAi) and hydroxylase inhibitors, respectively. These techniques are essential tools for the analysis of the function of the HIF hydroxylases, allowing the investigation and discovery of novel functions and substrates of these enzymes.

Key words HIF, Hydroxylase, PHD, Prolyl hydroxylation, FIH, Asparagine hydroxylation, Hypoxia, Inhibitor, Knockdown, siRNA, shRNA

1 Introduction

Oxygen-sensing hydroxylases are Fe^{2+}- and 2-oxoglutarate (2-OG)-dependent dioxygenases that regulate the α subunits of the dimeric transcription factor hypoxia-inducible factor (HIF) [1]. Hydroxylation of distinct proline residues by the prolyl hydroxylases 1–3 (PHD1–3) regulates HIF-α stability by targeting it for proteasomal degradation [1], whereas asparagine hydroxylation by factor-inhibiting HIF (FIH) prevents interaction with the transcriptional coactivator p300/CBP, inhibiting HIF transactivation activity [1]. Genetic knockdown and pharmacologic inhibition of the HIF hydroxylases have previously been used as powerful tools to

L. Eric Huang (ed.), *Hypoxia: Methods and Protocols*, Methods in Molecular Biology, vol. 1742,
https://doi.org/10.1007/978-1-4939-7665-2_1, © Springer Science+Business Media, LLC 2018

study the function of these enzymes and the molecular mechanisms involved in the regulation of HIF and other signaling pathways, such as nuclear factor κB (NF-κB) [2–9]. These two methods represent different strategies of either directly inhibiting hydroxylase activity or indirectly reducing it by interfering with hydroxylase mRNA and protein level, respectively. Knockdown of hydroxylases is mediated by RNA interference (RNAi) through transient or stable expression of short interfering RNAs (siRNAs) or short hairpin RNAs (shRNAs). siRNA/shRNA-mediated RNAi leads to the degradation of mRNA complementary to the guide strand of the siRNA/shRNA (Fig. 1) [10]. Therefore, the targeted protein is no longer translated, while protein degradation still occurs, leading to a reduced amount of the targeted protein. The specificity of siRNA/shRNA-mediated RNAi is provided by the sequence of the short RNAs utilized [10]. The reported half-life values in HeLa cells of the HIF hydroxylase isoforms are 10.3 h for PHD1, 13.1 h for PHD2, and 1.7 h for PHD3 and have not been determined for FIH [11]. These values are important for transient knockdown of hydroxylases, as the incubation time with siRNAs needs to sufficiently exceed the protein degradation rate to obtain an efficient knockdown. In contrast to RNAi, pharmacologic inhibitors abrogate the enzymatic activity of the protein, but the protein itself is still present. The two most common modes of action of hydroxylase inhibitors are competitive inhibition by 2-oxoglutarate analogues and iron chelation, which deplete the cofactor Fe^{2+} (Fig. 1) [12]. Of note, due to the different mode of action of both techniques, hydroxylase knockdown and pharmacologic inhibition can be combined to investigate enzymatic and nonenzymatic functions of hydroxylases [8, 9, 13]. In this chapter, we describe methods for reducing hydroxylase protein levels and/or activity by RNAi or different pharmacologic hydroxylase inhibitors.

2   Materials

Prepare all solutions with ultrapure water (such as double-distilled H_2O (ddH_2O)) and, where applicable, sterile solvents and cell culture grade reagents. Sterile filter or autoclave non-sterile solutions before applying them to cells. Unless otherwise indicated, prepare and store reagents at room temperature. Follow all waste disposal regulations, especially when working with biohazardous materials.

2.1   General Materials

1. Laminar flow hood.
2. CO_2- and temperature-controlled cell culture incubator with humidified atmosphere (5 % CO_2, 37 °C).
3. Cell line(s).

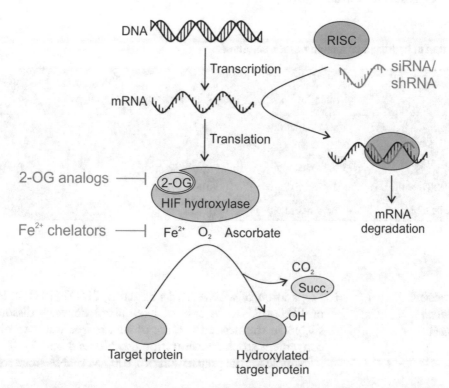

Fig. 1 Inhibition of HIF hydroxylase activity and protein expression by genetic knockdown or pharmacologic inhibitors. siRNA/shRNA-mediated knockdown of HIF hydroxylases leads to a RNA-induced silencing complex (RISC)-dependent degradation of the hydroxylase mRNA and a subsequent reduction of hydroxylase protein levels due to the diminished translation and ongoing protein degradation. Hydroxylase inhibitors prevent in the two main modes of action either the binding of the co-substrate 2-oxoglutarate (2-OG analogues) or sequester of the cofactor Fe^{2+} (Fe^{2+} chelators) so that hydroxylase activity is abolished. *Succ.,* succinate

4. Cell culture medium: choose adequate culture medium and supplements for your cell line.

5. Cell culture dishes or flasks of your choice (*see* **Note 1**).

6. Sterile filters (0.2 μm for chemical solutions; 0.45 μm for viral pseudoparticle solutions).

7. Syringes (10 mL, 20 mL).

2.2 Pharmacologic Hydroxylase Inhibition

1. If the inhibitor is not in liquid form, dissolve it in the appropriate solvent and desired volume to obtain the desired stock concentration (*see* **Notes 2** and **3**). Commonly used stock concentrations of different inhibitors as well as the amounts and volumes for the preparation of these are indicated in Table 1.

Aliquot the inhibitor stocks according to your needs, and store the aliquots at −20 °C, if not otherwise indicated in the data sheet.

Table 1
Preparation of hydroxylase inhibitor stock solutions

Inhibitor	Inhibitor class	Stock concentration	Compound (mg)	Solvent (μL)
Dimethyloxalylglycine (DMOG)	2-OG analogue	1 M in DMSO	10	57.11
Desferrioxamine (DFX)	Iron chelator	50 mM in ddH$_2$O	10	305.51
FG-4592 (roxadustat)	2-OG analogue	100 mM in DMSO	10	283.82
JNJ-42041935	2-OG competitor	100 mM in DMSO	10	288.36

2.3 Transient Transfection with siRNA

1. Preparation of siRNA stocks targeting PHD1, PHD2, PHD3, or FIH mRNA: in case of lyophilized delivery, dissolve the siRNA in the desired volume of RNase-free water to obtain a convenient stock concentration (*see* **Notes 4** and **5**).

   Example for the preparation of a 100 μM siRNA stock solution:

   siRNA amount n: 31.3 nmol

   Final concentration c: $100\ \mu mol/L = 100\ nmol/mL$

   Calculation of volume: $\text{volume } V = \dfrac{\text{siRNA amount } n}{\text{Final concentration } c}$

   $$V = \dfrac{31.3\ \text{nmol}}{100\ \text{nmol} / \text{mL}}$$
   $$= 0.313\ \text{mL}$$
   $$= 313\ \mu L$$

   Subsequently, determine the precise concentration of your stock solution using a spectrophotometer (*see* **Note 6**).

2. Spectrophotometer.

3. Transfection reagent: Lipofectamine 2000 (Invitrogen, Thermo Fisher Scientific, Waltham, MA, USA; store at 4 °C) or 1 mg/mL polyethylenimine (PEI) in sterile ddH$_2$O (25 kDa; Polysciences, Inc., Warrington, PA, USA; store at −20 °C) (*see* **Note 7**).

4. Opti-MEM (Gibco, Thermo Fisher Scientific).

2.4 Lentiviral Transduction with shRNA

1. Plasmid coding for shRNA (*see* **Note 8**).

2. Plasmid coding for lentiviral packaging components (e.g., ViraPower lentiviral expression vector system, Invitrogen, Thermo Fisher Scientific).

3. Transfection reagent: Lipofectamine 2000 or 1 mg/mL PEI in sterile ddH$_2$O (*see* **Note 7**).

4. Opti-MEM.

5. HEK293 or HEK293T cells for lentivirus production (*see* **Note 9**).

6. T75 cell culture flasks.

7. 50 mg/mL polybrene solution: weigh 2.5 g polybrene and dissolve it in 50 mL ddH$_2$O. Sterile filter the solution and store it at 4 °C.

8. Cell culture medium containing 12 µg/mL polybrene: mix 12 µL of 50 mg/mL polybrene solution with 50 mL of cell culture medium. Sterile filter and store it at 4 °C for up to several months.

3 Methods

All work is carried out in a laminar flow hood (sterile environment) unless otherwise indicated. Standard cell culture conditions refer to culture in a humidified atmosphere containing 5 % CO$_2$ and 20.9 % O$_2$ at 37 °C in a cell culture incubator.

3.1 Cell Treatment with Hydroxylase Inhibitors

1. Seed cells in 60 mm dishes 1 day prior to the treatment, and allow them to settle overnight in standard cell culture conditions (*see* **Notes 10** and **11**).

2. Prepare a stock solution of the inhibitor according to Table 1.

3. Dilute the inhibitor and the respective solvent control in cell culture medium (10% FCS, 1% Penicillin-Streptomycin) according to Table 2. Make sure to prepare sufficient amounts for the planned number of dishes (*see* **Note 12**).

4. Sterile filter the medium containing the inhibitor, if the inhibitor solution was not sterile.

5. Remove the cell culture medium from the dish using a vacuum pump, and replace it with 4 mL medium containing the inhibitor or the solvent as control.

6. Incubate cells in standard cell culture conditions.

7. Analyze the samples at the time point(s) of interest according to your research question (Fig. 2a) (*see* **Note 13**).

Table 2
Preparation of cell culture media containing hydroxylase inhibitor

Inhibitor	Stock conc.	Dilution in medium	Final conc.	References
DMOG	1 M in DMSO	1:1000 → e.g., 4 mL + 4 µL	1 mM	[2, 8, 9]
DFX	50 mM in ddH₂O	1:500 → e.g., 4 mL + 8 µL	100 µM	[2]
FG-4592	100 mM in DMSO	1:1000 → e.g., 4 mL + 4 µL	100 µM	[19]
JNJ-42041935	100 mM in DMSO	1:1000 → e.g., 4 mL + 4 µL	100 µM	[8, 20]

Conc., concentration

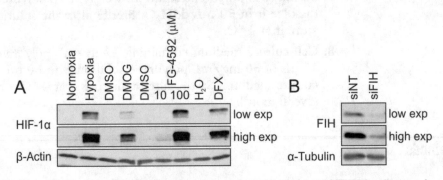

Fig. 2 Detection of hydroxylase inhibition and knockdown by immunoblot. (**a**) HEK293 cells were incubated in normoxia (21% O_2) or hypoxia (0.2% O_2) or treated with 1 mM DMOG, 10 or 100 µM FG-4592 (roxadustat), 100 µM DFX, or respective controls for 24 h. Cell lysates were analyzed for HIF-1α and β-actin protein levels by immunoblotting. HIF-1α was stabilized when HIF hydroxylases were inhibited in hypoxia or upon treatment with hydroxylase inhibitors. (**b**) HEK293 cells were transiently transfected with 50 nM nontarget siRNA (siNT) or siRNA targeting FIH (siFIH) for 48 h, and cell lysates were analyzed for FIH and α-tubulin protein levels by immunoblot. FIH protein levels were reduced upon siRNA knockdown. *Low exp.*, low exposure; *high exp.*, high exposure

3.2 Transient Transfection of Cells with siRNA Targeting the HIF Hydroxylases

1. Seed cells in 60 mm dishes 1 day prior to the transfection, and allow them to settle overnight in standard culture conditions (*see* **Notes 10** and **14**).

2. Dilute the siRNA stock to a working concentration of 10 µM by combining one part siRNA and nine parts of RNase-free water, e.g., 10 µL 100 µM siRNA stock and 90 µL RNase-free water (*see* **Note 15**).

3. Prepare Mix A containing (a) siRNA and (b) Opti-MEM separately for each siRNA and/or concentration (*see* **Notes 16–18**).

   (a) 7.5–15 µL siFIH (final concentration 25–50 nM)

   or 6 µL siPHD1 (final concentration 20 nM)

   or 1.5 µL siPHD2 (final concentration 5 nM)

   or 1.5 µL siPHD3 (final concentration 5 nM)

   (b) 150 µL Opti-MEM

4. Prepare Mix B containing transfection reagent and Opti-MEM (*see* **Note 16**).

   (a) 4.8 μL Lipofectamine 2000

   (b) 150 μL Opti-MEM

5. Incubate Mix A and B separately for 5 min at room temperature.

6. Combine Mix A and B (= Mix C), mix gently by pipetting up and down three to four times, and incubate for 20–45 min at room temperature.

7. In the meantime, aspirate the medium from the cells and add antibiotic-free medium. Adjust the volume of antibiotic-free medium to the volume of Mix C which will be added in the next step to obtain a final volume of 3 mL.

$$\begin{aligned} V_{\text{Mix C}} &= V_{\text{siFIH}} + V_{\text{Lipofectamine}} + \left(V_{\text{Opti-MEM}}\right) \times 2 \\ &= 7.5\,\mu\text{L} + 4.8\,\mu\text{L} + \left(150\,\mu\text{L}\right) \times 2 \\ &= 312.3\,\mu\text{L} \\ V_{\text{medium}} &= V_{\text{total}} - V_{\text{Mix C}} \\ &= 3000\,\mu\text{L} - 312.3\,\mu\text{L} \\ &= 2687.7\,\mu\text{L} \end{aligned}$$

   V, volume

8. Add Mix C dropwise to each dish containing cells and antibiotic-free medium (*see* **Note 19**).

9. Mix gently by rocking the plate back and forth in different directions, and then incubate the cells in standard cell culture conditions.

10. Change the culture medium 24 h after transfection, and carefully wash the cells with 5 mL PBS to remove residual transfection reagent. Incubate the cells in standard culture conditions for further 24 h (*see* **Notes 20** and **21**).

11. Harvest the cells at the desired time point(s), and analyze the samples according to your research question (Fig. 2b).

3.3 Stable Transduction of Cell Lines with shRNA Targeting the HIF Hydroxylases

3.3.1 Production of Lentiviral Pseudoparticles for Stable shRNA-Mediated Knockdown of PHD1/PHD2/PHD3/FIH

1. Seed 3×10^6 HEK293T cells per 75 cm² flask in 15 mL each.

2. Transfect HEK293T cells with plasmid coding for lentiviral packaging components and shRNAs targeting the hydroxylase of interest or control shRNA (*see* **Note 22**). Prepare Mix A containing plasmids and Opti-MEM.

   (a) 3.0 μg shRNA plasmid

   (b) 4.2 μg pLP1 (lentiviral packaging plasmid containing the HIV-1 gag and pol genes)

   (c) 2.0 μg pLP2 (lentiviral packaging plasmid containing the HIV-1 rev gene)

(d) 2.8 μg pVSV-G (lentiviral packaging plasmid containing the G glycoprotein gene from *Vesicular stomatitis virus*)

(e) Fill up to 200 μL with Opti-MEM.

3. Prepare Mix B containing transfection reagent and Opti-MEM (*see* **Note 7**).

(a) 5 μL PEI/μg plasmid = 60 μL PEI

(b) Fill up to 200 μL with Opti-MEM.

4. Incubate Mix A and B separately for 5 min at room temperature.

5. Add Mix B to A dropwise (= Mix C), mix by vortexing or flicking the tube, and incubate for 25–30 min at room temperature.

6. During the incubation time, aspirate the medium from the cells and add fresh medium. Adjust the volume of medium to the volume of Mix C, which will be added in the next step to obtain a total volume of 12 mL.

$$V_{\text{Mix C}} = V_{\text{Plasmids}} + V_{\text{PEI}} + V_{\text{Opti-MEM}}$$
$$= 400\,\mu L$$

$$V_{\text{medium}} = V_{\text{total}} - V_{\text{Mix C}}$$
$$= 12\,\text{mL} - 400\,\mu L$$
$$= 11.6\,\text{mL}$$

V, volume

7. Add Mix C dropwise to each flask containing cells and medium. Mix gently by rocking the flask back and forth (*see* **Note 19**).

8. Incubate the cells at standard cell culture conditions.

9. Remove the medium 24 h after the transfection, and carefully wash the cells with 10 mL PBS to remove residual transfection reagent.

10. Add 15 mL fresh standard medium, and culture the cells for 48–72 h at standard cell culture conditions for production of the lentiviral pseudoparticles (*see* **Note 23**).

11. Collect the cell culture supernatant and centrifuge it for 5 min at 220 x *g* to remove cell debris.

12. Filter the supernatant through a 0.45 μm filter for further removal of cell debris. The filtrate contains the lentiviral pseudoparticles.

13. Use the viral stock for infection of cells with the pseudoparticles containing the shRNA, or store the stocks at 4 °C (*see* **Note 24**).

3.3.2 Infection of Cells with Lentiviral Pseudoparticles for Stable Knockdown of Hydroxylases

1. Seed the cells of choice to reach approximately 50% confluency in a 6-well plate on the next day.

2. On the next day, prepare a 1:1 mixture of viral stock and cell culture medium containing 12 μg/mL polybrene in a volume

that is sufficient for the number of wells to be transduced with the shRNAs (final concentration of polybrene is 6 µg/mL; *see* **Note 25**).

3. Apply 1 mL of the mixture to each well, and incubate the cells in standard cell culture conditions for 24 h (*see* **Note 26**).

4. Change the medium 24 h after infection to remove the lentiviral pseudoparticles, and culture the cells under standard culture conditions for further 48 h to allow for stable integration of the shRNA into the host cell genome.

5. Start the selection of the infected cells 72 h after infection using respective antibiotics at a concentration sufficient for selection of transduced cells (*see* **Note 27**). Cells can be expanded to larger culture dishes during the selection period.

6. Analyze pools of clones for stable expression of the transfer vector and its respective shRNA following 1–2 weeks of selection by comparing the amount of target mRNA and protein between the control and knockdown cells (*see* **Note 28**).

7. Freeze pools of clones at early passages, and, if desired, perform single cell cloning from early passage pools of clones to obtain stable knockdown clones (*see* **Note 29**).

4   Notes

1. The sizes of cell culture dishes or flasks are chosen according to the needs and experimental setup. For reasons of simplification, the procedures are described for use of 60 mm cell culture dishes if not otherwise indicated. In order to adjust the volumes for different cell culture dishes or flasks, change the volumes proportional to the change of the cell culture dish/flask surface.

2. The solubility of a compound in different solvents is commonly indicated in the data sheet provided by the manufacturer. In case this information is not provided, the literature should be searched for further details. If no information is available, dissolve the compound stepwise in different solvents until it is completely in solution and calculate the achieved stock concentration.

3. Be aware of the effect that a solvent itself might have on the cells. Therefore, try to prepare the stock as highly concentrated as possible to reduce the amount of solvent on the cells later on and always include a solvent control in every experiment.

4. siRNA sequences/siRNA pools and final concentrations that were successfully used in our hands are listed in Table 3 [8].

Table 3
Recommended siRNA sequences and concentrations

siRNA	Sequence	Final conc. (nM)
siFIH[a]	5′-GUUGCGCAGUUAUAGCUUC-3′	25–50
siPHD1	Dharmacon ON-TARGETplus SMARTpool[b]	20
siPHD2	Dharmacon ON-TARGETplus SMARTpool[b]	5
siPHD3	Dharmacon ON-TARGETplus SMARTpool[b]	5

[a]The sequence for siFIH was previously reported by Cockman and colleagues (sequence F1) [5]. [b]GE Healthcare, Chalfont St. Giles, UK

5. Stock concentrations typically range from 20 to 100 μM; for reasons of simplification, we commonly use 100 μM stock solutions. siRNAs are sensitive to RNases; therefore, it is recommended to aliquot the siRNA stock solution to avoid contamination.

6. Photometric determination of the siRNA concentration usually leads to values with the units "ng/μL." To convert this into μmol/l, the molecular weight (MW) of the double-stranded siRNA has to be calculated. The average MW of one base pair of RNA is estimated as 660 g/mol. An exemplary calculation is shown below:

   Measured concentration
   (Conc.):

   $$1000\,\frac{ng}{\alpha L}$$

   siRNA length: 21 bp

   MWsiRNA:

   $$21\times660\,\frac{g}{mol}=13680\,\frac{g}{mol}$$
   $$=13.86\,\frac{ng}{pmol}$$

   siRNA concentration
   in μmol/L:

   $$\frac{Conc.}{MW_{siRNA}}=\frac{1000\,\frac{ng}{\mu L}}{13.86\,\frac{ng}{pmol}}$$
   $$=72.15\,\frac{pmol}{\mu L}$$
   $$=72.15\,\frac{\mu mol}{L}$$

7. Use a transfection reagent of choice according to the manufacturer's guidelines. In this chapter, we will describe the procedures using either Lipofectamine 2000 or PEI as a transfection reagent. In our experience, PEI transfection is sufficient for plasmids, whereas Lipofectamine 2000 is best for transfection of siRNAs.

8. Plasmids coding for shRNAs can be either cloned in the laboratory or ordered as purified circular DNA or bacterial glycerol stock. If a glycerol stock is used, the plasmid will need to be amplified and purified. High purity of the DNA is important to avoid contaminations and to ensure transduction of sufficient amounts of DNA.

9. HEK293T cells are commonly used to produce lentiviral pseudoparticles as they are easy to maintain and show a high transfectability with packaging and target plasmids. This is desirable in order to obtain high virus titers.

10. Seed cells to obtain maximum 80% confluency of evenly seeded cells at the time point of harvest to avoid local hypoxia, nutrient deprivation, and growth inhibition [14]. In case of working with a different cell line, optimize the seeding density prior to the experiment by seeding different numbers of cells and by observing the confluency when the cells are settled at the time point(s) of harvest.

11. Make sure to seed a sufficient number of dishes for all the treatments and time points you want to analyze. Remember to have a sample that is treated with the solvent only for every time point to be able to exclude any potential solvent effects in the experiment.

12. Adjust the volumes according to the number of dishes that will be treated with each inhibitor or respective control, and prepare a master mix containing an additional 10% of each volume (1.1×). This will allow treatment of all samples with the same mix of medium and inhibitor, which is important for subsequent comparisons of different samples and time points.

13. Following hydroxylase inactivation, HIF-1α is maximally stabilized after 6–8 h in most cell lines, whereas HIF-2α is maximally stabilized at 24 h or later [7, 15–17]. Depending on the research aim and readout, time points can be adjusted accordingly.

14. Seed one additional dish for every time point that will be transfected with an siRNA that does not have a target in the cell (nontarget siRNA (siNT)). This serves as a control for potential general effects of siRNA transfection.

15. Adjust these calculations to the siRNA concentration that was measured when preparing the stock solution. Prepare enough working solution for three independent experiments to avoid repeated freeze-thaw cycles of the 100 μM stock.

16. Adjust the volumes according to the number of dishes that will be transfected with each independent siRNA, and prepare a master mix containing an additional 10% of each volume (1.1×). This will allow to transfect all the samples with the same transfection mix, which is important for subsequent comparisons of different samples.

17. The optimum final concentration of an siRNA has to be titrated by transfecting different amounts and observing the effect either on mRNA or protein level. siRNA final concentrations usually lie between 10 and 100 nM. Generally, off-target effects are more likely to occur at higher siRNA concentrations. Therefore, the lowest concentration with sufficient knockdown efficiency should be used.

18. The siRNA concentrations listed in the table have worked in our hands in the cell lines HeLa and HEK293 [8, 9]. The concentration required for other cell lines might differ depending on different factors, such as hydroxylase expression level and transfectability.

19. Add the mix dropwise and distribute it evenly over the dish to avoid high local concentrations of transfection reagent that might be cytotoxic. Moreover, this facilitates an even distribution of the transfection mix over all cells.

20. The incubation time with siRNAs needs to exceed the protein half-life to a sufficient degree to obtain efficient knockdown on protein level. In our hands, the knockdown efficiency was optimal 48 h after transfection but might differ in different cell lines.

21. Repetitive transfection might be required to optimize the knockdown efficiency, especially in cells that are difficult to transfect. For this purpose, repeat the transfection procedure as before, and investigate the knockdown efficiency at different time points following the second round of transfection (e.g., at 24, 48, and 72 h).

22. Prior to the stable transfection of target cells, the cells can be transiently transfected with the shRNA plasmids, and the knockdown efficiency on mRNA or protein level can be compared with control cells in order to determine the efficiency of different shRNAs targeting the same mRNA. shRNAs leading to sufficient knockdown can then be chosen for the production of lentiviral pseudoparticles and generation of stable cell lines.

23. The transfection efficiency can be estimated and visualized by fluorescence microscopy approx. 24 h post transfection, if an extra flask was transfected with a GFP overexpression plasmid. Transfected cells show green fluorescence, whereas non-transfected cells are not fluorescent.

24. Storage at 4 °C is possible for up to several months.

25. Polybrene increases the transduction efficiency of lentiviral pseudoparticles [18].

26. Infection efficiency can be monitored by parallel infection of cells with pseudoparticles containing a construct for overexpression of GFP (e.g., pLenti6-EGFP, Thermo Fisher Scientific) and analysis by fluorescent microscopy 24 h post-infection.

The proportion of green cells will help to estimate the proportion of transduced cells.

27. shRNA vectors encode for a specific antibiotic resistance gene which can differ depending on the type of vector. The optimum antibiotic concentration for selection of transduced cells should be determined by a killing curve in which nonresistant parental cells are exposed to increasing concentrations of the antibiotic prior to the stable transduction. The lowest concentration at which all cells die within 24–48 h is to be chosen for selection of transduced cells. If the antibiotic concentration is too high, stably transduced, resistant cells can also die due to the antibiotic. Carrying along a nonresistant cell culture sample during the selection period will help to judge selection efficiency. Non-transduced cells of the same cell line should be used for this purpose. These nonresistant, non-transduced cells should die during the selection, whereas cells that were successfully transduced should survive.

28. Slowly dividing cells may need a longer selection period.

29. Pools of clones contain cells that integrated the shRNA in different locations and potentially different numbers within the genome and can therefore have different knockdown efficiencies. To overcome this effect, single cell cloning can be used to establish cell clones with a high, even knockdown efficiency. But these clones could at the same time show clonal effects that might not be observed when pools of clones are analyzed. Therefore, several clones should be established and included in subsequent investigations in order to exclude clonal effects.

Acknowledgments

This work was supported by ERACoSysMed (15/ERA-CSM/3267) and Science Foundation Ireland (11/PI/1005) awarded to Cormac T. Taylor, and by a Forschungskredit of the University of Zurich (FK-15-046) awarded to Carsten C. Scholz.

References

1. Kaelin WG Jr, Ratcliffe PJ (2008) Oxygen sensing by metazoans: the central role of the HIF hydroxylase pathway. Mol Cell 30(4):393–402

2. Jaakkola P, Mole DR, Tian YM, Wilson MI, Gielbert J, Gaskell SJ, Kriegsheim A, Hebestreit HF, Mukherji M, Schofield CJ, Maxwell PH, Pugh CW, Ratcliffe PJ (2001) Targeting of HIF-alpha to the von Hippel-Lindau ubiquity-lation complex by O2-regulated prolyl hydroxylation. Science 292(5516):468–472

3. Berra E, Benizri E, Ginouves A, Volmat V, Roux D, Pouyssegur J (2003) HIF prolyl-hydroxylase 2 is the key oxygen sensor setting low steady-state levels of HIF-1alpha in normoxia. EMBO J 22(16):4082–4090

4. Appelhoff RJ, Tian YM, Raval RR, Turley H, Harris AL, Pugh CW, Ratcliffe PJ, Gleadle

JM (2004) Differential function of the prolyl hydroxylases PHD1, PHD2, and PHD3 in the regulation of hypoxia-inducible factor. J Biol Chem 279(37):38458–38465

5. Cockman ME, Lancaster DE, Stolze IP, Hewitson KS, McDonough MA, Coleman ML, Coles CH, Yu X, Hay RT, Ley SC, Pugh CW, Oldham NJ, Masson N, Schofield CJ, Ratcliffe PJ (2006) Posttranslational hydroxylation of ankyrin repeats in IkappaB proteins by the hypoxia-inducible factor (HIF) asparaginyl hydroxylase, factor inhibiting HIF (FIH). Proc Natl Acad Sci U S A 103(40):14767–14772

6. Cummins EP, Berra E, Comerford KM, Ginouves A, Fitzgerald KT, Seeballuck F, Godson C, Nielsen JE, Moynagh P, Pouyssegur J, Taylor CT (2006) Prolyl hydroxylase-1 negatively regulates IkappaB kinase-beta, giving insight into hypoxia-induced NFkappaB activity. Proc Natl Acad Sci U S A 103(48):18154–18159

7. Stiehl DP, Wirthner R, Koditz J, Spielmann P, Camenisch G, Wenger RH (2006) Increased prolyl 4-hydroxylase domain proteins compensate for decreased oxygen levels. Evidence for an autoregulatory oxygen-sensing system. J Biol Chem 281(33):23482–23491

8. Scholz CC, Cavadas MA, Tambuwala MM, Hams E, Rodriguez J, von Kriegsheim A, Cotter P, Bruning U, Fallon PG, Cheong A, Cummins EP, Taylor CT (2013) Regulation of IL-1beta-induced NF-kappaB by hydroxylases links key hypoxic and inflammatory signaling pathways. Proc Natl Acad Sci U S A 110(46):18490–18495

9. Scholz CC, Rodriguez J, Pickel C, Burr S, Fabrizio JA, Nolan KA, Spielmann P, Cavadas MA, Crifo B, Halligan DN, Nathan JA, Peet DJ, Wenger RH, Von Kriegsheim A, Cummins EP, Taylor CT (2016) FIH regulates cellular metabolism through hydroxylation of the deubiquitinase OTUB1. PLoS Biol 14(1):e1002347

10. Carthew RW, Sontheimer EJ (2009) Origins and mechanisms of miRNAs and siRNAs. Cell 136(4):642–655

11. Bagnall J, Leedale J, Taylor SE, Spiller DG, White MR, Sharkey KJ, Bearon RN, See V (2014) Tight control of hypoxia-inducible factor-alpha transient dynamics is essential for cell survival in hypoxia. J Biol Chem 289(9):5549–5564

12. Scholz CC, Taylor CT (2013) Hydroxylase-dependent regulation of the NF-kappaB pathway. Biol Chem 394(4):479–493

13. Fu J, Taubman MB (2010) Prolyl hydroxylase EGLN3 regulates skeletal myoblast differentiation through an NF-{kappa}B-dependent pathway. J Biol Chem 285(12):8927–8935

14. Wenger RH, Kurtcuoglu V, Scholz CC, Marti HH, Hoogewijs D (2015) Frequently asked questions in hypoxia research. Hypoxia 3:35–43

15. Bruning U, Cerone L, Neufeld Z, Fitzpatrick SF, Cheong A, Scholz CC, Simpson DA, Leonard MO, Tambuwala MM, Cummins EP, Taylor CT (2011) MicroRNA-155 promotes resolution of hypoxia-inducible factor 1alpha activity during prolonged hypoxia. Mol Cell Biol 31(19):4087–4096

16. Stiehl DP, Bordoli MR, Abreu-Rodriguez I, Wollenick K, Schraml P, Gradin K, Poellinger L, Kristiansen G, Wenger RH (2012) Non-canonical HIF-2alpha function drives autonomous breast cancer cell growth via an AREG-EGFR/ErbB4 autocrine loop. Oncogene 31(18):2283–2297

17. Cavadas MA, Mesnieres M, Crifo B, Manresa MC, Selfridge AC, Scholz CC, Cummins EP, Cheong A, Taylor CT (2015) REST mediates resolution of HIF-dependent gene expression in prolonged hypoxia. Sci Rep 5:17851

18. Coelen RJ, Jose DG, May JT (1983) The effect of hexadimethrine bromide (polybrene) on the infection of the primate retroviruses SSV 1/SSAV 1 and BaEV. Arch Virol 75(4):307–311

19. Jain IH, Zazzeron L, Goli R, Alexa K, Schatzman-Bone S, Dhillon H, Goldberger O, Peng J, Shalem O, Sanjana NE, Zhang F, Goessling W, Zapol WM, Mootha VK (2016) Hypoxia as a therapy for mitochondrial disease. Science 352(6281):54–61

20. Barrett TD, Palomino HL, Brondstetter TI, Kanelakis KC, Wu X, Haug PV, Yan W, Young A, Hua H, Hart JC, Tran DT, Venkatesan H, Rosen MD, Peltier HM, Sepassi K, Rizzolio MC, Bembenek SD, Mirzadegan T, Rabinowitz MH, Shankley NP (2011) Pharmacological characterization of 1-(5-chloro-6-(trifluoromethoxy)-1H-benzoimidazol-2-yl)-1H-pyrazole-4-carboxylic acid (JNJ-42041935), a potent and selective hypoxia-inducible factor prolyl hydroxylase inhibitor. Mol Pharmacol 79(6):910–920

Chapter 2

Kinetic Analysis of HIF Prolyl Hydroxylases

Peppi Koivunen and Johanna Myllyharju

Abstract

Kinetic analyses of HIF prolyl 4-hydroxylases (HIF-P4Hs) allow determination of substrate, cosubstrate and cofactor requirements, analysis of the reaction rate, and inhibitory properties of the isoenzymes in vitro. Here we describe an assay measuring the substrate hydroxylation-coupled decarboxylation of radioactive 2-oxoglutarate to radioactive carbon dioxide as a fast, efficient, and diverse method to analyze the enzyme kinetics of HIF-P4Hs.

Key words HIF-P4H, 2-oxoglutarate, Substrate, Inhibitor, Kinetics

1 Introduction

Enzymes are required as catalysts to speed up most reactions in our body. HIF prolyl 4-hydroxylases (HIF-P4Hs, also known as PHDs and EglNs) catalyze the post-translational 4-hydroxylation of two prolyl residues in the HIFα subunit [1–3]. They use Fe^{2+} and molecular oxygen for the catalysis, while 2-oxoglutarate functions as a cosubstrate that is stoichiometrically decarboxylated to succinate and CO_2 during the reaction [1–4]. Ascorbate is needed to support the reaction but it is not a direct cofactor [4–6]. Three HIF-P4H isoenzymes, HIF-P4Hs 1–3, located in differing cellular compartments exist in human [1–3, 7]. Kinetic analyses of HIF-P4Hs have shown that they have a very low affinity and thus high demand for O_2, making them excellent oxygen sensors in the body [4, 8]. The analyses have also shown that HIF-P4Hs accept many different amino acids next to the target prolyl residue in their substrate [9, 10] and that substrate length has an effect on affinity—the longer the substrate the higher the affinity [8]. Inhibitory analyses have shown that some divalent metals, iron chelators, and naturally occurring and synthetic 2-oxoglutarate analogues act as competitive inhibitors for HIF-P4Hs with respect to iron and 2-oxoglutarate, respectively [11–14]. The kinetic analyses have

L. Eric Huang (ed.), *Hypoxia: Methods and Protocols*, Methods in Molecular Biology, vol. 1742,
https://doi.org/10.1007/978-1-4939-7665-2_2, © Springer Science+Business Media, LLC 2018

also revealed differences between the three HIF-P4Hs isoenzymes in their catalytic and inhibitory properties [4, 8, 11, 15].

Kinetic studies allow determination of the reaction rate and specific activity, substrate, cosubstrate, and cofactor requirements for an enzyme in vitro. In case of HIF-P4Hs, it has also enabled the discovery of enzyme inhibitors, some of which have advanced to clinical trials as potential novel therapeutics [16]. We describe here an assay measuring the HIFα substrate hydroxylation-coupled decarboxylation of ^{14}C–labeled 2-oxoglutarate to [14]CO_2 as a fast, efficient, and diverse method to study the enzyme kinetics of HIF-P4Hs [4]. This assay can also be used to study the inhibitory properties of HIF-P4Hs [4]. The assay was modified from an assay originally developed to determine the enzyme kinetics of collagen prolyl 4-hydroxylases and lysyl hydroxylases [17], and we have used the same principle to determine enzyme kinetics for several additional 2-oxoglutarate-dependent dioxygenases [18–20]. It has to be emphasized that this assay alone is not suitable for detecting potential novel substrates for HIF-P4Hs because it does not determine the actual formation of 4-hydroxyproline, and importantly some non-substrate peptides have been shown to lead to a high rate of uncoupled 2-oxoglutarate decarboxylation [9]. Therefore, hydroxylation of a proline in a putative novel substrate always needs to be verified by additional methods, such as radiochemical assay of 4-hydroxyproline or mass spectrometry [21–23].

2 Materials

2.1 Determination of Prolyl 4-Hydroxylase Activity Based on Decarboxylation of 2-Oxo[1-^{14}C] Glutarate

1. Purified recombinant HIF-P4Hs produced in *Escherichia coli* or insect cells [4, 11, 24].

2. Synthetic HIF-P4H peptide substrate DLDLEMLAPYIPMDDDFQL (*see* **Note 1**). Prepare a 500 µM stock solution in dimethyl sulfoxide.

3. 2-oxo[1-^{14}C]glutarate (PerkinElmer) mixed with unlabeled 2-oxoglutarate to give a final concentration of 2 mM with 1100 or 11,100 dpm/nmol (*see* **Note 2**).

4. Cofactor solution: 0.5 M Tris–HCl pH 7.8 (25 °C), 20 mg/mL bovine serum albumin (BSA) in H_2O, catalase (Sigma C-100), 0.01 M dithiothreitol (DTT) in H_2O, 20 mM ascorbate in H_2O, 100 µM $FeSO_4 \times 7\ H_2O$ in H_2O. BSA, DTT, ascorbate, and $FeSO_4 \times 7\ H_2O$ should be dissolved immediately before use in cold H_2O.

5. 1 M KH_2PO_4, pH 5.0.

6. Soluene-350 (PerkinElmer).

7. Scintillation solution: 15 g PPO (2,5-diphenyloxazole) + 50 mg POPOP equals 1,4-bis[2-(4-methyl-5-phenyloxazolyl)]benzene are dissolved in 1000 mL of toluene; 600 mL of Cellosolve (ethylene glycol monomethyl ether) is then added.

8. Other materials and equipment: test tubes, rubber stoppers with metal wire hooks (Fig. 1), filter paper, needles (23Gx1"), 1 ml syringes, liquid scintillation counter. For the determination of a K_m value for oxygen, a hypoxia workstation or gas mixtures with specific oxygen concentrations are needed.

3   Methods

3.1   Determination of Prolyl 4-Hydroxylase Activity Based on Decarboxylation of 2-Oxo[1-¹⁴C] Glutarate [4, 8, 9, 11, 12, 17]

1. Cut about 1 × 2 cm pieces of filter paper, attach them to the wire hooks of rubber stoppers (Fig. 1), wet the papers in Soluene-350, and let dry.

2. Put test tubes on ice and add cold H_2O to make a final volume of 1 mL after the addition of the substrate, cofactor solution, enzyme sample, and radioactive 2-oxoglutarate (*see* **Note 3**).

3. Prepare the cofactor solution. When determining K_m values for ascorbate or iron, leave them out from the cofactor solution and vary their concentration as described in Subheading 3.2. The amounts for preparing the cofactor solution are given here for a final reaction volume of 1 mL, multiply according to the number of reactions:

0.5 M Tris	100 µL
20 mg/mL BSA	100 µL (*see* **Note 4**)
Catalase	4 µL
0.01 M DTT	10 µL
20 mM ascorbate	100 µL
100 µM FeSO₄	50 µL
dH₂O	36 µL
	400 µL

4. Add 400 µL of the cofactor solution to each tube.

5. Add the recombinant HIF-P4H. Activities of the purified enzyme batches may vary. For kinetic analyses the amount of enzyme to be used needs to be determined so that it does not result in complete hydroxylation of the peptide substrate under the standard conditions. Examples of the obtained hydroxylation percentage with differing amounts of specific batches of recombinant human HIF-P4H-1 and HIF-P4H-2 purified from insect cells are shown in Fig. 2.

6. Add 100 µL of the DLDLEMLAPYIPMDDDFQL peptide substrate to a final concentration of 50 µM. When determining

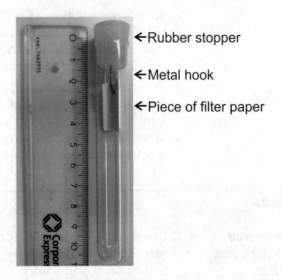

Fig. 1 Picture of a test tube and a rubber stopper with a metal wire hook to which the piece of filter paper catching CO_2 is attached

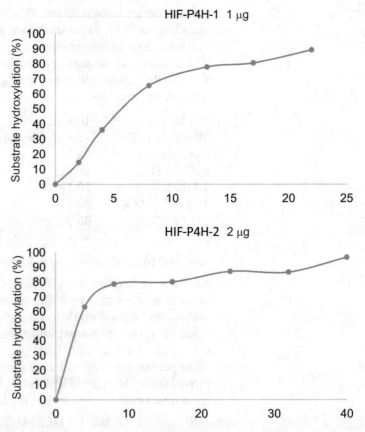

Fig. 2 Examples of peptide substrate hydroxylation percentages obtained with differing amounts of different batches of purified recombinant human HIF-P4H-1 and HIF-P4H-2

the K_m value for the substrate, vary its concentration as described in Subheading 3.2. In the presence of all cosubstrates and cofactors but in the absence of any peptide substrate, HIF-P4Hs catalyze an uncoupled decarboxylation of 2-oxoglutarate at a rate that is about 3% of that of the hydroxylation-coupled decarboxylation [11]. To obtain this blank value, omit the peptide substrate.

7. Add 60 μL of radioactive 2-oxoglutarate to a final concentration of 120 μM.

8. Seal the tubes immediately with the rubber stoppers with hanging filter papers, and mix.

9. Incubate in a 37°C water bath 20 min (with gentle shaking).

10. Transfer the tubes back to ice, and inject 1 mL of 1 M KH_2PO_4 through the rubber stoppers.

11. Shake tubes at room temperature for 30 min.

12. Place the filter papers into counting vials with 5 mL of the scintillation solution, and count the ^{14}C radioactivity in a liquid scintillation counter.

3.2 Determination of K_m and V_{max} Values

The K_m value for a substrate is determined in a simplified procedure by increasing its concentration while keeping the concentrations of other substrates saturating (*see* **Note 5**) and steady. When doable, it is important to increase the concentration of the substrate to a saturating concentration. For more detailed kinetic analyses, vary the concentration of one substrate in the presence of different fixed concentrations of the second substrate while keeping the concentrations of the other reaction components constant [25]. V_{max} values are determined at the saturating (*see* **Note 5**) concentrations of the reaction components. Determination of the K_m and V_{max} values of HIF-P4H-1 for oxygen is given below as an example:

1. Carry out the 2-oxoglutarate decarboxylation assay in the presence of increasing concentration of oxygen while keeping the concentrations of other reaction components saturating (*see* **Note 5**) and constant. The oxygen concentration can be varied in a hypoxia workstation or by using gas mixtures containing different oxygen concentrations. Prepare blank reactions without the peptide substrate for each oxygen concentration studied. When using the hypoxia workstation, allow adjustment of the reactions to the ambient oxygen tension for 20 min before initiation of the reaction. If using gas mixtures, bubble the reaction mixtures with the gas without enzyme and 2-oxo[1-^{14}C]glutarate for 10 min at room temperature. Close the tubes airtightly and equilibrate for 30 min by shaking. Then add enzyme and 2-oxo[1-^{14}C]glutarate and incubate at 37 °C.

2. Carry out the reactions as described in Subheading 3.1.

3. To calculate the K_m and V_{max} values, plot the substrate concentration (x-axis) vs. enzyme activity in dpm (y-axis) (Fig. 3). The plot should follow the Michaelis-Menten kinetics. The maximum rate of reaction, V_{max}, is the rate of the reaction when the enzyme is saturated with substrate (Fig. 3). The K_m value is the concentration of the substrate that gives ½ of the V_{max} value (Fig. 3). It is highly advisable to calculate the K_m and V_{max} values also by using the Lineweaver-Burk equation in which 1/ substrate concentration (x-axis) vs. 1/enzyme activity in dpm (y-axis) is plotted and which gives a line. The V_{max} value is the inverse of the intersection at y-axis, whereas when the line is extended to the negative side, the intersection of the x-axis equals $-1/K_m$. The Lineweaver-Burk equation typically gives more reliable values when saturation is difficult to reach (such as in the case of HIF-P4Hs for oxygen) (Fig. 3).

3.3 Determination of IC₅₀ Values

The IC_{50} value is determined over a range of inhibitor concentrations at one concentration of the substrate (or cosubstrate or cofactor) that the inhibitor competes with (e.g., 2-oxoglutarate or iron). For IC_{50} determination, use a constant concentration of the substrate (or cosubstrate or cofactor) of four times that of its K_m value, and increase the concentrations of the inhibitor up to a level where 100% inhibition is achieved. Below is given an example of the determination of the IC_{50} value of HIF-P4H-3 for cobalt chloride that is competitive with respect to iron:

1. Carry out the 2-oxoglutarate decarboxylation assay in the presence of 0.4 µM $FeSO_4$ concentration (equals 4× K_m of HIF-P4H-3 for iron [11]) and six increasing $CoCl_2$ concentrations (2–100 µM). Use saturating concentrations (*see* **Note 5**) of the other reaction components. Prepare control reactions without the inhibitor (0% inhibition) and a blank without the peptide substrate for each inhibitor concentration.

2. Carry out the reactions as described in Subheading 3.1.

3. To calculate the IC_{50} value, plot the inhibitor concentration (x-axis) vs. inhibition percentage (y-axis) (Fig. 4). Calculate the inhibition percentage by first subtracting the dpm of the blank from all samples with the corresponding inhibitor concentration. Then, set the reaction without inhibitor as 100% active, and calculate the activity percentage of each reaction containing the inhibitor by dividing its activity in dpm with that of the 100% active reaction and multiply by 100. Finally, convert activity percentages to inhibition percentages by subtracting 100 from all values. The IC_{50} value is the concentration of the inhibitor that results in 50% inhibition of the enzyme activity (Fig. 4).

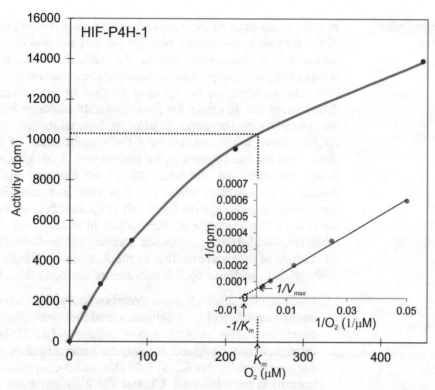

Fig. 3 Example of the Michaelis-Menten (big picture) and Lineweaver-Burk (inset) plots used to determine the K_m and V_{max} values of HIF-P4H-1 for oxygen. The source of enzyme was a crude cell lysate of insect cells expressing recombinant human HIF-P4H-1

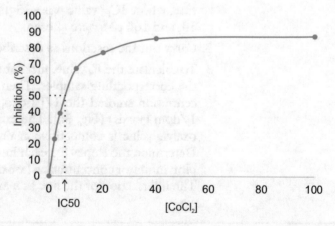

Fig. 4 Example of a plot used to determine the IC_{50} value of HIF-P4H-3 for cobalt chloride. The source of enzyme was recombinant human HIF-P4H-3 purified from insect cells

3.4 Determination of K_i Values

K_i value is determined by independently varying the concentration of the substrate, cosubstrate or cofactor, and the concentration of inhibitor. The nature of inhibition of a particular inhibitor may be competitive, noncompetitive, or uncompetitive with respect to the substrate, cosubstrate, or cofactor. In case of HIF-P4Hs when determining the K_i values for 2-oxoglutarate analogue inhibitors, we typically vary the concentration of 2-oxoglutarate, while for divalent metal inhibitors, we vary the concentration of iron. In these cases the inhibitors tested compete with 2-oxoglutarate and iron, respectively, for the active site of the enzyme, resulting in double-reciprocal plots where the lines with increasing inhibitor concentration intercept on the y-axis (Fig. 5a). Prepare reactions for at least four inhibitor concentrations in at least four different substrate, cosubstrate, or cofactor concentrations. Below is given an example of the determination of the K_i value of HIF-P4H-2 for oxalylglycine (see **Note 6**), a 2-oxolutarate analogue (Fig. 5):

1. Carry out the 2-oxoglutarate decarboxylation assay in the presence of four 2-oxo[1-^{14}C]glutarate and four oxalylglycine concentrations. Use saturating concentrations (see **Note 5**) of other reaction components. Prepare blank reactions without the peptide substrate for all inhibitor and 2-oxoglutarate concentration combinations. Choose the 2-oxoglutarate concentrations to be used around its K_m value and the inhibitor concentrations around its IC_{50} value (if known). For a crude human HIF-P4H-2 sample from insect cells, whose K_m for 2-oxoglutarate is 60 μM [4], concentrations 20, 60, 120, and 240 μM of 2-oxo[1-^{14}C]glutarate were chosen. For oxalylglycine, whose IC_{50} value was ~35 μM, concentrations of 10, 20, 50, and 100 μM were chosen.

2. Carry out the reactions as described in Subheading 3.1.

3. To calculate the K_i value, first subtract dpm of the blanks from the corresponding samples. Then plot for each inhibitor concentration studied the $1/[$2-oxo[1-^{14}C]glutarate$]$ (x-axis) vs. $1/$dpm (y-axis) (Fig. 5a). The extended lines cross at y-axis as oxalylglycine is competitive against 2-oxoglutarate (Fig. 5a). Determine the slopes for each line and plot them to a second plot: inhibitor concentration (x-axis) vs. slope (y-axis) (Fig. 5b). The intersection of the line at x-axis equals $-K_i$ (Fig. 5b).

4   Notes

1. We typically use the 19-residue HIF1α peptide DLDLEMLAPYIPMDDDFQL for HIF-P4H activity assays, but shorter or longer variants and peptides representing the ODDD of HIF2α or HIF3α can also be used [8, 9].

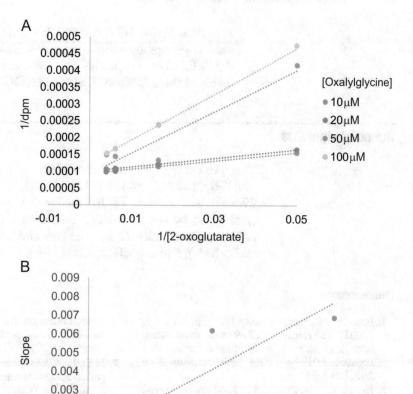

Fig. 5 Example of plots (**a**, **b**) used to determine the K_i value of HIF-P4H-2 for oxalylglycine. The source of enzyme was a crude cell lysate of insect cells expressing recombinant human HIF-P4H-2

2. To increase the sensitivity of the assay, we have used also 2-oxoglutarate with 55,500 dpm/nmol [8], but in such reactions the background radioactivity is also markedly higher and needs to be carefully controlled.

3. To save the reaction components, the reaction volumes can be scaled down from 1 mL. In such case adjust the volumes of the reaction components accordingly. Nowadays we mainly use 250 μL reaction volumes.

4. Some 2-oxoglutarate analogue inhibitors bind tightly to BSA. Omit BSA in such reactions.

5. Generally, a concentration of $10 \times K_m$ value is considered as a saturating concentration of a substrate.

6. Note that the cell membrane-permeable forms of certain inhibitors, such as dimethyloxalylglycine (DMOG), do not act

as such as HIF-P4H inhibitors, but require the cleavage of the methyl groups by cellular esterases. Thus, when determining the inhibitory properties in vitro, use the non-methylated forms of the inhibitors, i.e., for DMOG use oxalylglycine.

Acknowledgments

This work was supported by Academy of Finland through Grants 120156, 140765 and 218129 (P.K.), 200471, 202469, and 296498 and Center of Excellence 2012–2017 Grant 251314 (J.M.) and by the S. Jusélius Foundation (P.K., J.M.), the Emil Aaltonen Foundation (P.K.), the Jane and Aatos Erkko Foundation (P.K., J.M.), and FibroGen, Inc. (J.M.).

References

1. Epstein AC, Gleadle JM, McNeill LA et al (2001) C. elegans EGL-9 and mammalian homologs define a family of dioxygenases that regulate HIF by prolyl hydroxylation. Cell 107(1):43–54

2. Bruick RK, McKnight SL (2001) A conserved family of prolyl-4-hydroxylases that modify HIF. Science 294(5545):1337–1340

3. Ivan M, Haberberger T, Gervasi DC et al (2002) Biochemical purification and pharmacological inhibition of a mammalian prolyl hydroxylase acting on hypoxia-inducible factor. Proc Natl Acad Sci U S A 99(21):13459–13464

4. Hirsilä M, Koivunen P, Gunzler V et al (2003) Characterization of the human prolyl 4-hydroxylases that modify the hypoxia-inducible factor. J Biol Chem 278(33):30772–30780

5. Flashman E, Davies SL, Yeoh KK et al (2010) Investigating the dependence of the hypoxia-inducible factor hydroxylases (factor inhibiting HIF and prolyl hydroxylase domain 2) on ascorbate and other reducing agents. Biochem J 427(1):135–142

6. Briggs KJ, Koivunen P, Cao S et al (2016) Paracrine induction of HIF by glutamate in breast cancer: EglN1 senses cysteine. Cell 166(1):126–139

7. Metzen E, Berchner-Pfannschmidt U, Stengel P et al (2003) Intracellular localisation of human HIF-1 alpha hydroxylases: implications for oxygen sensing. J Cell Sci 116(Pt 7):1319–1326

8. Koivunen P, Hirsilä M, Kivirikko KI et al (2006) The length of peptide substrates has a marked effect on hydroxylation by the hypoxia-inducible factor prolyl 4-hydroxylases. J Biol Chem 281(39):28712–28720

9. Li D, Hirsilä M, Koivunen P et al (2004) Many amino acid substitutions in a hypoxia-inducible transcription factor (HIF)-1alpha-like peptide cause only minor changes in its hydroxylation by the HIF prolyl 4-hydroxylases: substitution of 3,4-dehydroproline or azetidine-2-carboxylic acid for the proline leads to a high rate of uncoupled 2-oxoglutarate decarboxylation. J Biol Chem 279(53):55051–55059

10. Huang LE, Pete EA, Schau M et al (2002) Leu-574 of HIF-1alpha is essential for the von Hippel-Lindau (VHL)-mediated degradation pathway. J Biol Chem 277(44):41750–41755

11. Hirsilä M, Koivunen P, Xu L et al (2005) Effect of desferrioxamine and metals on the hydroxylases in the oxygen sensing pathway. FASEB J 19(10):1308–1310

12. Koivunen P, Hirsilä M, Remes AM et al (2007) Inhibition of hypoxia-inducible factor (HIF) hydroxylases by citric acid cycle intermediates: possible links between cell metabolism and stabilization of HIF. J Biol Chem 282(7):4524–4532

13. Isaacs JS, Jung YJ, Mole DR et al (2005) HIF overexpression correlates with biallelic loss of fumarate hydratase in renal cancer: novel role of fumarate in regulation of HIF stability. Cancer Cell 8(2):143–153

14. Selak MA, Armour SM, MacKenzie ED et al (2005) Succinate links TCA cycle dysfunction to oncogenesis by inhibiting HIF-alpha prolyl hydroxylase. Cancer Cell 7(1):77–85

15. McNeill LA, Flashman E, Buck MR et al (2005) Hypoxia-inducible factor prolyl hydroxylase 2 has a high affinity for ferrous iron and 2-oxoglutarate. Mol Biosyst 1(4):321–324

16. Myllyharju J (2013) Prolyl 4-hydroxylases, master regulators of the hypoxia response. Acta Physiol (Oxf) 208(2):148–165

17. Kivirikko KI, Myllylä R (1982) Posttranslational enzymes in the biosynthesis of collagen: intracellular enzymes. Methods Enzymol 82(Pt A):245–304

18. Koivunen P, Hirsilä M, Gunzler V et al (2004) Catalytic properties of the asparaginyl hydroxylase (FIH) in the oxygen sensing pathway are distinct from those of its prolyl 4-hydroxylases. J Biol Chem 279(11):9899–9904

19. Laukka T, Mariani CJ, Ihantola T et al (2016) Fumarate and succinate regulate expression of hypoxia-inducible genes via TET enzymes. J Biol Chem 291(8):4256–4265

20. Tiainen P, Pasanen A, Sormunen R et al (2008) Characterization of recombinant human prolyl 3-hydroxylase isoenzyme 2, an enzyme modifying the basement membrane collagen IV. J Biol Chem 283(28):19432–19439

21. Juva K, Prockop DJ (1966) Modified procedure for the assay of H-3-or C-14-labeled hydroxyproline. Anal Biochem 15(1):77–83

22. Jaakkola P, Mole DR, Tian YM et al (2001) Targeting of HIF-alpha to the von Hippel-Lindau ubiquitylation complex by O2-regulated prolyl hydroxylation. Science 292(5516):468–472

23. Ivan M, Kondo K, Yang H et al (2001) HIFalpha targeted for VHL-mediated destruction by proline hydroxylation: implications for O2 sensing. Science 292(5516):464–468

24. Hewitson KS, Schofield CJ, Ratcliffe PJ (2007) Hypoxia-inducible factor prolyl-hydroxylase: purification and assays of PHD2. Methods Enzymol 435:25–42

25. Myllylä R, Tuderman L, Kivirikko KI (1977) Mechanism of the prolyl hydroxylase reaction. 2. Kinetic analysis of the reaction sequence. Eur J Biochem 80(2):349–357

Chapter 3

Mass Spectrometry and Bioinformatic Analysis of Hydroxylation-Dependent Protein-Protein Interactions

Javier Rodriguez and Alex von Kriegsheim

Abstract

Characterization of how a stimulus regulates the dynamics of protein-protein interaction is critical for understanding how a particular protein is regulated in an intracellular signaling network. Protein hydroxylation, which is a posttranslational modification catalyzed by oxygen-dependent enzymes, is a crucial regulator of protein-protein interactions. Under low oxygen conditions, the activity of many hydroxylases is inhibited, which results in a reduction of substrate hydroxylation. These changes alter the interactome of the substrate, and this dynamic rewiring of signaling networks explains crucial aspects of the adaptive response to hypoxia. In order to fully understand the systemic role of hydroxylation, it is necessary to identify a comprehensive set of substrates, as well as to determine which residues are hydroxylated. In addition, hydroxylation-dependent changes in the interactome of the substrates are indicative of the molecular function of the modification. To identify new substrates of hydroxylases, we have developed an approach involving the use of a pharmacological substrate-trap strategy followed by label-free quantitative mass spectrometry. An overview is provided for the sample preparation, mass spectrometry techniques, and statistical analysis used for detection of new substrates, hydroxylated residue, and hydroxylation-dependent protein-protein interaction changes.

Key words Hydroxylation, Mass spectrometry, Quantitative interaction proteomics.

1 Introduction

Cell signaling is the transduction mechanism that allows cells to respond to disparate extrinsic and intrinsic stimuli including mitogenic or stress signals such as oxygen deprivation or hypoxia. Intrinsically, cell signal transduction is based on the formation and dissolution of protein-protein interactions, which transmit, translate, and transform stimuli into appropriate biological responses. This in turn induces the adaptation of the organism to the environment. Characterization of how protein-protein interactions are dynamically regulated by a stimulus can produce a better insight of the function and regulatory network of a number of signaling proteins.

L. Eric Huang (ed.), *Hypoxia: Methods and Protocols*, Methods in Molecular Biology, vol. 1742, https://doi.org/10.1007/978-1-4939-7665-2_3, © Springer Science+Business Media, LLC 2018

Low oxygen, or hypoxia, is one example of how environmental cues shape cell fate decisions through the dynamic regulation of protein-protein interactions and posttranslational modifications. The changes in the interactome are mediated by hydroxylases, which are a family of enzymes that act as cellular oxygen sensors. These enzymes require molecular oxygen to catalyze the hydroxylation of specific amino acid residues. The best characterized substrates of this family of enzymes are the alpha subunits of the transcription factor HIF (hypoxia-inducible factor 1). In the presence of molecular oxygen, a family of prolyl (PHD1-3) and asparagine (FIH) hydroxylases hydroxylate HIF1-3α on specific residues. These hydroxylations regulate the HIF interactome promoting the proteasomal degradation and inactivation of the transcription factor and consequently the reduction in its expression and activity [1–3].

In addition to HIFs, several other PHD and FIH substrates, such as Foxo3a [4], actin [5], PKM2 [6], IκBβ [7], CEP192 [8], OTUB1 [9], MAPK6, and RIPK4 [10], have been described over the years, suggesting that hypoxia and hydroxylases regulate additional aspects of the cellular signaling machinery. Many novel substrates have been described recently, and there has been an acceptance in the field that additional substrates are still to be identified. One way of identifying new substrates is to exploit the fact that an enzyme-substrate complex has to be formed for the reaction to take place and that this complex is stabilized when the hydroxylation reaction is inhibited. Thus, hydroxylase interactors that are enriched in the presence of an inhibitor can lead to the identification of likely substrates. Usually, quantitative mass spectrometry is used for detection and quantification of the interacting protein/substrates.

We employed a pharmacological substrate-trap strategy in conjunction with label-free quantitative mass spectrometry to detect HIF hydroxylase substrates. To trap the complex, the cell permeable pan-hydroxylase inhibitor dimethyloxalylglycine (DMOG) was used. DMOG is a synthetic analogue of α-ketoglutarate and is commonly used to inhibit hydroxylases [11]. DMOG blocks the resolution of the active enzyme-substrate complex by preventing the hydroxylation from taking place (Fig. 1). Therefore, hydroxylase interactors, whose abundance increases upon DMOG treatment, can be considered potential substrates of the hydroxylases.

2   Materials

1. Sample: 3×10 cm plate per condition (DMOG or DMSO) of HEK 293T (1×10^6 cells) cells expressing FLAG−/V5-tagged protein of interest (hydroxylase or substrate, depending if a

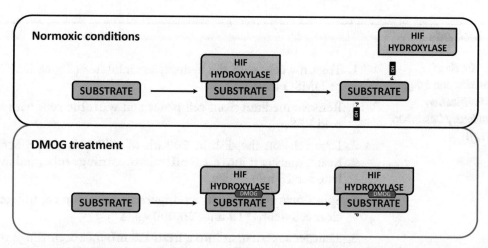

Fig. 1 Cartoon of substrate-trap mechanism: in the absence of DMOG, the hydroxylases bind to the substrate and release them upon completed hydroxylation. In the presence of DMOG, the hydroxylation is inhibited, and the enzyme-substrate complex is trapped

substrate screen or hydroxylation/interaction screen is planned) or empty vector (*see* **Note 1**).

2. Inhibitor of hydroxylases: dimethyloxallyl glycine (DMOG).

3. Lysis buffer: 1% Triton-x100, 20 mM Tris–HCl (pH 7.5), 150 mM NaCl, supplemented with protease (5 μg/mL leupeptin, 2,2 μg/mL aprotinin) and phosphatase (20 mM β-glycerophosphate) inhibitors.

4. Affinity media: anti-V5 beads and anti-FLAG M2 beads.

5. Wash buffer: 20 mM Tris–HCl (pH 7.5), 150 mM NaCl, and 1 mM EDTA.

6. Elution buffer 1: 2 M urea, 50 mM Tris–HCl (pH 7.5), and 5 μg/mL trypsin sequencing grade (V5111 Promega) (see **Note 2**).

7. Elution buffer 2: elution buffer 1 without trypsin and 1 mM DTT (*see* **Note 2**).

8. Alkylation buffer: Iodoacetamide (5 mg/mL) in H_2O (HPLC grade) (*see* **Note 2**).

9. Digestion termination: trifluoroacetic acid (LC/MS grade).

10. Desalting media: C-18 cartridge Empore material (*see* **Note 3**).

11. Solution A: H_2O (HPLC grade), 0.1% trifluoroacetic acid.

12. Solution B: 50% acetonitrile, 0.1% trifluoroacetic acid.

3   Methods

3.1   On Bead Digestion and Stage Tip Purification (Previously Described in [12])

1. Treat the cells with the hydroxylase inhibitor (2 mM DMOG) or DMSO for 4 h.

2. Remove medium from cell plates and wash the cells with ice-cold PBS.

3. Lyse cells on the dish in 500 μL of cold lysis buffer. Scrape lysate, transfer it into a 1.5 mL microcentrifuge tube, and incubate for 15 min on ice.

4. Spin down the lysate in a refrigerated benchtop centrifuge to clear cell debris (10 min, $20,000 \times g$, 4 °C).

5. Transfer supernatant into a fresh 1.5 mL microcentrifuge tube and discard the pelleted cell debris.

6. Add 5–7 μL of anti-V5/FLAG M2 bead slurry to the cleared lysates, and incubate with end-to-end rotation at 4 °C for 2 h.

7. Give a pulse of centrifugation to pellet the beads and remove the supernatant. Wash with 500 μL of lysis buffer, centrifuge, and discard the supernatant (three times) (*see* **Note 4**).

8. Repeat the previous step but using the wash buffer (twice). Transfer the beads to a new tube in the final wash.

9. Resuspend the beads in 60 μL of elution buffer 1. Incubate at 27 °C for 30 m in a thermomixer (with shaking at 800 rpm).

10. Centrifuge and collect the supernatant into a fresh tube.

11. Wash the beads in 25 μL of elution buffer 2. Centrifuge, collect the supernatant, and pool with the previous eluate (twice).

12. Incubate at room temperature (RT) overnight to allow the digestion.

13. Add 20 μL of alkylation buffer and keep for 30 min in the dark (RT).

14. Add trifluoroacetic acid (1% final concentration) to terminate the digestion and alkylation reaction.

15. Activate the C18 stage tip with 50 μL of Solution B (*see* **Note 5**).

16. Wash the C18 material with 50 μL of Solution A.

17. Load the sample into the C18 stage tip and pass the sample through it.

18. Wash two times with 50 μL of solution A (discard eluate).

19. Elute two times with 25 μL of Solution B and pool together the eluents into a new tube.

20. Lyophilize the eluents in a SpeedVac concentrator (a small drop of about 5 μL should be left).

21. Resuspend in 12 μL of Solution A before analysis by liquid chromatography-mass spectrometry (LC-MS) (*see* **Note 6**).

3.2 Bioinformatic Search for DMOG-Induced Hydroxylase/ Substrate Interactome

In the following section, we describe the use of the MaxQuant software package [13] as well as the selection criteria for detection of the hydroxylase substrates using the "DMOG-trap approach" (Fig. 2):

1. These are the main parameters required for the analysis: in the experimental design template, treat each single file as a separate sample for the subsequent statistical analysis. On the group-specific parameter sheet, select the enzyme that has been used for the digestion (i.e., trypsin). Specify methionine oxidation and N-terminal acetylation as variable modifications, and select LFQ as mechanism of quantification (LFQ min. Ratio count 1). On the global parameter sheet, select a human database (i.e., Uniprot HUMAN), and specify carbamidomethylation as fixed modification.

2. After the completion of the search, open the proteingroups.txt in the result folder file with a spread-sheep program, and replace missing LFQ intensity values (0) by a constant (1) to allow statistical analysis and calculation of ratios.

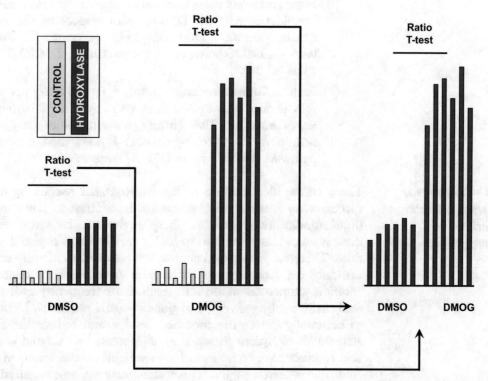

Fig. 2 Illustration of data analysis. The LFQ intensity values were averaged and filtered via a *t*-test and ratio cutoff versus the respective negative controls. All significant hits were then additionally compared to each other after the hydroxylase input was normalized. The proteins whose binding was significantly increased by DMOG were deemed to be potential substrates

3. Remove "reverse" and "identified only with modified peptide" entries.

4. Generate the average LFQ intensity for each condition (in total there are six values, three biological with two technical replicates for each one), and generate the ratio by dividing it by the average of the respective negative control (DMSO or DMOG). In addition, use a t-test to calculate the p-value of each experimental group when compared to the respective negative control.

5. To identify specific interactions enriched upon DMOG treatment, the potential substrates, we used the following criteria:

   (a) Select specific interactions for each hydroxylase with respect of its negative control (ratio LFQ Intensity (average hydroxylase/average negative control) >2, t-test <0.05).

   (b) Combine specific interactions for each hydroxylase in both conditions (DMSO-DMOG).

   (c) Normalize the LFQ values of the specific interactors against the average of the LFQ values of the bait (see **Note 7**).

   (d) Filter and select these interactions that are increased in the sample treated with DMOG with respect of the non-treated sample (ratio MaxLFQ (average [hydroxylase + DMOG]/average [hydroxylase + DMSO]) >2, t-test <0.05).

   (e) Reduce Uniprot accession numbers to one entry per protein group, and upload into pathway analysis software suites such as IPA (http://www.ingenuity.com∕) or StringDB (http://string-db.org). IPA was used to identify pathways enriched upon DMOG treatment.

3.3 Bioinformatic Analysis to Identify Specific Hydroxylations

The accurate identification of the hydroxylated residue by mass spectrometry is not a trivial endeavor. In contrast to posttranslational modifications such as phosphorylation, acetylation, etc., there is no unbiased method to highly enriched hydroxylated peptides. Therefore, a targeted approach is required to identify enzymatically regulated hydroxylation sites of a selected target. In addition, numerous amino acid residues are frequently hydroxylated/oxidated by enzymatic or nonenzymatic processes. In order to accurately detect the modified residue and reduce the false assignment of hydroxylation sites, all potentially oxidated amino acid residues have to be included as variable modifications in the Andromeda search engine. This will increase the time required for searching but is absolutely required for the correct assignment of the modified residue. In addition, it may be necessary to use several endopeptidase, or even combinations, to cover the entire length of a potential substrate. We commonly use three enzymes with orthogonal specificities for the initial screen (GluC, trypsin, and chymotrypsin).

In the following steps, we describe the use of MaxQuant software package and the selection criteria for detection of enzymatic-related hydroxylation sites (*see* **Note 8**):

1. Generate in Andromeda configuration the modification entry that includes the hydroxylation/oxidation of methionine, tryptophan, tyrosine, proline, asparagine, lysine, histidine, cysteine, and aspartate. Configure the addition of an oxygen (mass 15.994914) to these amino acids (position, anywhere; type, standard; and new terminus, none) (*see* **Note 9**).

2. Analyze raw files against a human database (Uniprot HUMAN). On the group-specific parameter sheet, select the enzyme that has been used for the digestion (i.e., trypsin), and select carbamidomethylation as fixed modification and as variable the newly coded modification specified under 1 (oxidation (MWYPNKHCD)) and N-terminal acetylation. Limit the maximal amount of modifications to 4. This not only reduces the analysis time but also reduces the false-positive assignment of distinct PTMs as multiple hydroxylated peptides. Phosphorylation and 5× hydroxylation induce a nearly identical mass shift. In the experimental design template, consider each single file as separate for the subsequent statistical analysis.

3. After the completion of the search, open the Oxidation (MWYPNKHCD).txt file in the result folder, and filter the gene name with the ID of interest. Filter again for the amino acid that is expected to be hydroxylated (i.e., in the case of working with FIH as hydroxylase, filter for asparagines).

4. For the selection of the specific hydroxylation sites, we used the following criteria:

   (a) Filter the sites with a localization higher than 0.9.

   (b) Flter and select those sites which ratio modified/unmodified are reduced in the sample treated with DMOG treatment (ratio (average ratio modified/unmodified hydroxylase)/(average ratio modified/unmodified +DMOG) >1, *t*-test <0.05) (*see* **Note 10**).

4   Notes

1. For the experimental workflow, the samples of the different experimental conditions were performed as biological triplicates and technical duplicates. We generally transfected 1 μg of

DNA in one 10 cm dish, in order not to overexpress the hydroxylase/substrate excessively.

2. Elution buffer 1–2 and the alkylation buffer should be made fresh.

3. For stage tip preparation, a small disk of Empore material 3 M was placed in an ordinary pipette tip [14].

4. Remove the supernatant carefully using a gel-loader tip connected to a vacuum pump. Be careful not to remove pelleted beads. Add the washing buffer and mix the beads by using a vortex.

5. Liquid was passed through the tips with a syringe or by light centrifugation ($0.9 \times g$).

6. With this volume we are able to analyze each sample at least two times (5 µL injected per run). Q-Exactive mass spectrometer was connected to an Ultimate Ultra3000 uHPLC system (both Thermo Scientific, Germany) incorporating an autosampler. 5 µL of each sample was loaded on a homemade column (100 mm length, 75 µm inside diameter [i.d.]) packed with 1.9 µm ReprosilAQ C18 (Dr. Maisch, Germany) and separated by an increasing acetonitrile gradient, using a 40 min reverse-phase gradient (from 3 to 32% acetonitrile) at a flow rate of 250 nL/min. The mass spectrometer was operated in positive ion mode with a capillary temperature of 220 °C, with a potential of 2000 V applied to the column.

7. The normalization is only required in the case that the LFQ values of the bait are affected by the distinct treatments.

8. For the detection of the hydroxylation, the experimental plan varies. The substrate is overexpressed and specifically immunoprecipitated under two conditions (DMSO and overexpression of hydroxylase/DMOG treatment). Hydroxylations can affect the stability of the protein. Thus, if the substrate is regulated in such manner, pretreatment with proteasomal inhibitors such as MG132 may be required to detect the hydroxylated residues.

9. For the detection and differentiation of these different oxidated amino acids, a high-resolution HPLC is required. Generally, hydroxylation of amino acids promotes a slight reduction of the elution time with respect of the unmodified peptide, whereas oxidation of amino acid side chains which include sulfur (methionine and cysteine) promotes a large shift in the elution time (Fig. 3).

10. If the hydroxylation is stoichiometric, the unmodified peptide may be difficult to detect. If this is the case, the value of the ratio modified/unmodified will be Na/N (due to absence of an unmodified intensity). Under these circumstances, the ratio of the intensity of the hydroxylated peptide over the intensity of the full protein can be used.

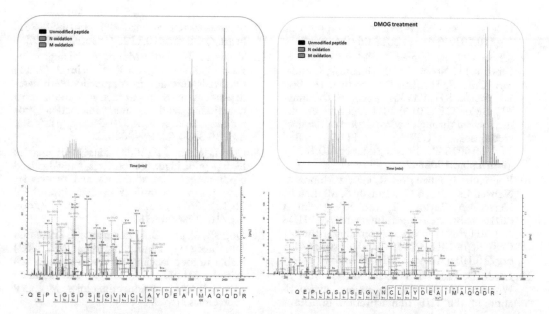

Fig. 3 (**a**) Extracted ion chromatogram (XIC) of asparagine hydroxylation, methionine oxidation, and the corresponding unmodified tryptic peptide of OTUB1 (_QEPLGSDSEGVNCLAYDEAIMAQQDR_). Methionine oxidation promotes a dramatic reduction in the elution time with respect to the unmodified peptide. In the case of the asparagine hydroxylation, this shift is smaller, with the elution time close to the unmodified peptide. (**b**) Representative fragmentation spectra of N(ox) and M(ox) of the previous described OTUB1 peptide

References

1. Bruick RK, McKnight SL (2001) A conserved family of prolyl-4-hydroxylases that modify HIF. Science 294(5545):1337–1340. https://doi.org/10.1126/science.1066373

2. Jaakkola P, Mole DR, Tian YM, Wilson MI, Gielbert J, Gaskell SJ, von Kriegsheim A, Hebestreit HF, Mukherji M, Schofield CJ, Maxwell PH, Pugh CW, Ratcliffe PJ (2001) Targeting of HIF-alpha to the von Hippel-Lindau ubiquitylation complex by O2-regulated prolyl hydroxylation. Science 292(5516):468–472. https://doi.org/10.1126/science.1059796

3. Huang LE, Gu J, Schau M, Bunn HF (1998) Regulation of hypoxia-inducible factor 1alpha is mediated by an O2-dependent degradation domain via the ubiquitin-proteasome pathway. Proc Natl Acad Sci U S A 95(14):7987–7992

4. Zheng X, Zhai B, Koivunen P, Shin SJ, Lu G, Liu J, Geisen C, Chakraborty AA, Moslehi JJ, Smalley DM, Wei X, Chen X, Chen Z, Beres JM, Zhang J, Tsao JL, Brenner MC, Zhang Y, Fan C, DePinho RA, Paik J, Gygi SP, Kaelin WG Jr, Zhang Q (2014) Prolyl hydroxylation by EglN2 destabilizes FOXO3a by blocking its interaction with the USP9x deubiquitinase. Genes Dev 28(13):1429–1444. https://doi.org/10.1101/gad.242131.114

5. Luo W, Lin B, Wang Y, Zhong J, O'Meally R, Cole RN, Pandey A, Levchenko A, Semenza GL (2014) PHD3-mediated prolyl hydroxylation of nonmuscle actin impairs polymerization and cell motility. Mol Biol Cell 25(18):2788–2796. https://doi.org/10.1091/mbc.E14-02-0775

6. Luo W, Hu H, Chang R, Zhong J, Knabel M, O'Meally R, Cole RN, Pandey A, Semenza GL (2011) Pyruvate kinase M2 is a PHD3-stimulated coactivator for hypoxia-inducible factor 1. Cell 145(5):732–744. https://doi.org/10.1016/j.cell.2011.03.054

7. Cummins EP, Berra E, Comerford KM, Ginouves A, Fitzgerald KT, Seeballuck F, Godson C, Nielsen JE, Moynagh P, Pouyssegur J, Taylor CT (2006) Prolyl hydroxylase-1 negatively regulates IkappaB kinase-beta, giving insight into hypoxia-induced NFkappaB activity. Proc Natl Acad Sci U S A 103(48):18154–18159. https://doi.org/10.1073/pnas.0602235103

8. Moser SC, Bensaddek D, Ortmann B, Maure JF, Mudie S, Blow JJ, Lamond AI, Swedlow JR, Rocha S (2013) PHD1 links cell-cycle progression to oxygen sensing through hydroxylation of the centrosomal protein Cep192. Dev

Cell   26(4):381–392.   https://doi.
org/10.1016/j.devcel.2013.06.014

9. Scholz CC, Rodriguez J, Pickel C, Burr S, Fabrizio JA, Nolan KA, Spielmann P, Cavadas MA, Crifo B, Halligan DN, Nathan JA, Peet DJ, Wenger RH, Von Kriegsheim A, Cummins EP, Taylor CT (2016) FIH regulates cellular metabolism through hydroxylation of the deubiquitinase OTUB1. PLoS Biol 14(1):e1002347. https://doi.org/10.1371/journal.pbio.1002347

10. Rodriguez J, Pilkington R, Garcia Munoz A, Nguyen LK, Rauch N, Kennedy S, Monsefi N, Herrero A, Taylor CT, von Kriegsheim A (2016) Substrate-trapped Interactors of PHD3 and FIH cluster in distinct signaling pathways. Cell Rep 14(11):2745–2760. https://doi.org/10.1016/j.celrep.2016.02.043

11. Lando D, Peet DJ, Whelan DA, Gorman JJ, Whitelaw ML (2002) Asparagine hydroxylation of the HIF transactivation domain a

hypoxic switch. Science 295(5556):858–861. https://doi.org/10.1126/science.1068592

12. Turriziani B, Garcia-Munoz A, Pilkington R, Raso C, Kolch W, von Kriegsheim A (2014) On-beads digestion in conjunction with data-dependent mass spectrometry: a shortcut to quantitative and dynamic interaction proteomics. Biology 3(2):320–332. https://doi.org/10.3390/biology3020320

13. Cox J, Mann M (2008) MaxQuant enables high peptide identification rates, individualized p.p.b.-range mass accuracies and proteome-wide protein quantification. Nat Biotechnol 26(12):1367–1372.   https://doi.org/10.1038/nbt.1511

14. Ong SE, Blagoev B, Kratchmarova I, Kristensen DB, Steen H, Pandey A, Mann M (2002) Stable isotope labeling by amino acids in cell culture, SILAC, as a simple and accurate approach to expression proteomics. Mol Cell Proteomics 1(5):376–386

Chapter 4

Acquisition of Temporal HIF Transcriptional Activity Using a Secreted Luciferase Assay

Miguel A.S. Cavadas, Cormac T. Taylor, and Alex Cheong

Abstract

Here we describe a simple method based on secreted luciferase driven by a hypoxia-inducible factor (HIF) response element (HRE) that allows the acquisition of dynamic and high-throughput data on HIF transcriptional activity during hypoxia and pharmacological activation of HIF. The sensitivity of the assay allows for the secreted luciferase to be consecutively sampled (as little as 1% of the total supernatant) over an extended time period, thus allowing the acquisition of time-resolved HIF transcriptional activity.

Key words Hypoxia, HIF, Luciferase, Transcription, Activity, Time-resolved

1 Introduction

Transcription factors promote global changes in gene expression allowing cells to adapt in time and space to a particular challenge. A prime example is the hypoxia-inducible factor (HIF), which induces hundreds of genes in response to oxygen deprivation [1, 2]. To investigate the full scope of regulatory mechanisms that modulate HIF activity during its induction and resolution, it is essential to have experimental methods that efficiently monitor transcription factor activity in a dynamic and high-throughput manner.

Commonly used promoter-luciferase reporter constructs rely on firefly [3] and the renilla [4] luciferases, which require lysis of the cultured cells at a given time point to obtain a measure of promoter activity. A novel generation of luciferase reporters based on *Gaussia* and *Cypridina* [5, 6] has the characteristic of being secreted out of the cell. This allows for repeated *sampling* of the culture medium on the same population of cells *over time*, using as little as 1% of the supernatant (e.g., 10 μL out of 1000 μL of medium on a 6-well plate) (Fig. 1).

Here we describe a simple method based on HRE-driven secreted luciferase, which allows the acquisition of dynamic and

L. Eric Huang (ed.), *Hypoxia: Methods and Protocols*, Methods in Molecular Biology, vol. 1742,
https://doi.org/10.1007/978-1-4939-7665-2_4, © Springer Science+Business Media, LLC 2018

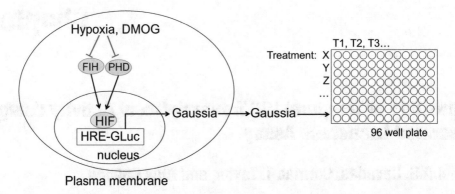

Fig. 1 HRE-GLuc assay for the measurement of HIF transcriptional activity. Cells are transfected with a Gaussia vector driven by the HRE promoter (HRE-GLuc). Cells are then subjected to the desired treatments (X, Y, Z), and if HIF expression is stabilized (e.g., by hypoxia or DMOG) it will bind to the HRE on the vector, Gaussia luciferase is produced and secreted out of the cell into the culture medium. The medium is collected at desired time interval (T1, T2, T3) into a 96-well plate. Gaussia substrate is added to the plate, and bioluminescence is measured, giving a readout of HIF transcriptional activity

high-throughput [7, 8] data on HIF transcriptional activity both in hypoxia and pharmacological activation of HIF (Fig. 1).

This method has allowed us to study how the inhibition of the PHD and FIH hydroxylases is required for maximal induction of the HIF transcriptional activity [8, 9] and how the HIF response is resolved with contributions from the microRNA (miR-155) [10] and transcriptional repressors [11]. The method has also been successfully adapted to measure the transcriptional activity of other transcription factors including NF-κB [12, 13] and the repressor element-1 silencing transcription factor (REST) [1].

2 Materials

2.1 Generation of Reporter Constructs

1. pGLuc-Mp vector (made from pGLuc-basic vector (NEB)) with customized minimal promoter inserted (*see* **Note 1**).

2. pTK-GLuc vector (NEB).

3. pCMV-CLuc vector (NEB).

4. Synthesized DNA sequence containing binding sites for HIF with restriction recognition sites for BglII and EcoRI 54 (*see* **Note 2**).

5. Restriction enzymes BglII and EcoRI.

6. Agarose.

7. Ethidium bromide (or Safe Red).

8. Gel purification kit.

9. T4 DNA ligase.

10. *E. coli* competent cells for subcloning.

11. Luria broth agar.

12. Luria broth.

13. Plasmid DNA extraction kit.

14. Spectrophotometer.

2.2 Cell Culture and Transfection

1. Human embryonic kidney cells (HEK293) (*see* **Note 3**).

2. Dulbecco's Modified Eagle's Medium (DMEM) including 10% fetal calf serum and 1% penicillin and streptomycin.

3. Optimem medium (Invitrogen).

4. Lipofectamine 2000 (Invitrogen).

2.3 Cell Treatments

1. Hypoxic chamber.

2. DMOG.

3. JNJ1935 (JNJ-42041935, Johnson & Johnson).

4. DMSO.

2.4 Measuring Gaussia Luciferase

1. 96-well white plates.

2. Plate reader able to read luminescence.

3. *Gaussia* luciferase substrate (coelenterazine).

4. *Cypridina* luciferase substrate (luciferin).

3   Methods

3.1 Making the HRE-GLuc Reporter Construct

1. Digest the DNA sequences and the pGLuc-Mp vector with the restriction enzymes BglII and EcoRI.

2. Run the digested DNA sequences in a 2% agarose gel; cut and gel purify the correct DNA fragments with the gel purification kit.

3. Ligate the DNA sequences to the opened pGLuc-Mp vector using T4 DNA ligase.

4. Transform the ligated DNA into *E. coli* competent cells.

5. Pick colony from agar plate (*see* **Note 4**).

6. Amplify the bacterial clone in Luria broth overnight.

7. Extract the plasmid DNA with the plasmid DNA extraction kit.

8. Measure the amount of DNA obtained using a spectrophotometer (*see* **Note 5**).

9. Have the DNA sequenced to verify correct insertion (*see* **Note 6**).

3.2 Transfection of HRE-GLuc Reporter

1. Seed 1×10^5 HEK293 cells per well of a 24-well culture plates (*see* **Note 7**) the day before transfection (*see* **Note 8**).

2. For transfection of 1 well, mix 100 ng of HRE-GLuc into 100 μL of Optimem (*see* **Note 9**); then add 0.5 μL of Lipofectamine 2000 (*see* **Note 10**).

3. Leave for 30 min at room temperature before adding to the well.

4. Cells should be transfected and ready to use the day after.

3.3 HIF Activation

1. Activation can be achieved physiologically by varying the oxygen tension inside the hypoxic chamber. Inside the hypoxic chamber cells are kept at a normal atmospheric pressure with 5% CO_2 and the desired O_2 and N_2 (e.g., 1% O_2 and 94% N_2). We have successfully monitored coherent changes in HIF transcriptional activity over oxygen tensions ranging from 0.5 to 10% oxygen (unpublished).

2. The culture medium must be pre-equilibrated in the hypoxia chamber overnight to the desired oxygen tension. At the start of the experiment in the hypoxia chamber, remove culture medium, and replace with fresh hypoxia-equilibrated medium (*see* **Note 11**).

3. Alternatively, HIF can be activated pharmacologically. We have used the pan-hydroxylase inhibitor dimethyl-oxaloylglycine (DMOG) and the PHD-selective inhibitor JNJ1935, to show the contribution of the FIH hydroxylase to the dynamics of HIF transcriptional activity (Fig. 2).

4. Pipette 10 μL of the culture medium from each well every 2 h into a 96-well white plate (*see* **Note 12**).

3.4 Data Acquisition

1. Reconstitute the luciferase substrate as recommended by the manufacturer (*see* **Note 13**).

2. Add the substrate to the wells, and read the luminescence in a plate reader (*see* **Note 14**).

3. The values obtained from the plate reader can be transferred to a spreadsheet (such as Microsoft Excel) for data analysis.

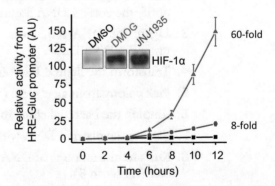

Fig. 2 Effect of hydroxylases inhibition by DMOG or JNJ1935 on HIF-1α transcriptional activity. Relative luciferase activity from HEK293 cells transfected with pHRE-GLuc plasmid and treated with DMOG (1 mM), JNJ1935 (100 μM), or vehicle control, DMSO, after 12 h. Compared to the pan-hydroxylase inhibitor DMOG, the PHD-specific inhibitor JNJ can induce similar HIF stabilization (inset) but much lower HIF transcriptional activity (8-fold vs. 60-fold), thus demonstrating that FIH (inhibited by DMOG but not by JNJ) is a key gatekeeper of HIF transcriptional activity. Data shown is adapted from [8]

3.5 Data Processing and Presentation

1. In order to compare among experiments, the raw data should be normalized by one of the following methods:

   (a) Mathematically by dividing by the sum of the values for the control condition (*see* **Note 15**).

   (b) By dividing by the total protein in each well at the end of 126 the experiment (*see* **Note 16**).

   (c) By dividing by a constitutively active luciferase reporter which does not use the same type of substrate as *Gaussia* luciferase (e.g., firefly or *Cypridina*) (*see* **Note 17**).

2. The data can be represented as either a plot of the relative luciferase that accumulates in the medium over time (Fig. 3a) or as a transcription rate (Fig. 3b). For the latter, the data needs to undergo either a first-order derivation or a linear regression to obtain the rate of transcription (*see* **Note 18**). The luciferase activity that accumulates in the culture medium over time (Fig. 3a) does not reflect what is known about the protein response of HIF-1α, which is transient in nature. The transcription rate (Fig. 3b) provides a better indication of the "activation" state of HIF at a given time point.

4 Notes

1. In this paper, we have customized the basic *Gaussia* luciferase vector by inserting a minimal promoter to create the pGLuc-Mp vector. Alternatively, it is also possible to purchase *Gaussia* vectors with a range of minimal promoters such as mini-TK (from the Herpes simplex virus thymidine kinase promoter).

2. This sequence corresponds to four copies of the promoter sequence of the erythropoietin gene previously shown to bind to HIF [14].

3. We have successfully used the *Gaussia* luciferase reporter assay in a variety of cell types. We here show data from HEK293 cells as an example. Consult the datasheet of the transfection reagents to verify that the desired cells can be efficiently transfected.

4. To rapidly screen a multitude of potential positive clones on the agar plate, we have used part of the bacterial clones as DNA template for RT-PCR. The primers are designed to amplify the region before and after the cloning site. Thus only the DNA from the positive clones will be amplified by the primers.

5. We use the NanoDrop spectrophotometer as it uses a small volume (1 μL) for measuring DNA concentration.

6. There are many free software for analyzing and archiving DNA sequences. We personally use ApE (A plasmid Editor; http://biologylabs.utah.edu/jorgensen/wayned/ape/ or SnapGene (http://www.snapgene.com).

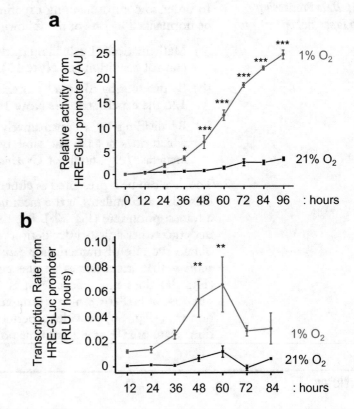

Fig. 3 The HIF-1α transcriptional activity is transiently increased in prolonged hypoxia. HEK293 cells were exposed 21 or 1% oxygen for the indicated time points after transfection with HRE-GLuc construct. (**a**) Relative activity from. HRE-GLuc promoter. (**b**) Transcription rate from HRE-GLuc promoter estimated by calculating the time derivatives of the data presented in (**a**), as described in **Note 18**. *AU* arbitrary units, *RLU* relative light units. Data are represented as mean ± SEM, $n = 5$ independent experiments. *$p < 0.05$, **$p < 0.01$, ***$p < 0.001$

7. If the transcriptional activity needs to be measured for longer than 12–24 h, less cells need to be seeded in the culture dish (with the transfection reagents adjusted accordingly), otherwise the cells will become over-confluent, and *Gaussia* will be released from cells that are dying or have their membrane integrity compromised, leading to a false-positive signal. In addition, the medium will become acidic from the accumulation of metabolites and can compromise the luciferase signal, as the reaction depends substantially on the pH of the medium [12].

8. We normally ensure that there are two wells for each of the conditions to be measured. The measurements from the two wells will be averaged to give a mean value.

9. If using CMV-CLuc as a control, co-transfect 100 ng of HRE-GLuc and 25 ng of CMV-CLuc, and use 0.6 μL Lipofectamine 2000.

10. Other transfection reagents have also been used successfully, including 0.5 μL FuGENE (Roche) for mouse embryonic fibroblasts and 2.5 μL branched polyethylenimine (PEI) for HEK293 cells. PEI used at 1 mg/mL working solution is very inexpensive (Sigma, 408727).

11. The volume we use is 400 μL for each well. This gives good coverage of the cells and is not drastically reduced by repeated sampling over a 12- to 24-h time period. For longer studies up to 72–96 h, one can start with larger volumes, e.g., start with 800–1000 μL, and decrease the initial seeding density, otherwise cell will become over-confluent and acidic.

12. We collect the medium at different time points directly into a 96-well plate and freeze it for later analysis. Activity will be stable for up to a week.

13. The *Gaussia* substrate (coelenterazine) is stored in the freezer. The working solution is quite stable, especially after the inclusion of the stabilizing agent (included in the NEB *Gaussia* substrate pack) and can be stored in the freezer for later use, being stable for a couple of weeks. The *Cypridina* substrate (luciferin) needs to be stored in the −80 °C, and working solutions need to be made fresh for each measurement.

14. If the samples from the experiment span more than one more plate, there should be an inter-plate control to be included in each plate. The values of the inter-plate control should be similar. If they are not, the values need to be adjusted, and the adjustment ratio is also applied to all the measurements from the plate.

15. This is a mathematical method to normalize a set of data [7]. Simply add all the values of the control condition, and use it to divide all the values from the experiment.

16. Wash cells in PBS, lyse with standard buffers (e.g., RIPA or Triton X-100), and perform protein quantification (e.g., detergent compatible DC protein assay, Bio-Rad). Lysates can also be used to measure the expression of HIF-1α and other proteins by Western blot. If hypoxia was used to stabilize HIF-1α, cells must be lysed inside the hypoxic chamber.

17. The constitutively active luciferase reporter must be transfected at the same time as the *Gaussia* luciferase construct. *See* **Note 9** for co-transfection of *Cypridina* and *Gaussia* luciferases. The principle is similar to the standard dual luciferase assay which uses firefly and renilla luciferases.

18. *Gaussia* luciferase accumulates in the culture medium. Thus the rate of *Gaussia* accumulating in the medium is considered to be proportional to the rate of transcription. This rate is obtained by first-order derivation, i.e., calculating the difference between two data points divided by the time interval between the two points. Linear regression is similar, except that this assumes the transcription rate to be linear over the range of time measured.

Acknowledgments

This work was supported by Science Foundation Ireland (grant number 06/CE/B1129).

References

1. Cavadas MAS, Mesnieres M, Crifo B, Manresa MC, Selfridge AC, Keogh CE, Fabian Z, Scholz CC, Nolan KA, Rocha LMA, Tambuwala MM, Brown S, Wdowicz A, Corbett D, Murphy KJ, Godson C, Cummins EP, Taylor CT, Cheong A (2016) REST is a hypoxia-responsive transcriptional repressor. Sci Rep 6:31355. https://doi.org/10.1038/srep31355

2. Elvidge GP, Glenny L, Appelhoff RJ, Ratcliffe PJ, Ragoussis J, Gleadle JM (2006) Concordant regulation of gene expression by hypoxia and 2-oxoglutarate-dependent dioxygenase inhibition—the role of HIF-1 alpha, HIF-2 alpha, and other pathways. J Biol Chem 281(22):15215–15226. https://doi.org/10.1074/jbc.M511408200

3. Dewet JR, Wood KV, Deluca M, Helinski DR, Subramani S (1987) Firefly luciferase gene: structure and expression in mammalian cells. Mol Cell Biol 7(2):725–737

4. Lorenz WW, McCann RO, Longiaru M, Cormier MJ (1991) Isolation and expression of a cDNA encoding Renilla reniformis luciferase. Proc Natl Acad Sci U S A 88(10):4438–4442. https://doi.org/10.1073/pnas.88.10.4438

5. Tannous BA, Kim DE, Fernandez JL, Weissleder R, Breakefield XO (2005) Codon-optimized Gaussia luciferase cDNA for mammalian gene expression in culture and in vivo. Mol Ther 11(3):435–443. https://doi.org/10.1016/j.ymthe.2004.10.016

6. Nakajima Y, Kobayashi K, Yamagishi K, Enomoto T, Ohmiya Y (2004) CDNA cloning and characterization of a secreted luciferase from the luminous Japanese ostracod, Cypridina noctiluca. Biosci Biotech Bioch 68(3):565–570. https://doi.org/10.1271/bbb.68.565

7. Bruning U, Fitzpatrick SF, Frank T, Birtwistle M, Taylor CT, Cheong A (2012) NF kappa B and HIF display synergistic behaviour during hypoxic inflammation. Cell Mol Life Sci 69(8):1319–1329. https://doi.org/10.1007/s00018-011-0876-2

8. Nguyen LK, Cavadas MA, Scholz CC, Fitzpatrick SF, Bruning U, Cummins EP, Tambuwala MM, Manresa MC, Kholodenko BN, Taylor CT, Cheong A (2013) A dynamic model of the hypoxia-inducible factor 1α (HIF-1α) network. J Cell Sci 126(Pt 6):1454–1463. https://doi.org/10.1242/jcs.119974

9. Cavadas MA, Nguyen LK, Cheong A (2013) Hypoxia-inducible factor (HIF) network: insights from mathematical models. Cell Commun Signal 11(1):42. https://doi.org/10.1186/1478-811X-11-42

10. Bruning U, Cerone L, Neufeld Z, Fitzpatrick SF, Cheong A, Scholz CC, Simpson DA, Leonard MO, Tambuwala MM, Cummins EP, Taylor CT (2011) MicroRNA-155 promotes resolution of hypoxia-inducible factor 1 alpha activity during prolonged hypoxia. Mol Cell Biol 31(19):4087–4096. https://doi.org/10.1128/mcb.01276-10

11. Cavadas MA, Mesnieres M, Crifo B, Manresa MC, Selfridge AC, Scholz CC, Cummins EP, Cheong A, Taylor CT (2015) REST mediates resolution of HIF-dependent gene expression in prolonged hypoxia. Sci Rep 5:17851. https://doi.org/10.1038/srep17851

12. Cavadas MA, Cheong A (2014) Monitoring of transcriptional dynamics of HIF and NFκB activities. Methods Mol Biol 1098:97–105. https://doi.org/10.1007/978-1-62703-718-1_8

13. Nguyen LK, Cavadas MA, Kholodenko BN, Frank TD, Cheong A (2015) Species differential regulation of COX2 can be described by an NFκB-dependent logic AND gate. Cell Mol Life Sci 72(12):2431–2443. https://doi.org/10.1007/s00018-015-1850-1

14. Wang GL, Semenza GL (1993) General involvement of hypoxia-inducible factor-1 in transcriptional response to hypoxia. Proc Natl Acad Sci U S A 90(9):4304–4308. https://doi.org/10.1073/pnas.90.9.4304

Chapter 5

Fluorescence Lifetime Imaging Microscopy (FLIM) as a Tool to Investigate Hypoxia-Induced Protein-Protein Interaction in Living Cells

Vera Schützhold, Joachim Fandrey, and Katrin Prost-Fingerle

Abstract

Fluorescence resonance energy transfer (FRET) is widely used as a method to investigate protein-protein interactions in living cells. A FRET pair donor fluorophore in close proximity to an appropriate acceptor fluorophore transfers emission energy to the acceptor, resulting in a shorter lifetime of the donor fluorescence. When the respective FRET donor and acceptor are fused with two proteins of interest, a reduction in donor lifetime, as detected by fluorescence lifetime imaging microscopy (FLIM), can be taken as proof of close proximity between the fluorophores and therefore interaction between the proteins of interest. Here, we describe the usage of time-domain FLIM-FRET in hypoxia-related research when we record the interaction of the hypoxia-inducible factor-1 (HIF-1) subunits HIF-1α and HIF-1β in living cells in a temperature- and CO_2-controlled environment under the microscope.

Key words Fluorescence lifetime imaging microscopy, FLIM, Fluorescence resonance energy transfer, FRET, Live cell imaging, Protein-protein interaction, Hypoxia-inducible factor-1, HIF-1

1 Introduction

Fluorescence lifetime imaging microscopy (FLIM) is a highly useful tool to investigate protein-protein interactions in living cells. Here, the principles of fluorescence resonance energy transfer (FRET) are combined with the measurement of the fluorophore lifetime [1]. If an excited donor fluorophore is spatially close to an acceptor, emission energy is transferred radiation-free and excites the acceptor fluorophore. The fluorescence lifetime of the donor fluorophore (τ_D), which is derived from the exponential decay of fluorescence intensity after excitation with a pulsed laser, decreases with FRET by addition of this non-radiative process to the sum of transition rates which determine fluorescence lifetime [2]. FLIM is a method to quantitatively analyze FRET. Via time-correlated single-photon counting (TCSPC), the fluorophore lifetimes are

L. Eric Huang (ed.), *Hypoxia: Methods and Protocols*, Methods in Molecular Biology, vol. 1742,
https://doi.org/10.1007/978-1-4939-7665-2_5, © Springer Science+Business Media, LLC 2018

detected, and with these information, a FRET fraction α can be calculated [3]. The FRET fraction α represents the relative amount of donor molecules interacting with an acceptor resulting in FRET.

With FLIM, also hypoxia-related protein interactions can be measured. Gene expression in hypoxia is mainly controlled by hypoxia-inducible factor-1 (HIF-1), which is a transcription factor composed of two subunits, HIF-1α and HIF-1β. Under normoxia, HIF-1α is hydroxylated by oxygen-dependent hydroxylases and thereby marked for von Hippel-Lindau protein (pVHL)-mediated proteasomal degradation [4]. Under hypoxia, the hydroxylases become inactive, and HIF-1α can accumulate and migrate into the nucleus where it forms a dimer with the β-subunit to work as a transcription factor [5]. This HIF complex assembly critically depends on protein-protein interaction. Here, we demonstrate and quantify the interaction of HIF-1α and HIF-1β via FLIM-FRET. HIF-1α and HIF-1β were fused with appropriate fluorophores. As the emission spectrum of the donor has to overlap with the excitation spectrum of the acceptor, mCitrine and mCherry are the fluorescent proteins of our choice (Fig. 1a). The selection of the most suitable fluorophores is dependent on their spectral properties, the microscopic setup, and available vectors [6] (*see* **Note 1**). The appropriate cell line and transfection reagent can be chosen depending on the question that is to be solved. In the following, the transfection of the human embryonic kidney cell line HEK293T using Lipofectamine® 3000 with the fluorophore constructs mCitrine-HIF-1α (donor) and mCherry-HIF-1β (acceptor) is described.

2   Materials

Store all reagents and media at 4 °C. All reagents and materials that get in contact with the cells need to be sterile/autoclaved.

2.1   Cell Culturing

1. Cell culture flask (75 cm² bottom area).

2. Full-growth medium/Dulbecco's Modified Eagle's Medium (DMEM; 4.5 g/L D-glucose, 4.5 g/L L-glutamine) supplemented with 100 U/mL penicillin, 100 µg/mL streptomycin, 10% fetal calf serum (FCS), 1 mM pyruvate.

3. Phosphate buffered saline (PBS): 137 mM NaCl, 2.7 mM KCl, 100 mM Na_2HPO_4, 18 mM KH_2PO_4, pH 7.4 (adjusted with HCl).

4. Solution of trypsin: 0.25% trypsin, 1 mM ethylenediaminetetraacetic acid (EDTA) in PBS.

2.2   Transfection

1. Glass-bottom microscopic (multi-well) dishes (thickness, 180 µm).

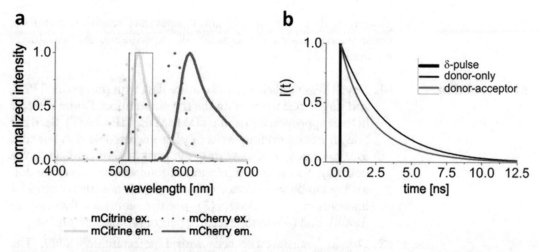

Fig. 1 Spectra of FRET pair and donor lifetime histogram. (**a**) Spectra of a typical FRET pair (peak-normalized): mCitrine as donor and mCherry as acceptor fluorophore, excitation, and emission. As indicated, in our approach, the laser excites the donor at 514 nm (δ-pulse), and the emission was recorded at 524–551 nm. (**b**) Exemplary TCSPC fit for the donor-only (3 ns) and the donor-acceptor lifetime (1.2 ns) resulting from an interaction fraction of 0.6 (Reproduced from [13] with permission from Elsevier)

2. Serum-free medium: full-growth medium/DMEM without FCS.

3. Lipofectamine® 3000 or transfection reagent of choice.

4. Fluorophore-protein constructs: mCitrine-HIF-1α (mCitrine-C1 was a gift from R. Campbell, M. Davidson, O. Griesbeck & R. Tsien; Addgene plasmid #54587), mCherry empty vector (#632524, Clontech, Heidelberg, Germany), mCherry-HIF-1β, mCitrine-mCherry fusion.

2.3 tdFLIM

1. Live cell fluorescence imaging medium (high glucose, 3.7 g/L sodium bicarbonate) supplemented with 100 U/mL penicillin, 100 μg/mL streptomycin, 4 mM L-glutamine.

2. Fluorescein (10 μM).

3. Pulsed laser (40–80 MHz).

4. Laser scanning microscope.

5. TCSPC detector.

6. SymPhoTime software from PicoQuant.

7. Fiji/ImageJ software (http://rsbweb.nih.gov/ij).

3 Methods

Carry out all procedures at 37 °C, if possible. Perform the measurements at 37 °C and 5% CO_2 (a heating system and a gas-mixing device at the microscope are required). Lifetimes of fluorescent protein excited states are pH and temperature sensitive; therefore, constant conditions throughout all experiments are absolutely mandatory [7].

3.1 Transfection

1. Day 1: Wash cells in the cell culture flask with pre-warmed PBS, and singularize them using the trypsin solution. Dilute the cells in their appropriate medium (DMEM for HEK293T). Seed the cells in minimum four wells of your microscopic dish (or seed four microscopic dishes) to get at least 80% of confluence the next day. You need a minimum of four wells to cover the following samples: (1) donor-only sample, (2) negative control for fluorophore interaction, (3) positive control (fluorophore fusion), and (4) donor-acceptor pair of proteins of interest.

2. Day 2: Transfect the cells with Lipofectamine® 3000. The appropriate amount of transfected DNA needs to be determined in preexperiments and depends on the size of the proteins of interest and the expression level in the cells. We transfect about 50 ng of the fluorophore only and fusion constructs and 200 ng of the mCitrine-HIF-1α and the mCherry-HIF-1β constructs in 250 μL medium per well in an 8-well dish (growth area per well, 1.0 cm²). Follow the instructions of the chosen transfection reagent: you will need to dilute the constructs in serum-free medium and incubate them with the transfection reagents.

3. Optional: About 6 h after transfection, replace with fresh full-growth medium (*see* **Note 2**).

3.2 tdFLIM

1. Day 3: Check the cells under a fluorescence microscope. If the fluorophore-protein constructs are already well expressed, you can measure the cells at the FLIM setup (*see* **Note 3**). Otherwise wait for another 12–24 h.

2. Replace the medium in the microscopic dish with pre-warmed imaging medium (*see* **Note 4**).

3. Use a microscope capable of FLIM, like the Leica TCS SP8, with the following settings: resolution of 16 bit and 512×512 dpi, 40× objective, pulsed white light laser excitation at 514 nm for mCitrine and 561 nm for mCherry, detection range of 524–551 nm for mCitrine and 571–671 nm for mCherry emission (*see* **Note 5**), heating system at 37 °C, gas-mixing device at 5% CO_2.

4. First of all, measure the lifetime of the instrumental response function (IRF) with a reference dye as a standard (*see* **Note 6**). You can use fluorescein (about 10 μM) or another fluorophore that fits to your setup. Test the settings in SymPhoTime. Adjust the intensity to approximately 80% of the detector

maximum: change the number of airy disks for the pinhole between 1 and approximately 4 or the laser power (start low in a range of 1–10%).

5. Run the FLIM measurement until 200 counts in the brightest pixel are reached (this number should be sufficient in a homogeneous liquid).

6. Put your donor-only sample under the microscope, and check the heating and gas-mixing devices. Temperature and CO_2 concentration should be stable at the adjusted settings. Focus to the cell layer with the desired objective (*see* **Note 7**).

7. Find an area with some well-transfected cells (*see* **Note 8**). Run a FLIM test (*see* **Note 9**) and potentially zoom into your cells. If the pixel width gets smaller than half of the wavelength of the laser pulse (here smaller than about 200 nm) due to the zooming in, decrease the aspect ratio to 256 × 256 dpi (*see* **Note 10**). Adjust the settings to get 80% of the detector maximum to your image section (change number of airy disks and the laser power, *see* **Note 11**). Try to run the same settings for all cells that you image in one sample. If the intensity varies a lot, change the laser power as an approximately linear setting instead of the airy number.

8. Run the FLIM measurement with a higher speed than the test and an end point of 500 counts for the brightest pixel. If you measure several cells at once and they have a big difference in their brightness, set 1000 counts for the brightest pixel as the end point to make sure that all cells will have enough counts per pixel for single-cell analysis.

9. Perform a sequential scan to record donor and acceptor fluorescence of the measured section. Avoid overexposure and change laser power if necessary. These images are necessary for intensity weighted interaction maps, since the count images from the FLIM analysis cannot be compared quantitatively.

10. Collect data until you have at least ten cells for your analysis. This number depends strongly on the experiment and might be much higher.

11. Perform measurements for all four conditions (donor-only, negative control, positive control, interaction sample) with at least ten cells for the analysis each (*see* **Note 12**).

3.3 Analysis

1. Save the data from the SymPhoTime software. You will have three files of each measurement that contain the information for the fluorophore lifetimes. The lifetime τ_D can be derived from the TCSPC data using a mono-exponential fit on donor-only sample histograms and τ_{DA} using a double-exponential decay

model for the donor-acceptor samples (Fig. 1b). This calculation can be done by, e.g., the SymphoTime software. You can use these results to calculate the FRET efficiency E (Eq. 1), which can be derived from the lifetime of the donor alone and the lifetime of donor and acceptor in complex (τ_{DA}) [8].

$$E = 1 - \frac{\tau_{DA}}{\tau_D} \qquad (1)$$

As the lifetimes of the donor-only and the donor-acceptor pair are globally constant, we continue with calculating the FRET fraction α for each pixel (Eq. 2) (*see* **Note 13**). Thus, we can analyze the results quantitatively, which means that α represents the actual fraction of interacting molecules.

$$I(t) = I_0\, e^{-\frac{t}{\tau_{AVG}}} = (1-\alpha)\, e^{-\frac{t}{\tau_D}} + \alpha\, e^{-\frac{t}{\tau_{DA}}} \qquad (2)$$

2. Apply a threshold on the counts per pixel map (*see* **Note 14**) to create a mask for the image analysis. Afterward, all background pixels should have no value (not a number, NaN).

3. Draw your ROI containing one cell for analysis, and calculate the mean α-value for this particular cell (*see* **Note 15**).

4. Calculate the mean α-values of all cells of one condition, and compare donor-only, negative control, positive control, and your interaction sample. You will realize that the highest α-fraction that can be obtained is about 0.6 (positive control, *see* **Note 16**). According to this, positive interaction results for your proteins of interest can result in an α-value range of approximately 0.1–0.6 (Fig. 2b).

4   Notes

1. There are several spectra viewers available online to help you find the most appropriate FRET pair for your experiment. Consider the efficiency of excitation using your available laser lines and emission filters, as well as fluorescent protein brightness, stability, and availability. In any case, the donor fluorophore needs to be monomeric.

2. The transfection reagent may be aggressive against your cells and needs to be removed after the transfection is completed. In addition, removal of cell debris reduces noise in your experiment.

3. Normally you can measure 24–72 h after transfection.

4. Imaging medium needs to lack phenol red that could influence your measurements due to its fluorescent properties [9].

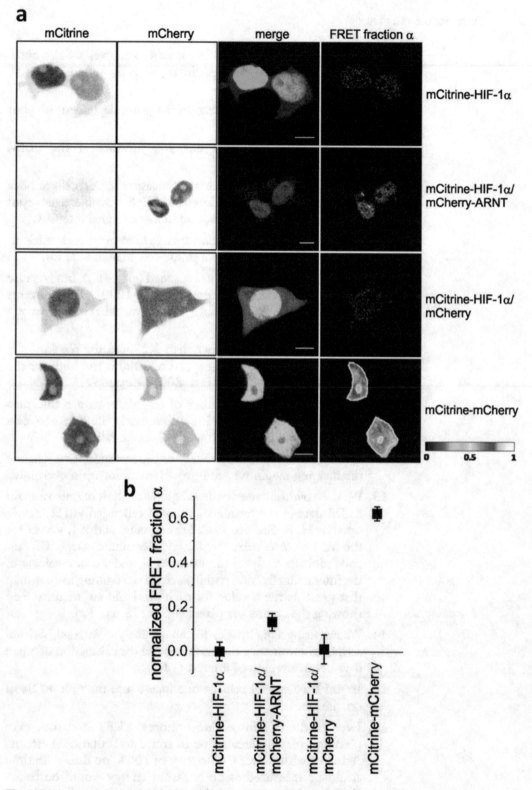

Fig. 2 FLIM measurement to visualize HIF-1α and HIF-1β (ARNT) interaction. (**a**) Intensity images and alpha maps of the donor-only mCitrine-HIF-1α, the co-transfection of mCitrine-HIF-1α and mCherry-HIF-1β, the donor and the acceptor only as negative control, and the direct donor-acceptor fusion as positive control. Scale bar 10 μm. (**b**) Resulting FRET fractions α for all four samples from (**a**) (Reproduced from [13] with permission from Elsevier)

Keep in mind that FCS, BSA, or other additives are also auto-fluorescent. If you need additives, keep them constant in all samples.

5. Laser wavelength and detection range are dependent on your chosen fluorophores.

6. Measure your IRF each day you measure at the FLIM microscope.

7. The 40× objective is sufficient to measure several cells at once without losing too much intensity. High NA objectives result in longer acquisition times, due to loss of signal intensity.

8. Measuring multiple cells at once saves time and is cost-efficient. You are still able to perform single-cell analysis later on.

9. Run the FLIM test with a lower speed (e.g., 400), but increase the speed for the measurement (e.g., 1000). This prevents pileup effects at the detector, which would result in wrong lifetime determination.

10. If the zoom factor is higher than 2.25 and the resolution is 512 × 512 dpi, a scanning speed of 1000 is too high for the Leica SP8 scanner. Adjust it to 800, for example.

11. Do not increase the number of airy disks to a z-thickness higher than 5 μm, which is approximately the thickness of a cell. This would result in low signal/noise ratio.

12. If two or more days are needed for the measurements, change the imaging medium to full-growth medium again overnight.

13. We use a python-based code for global analysis of time-domain FLIM data in the frequency domain called jediFLIM, developed by H. E. Grecco, K. C. Schuermann, and P. J. Verveer at the MPI of Molecular Physiology, Dortmund, Dept. II. This code globally analyzes the lifetimes τ_D and τ_{DA} and calculates the interacting fraction within each pixel, resulting in an α-map that presents the α-value for each pixel and an intensity map showing the counts per pixel (Fig. 2a) [3, 10–12].

14. We are using Fiji/ImageJ for image analysis with self-written scripts to automatize thresholding and the calculation of mean α-values in a region of interest (ROI).

15. If you have several cells in one image, use multiple ROIs in one image.

16. Even in directly fused fluorophores, FRET does not take place in every molecule due to steric and rotational effects. Actually, with a FRET efficiency of 100%, no donor lifetime could be measured as all emission energy would be transferred to the acceptor, and we derive the lifetime from donor emission only.

References

1. Sun Y, Periasamy A (2015) Localizing protein-protein interactions in living cells using fluorescence lifetime imaging microscopy. Methods Mol Biol 1251:83–107. https://doi.org/10.1007/978-1-4939-2080-8_6

2. Lakowicz JR (2013) Principles of fluorescence spectroscopy, 3rd edn. Springer, New York

3. Digman MA, Caiolfa VR, Zamai M, Gratton E (2008) The phasor approach to fluorescence lifetime imaging analysis. Biophys J 94(2):L14–L16. https://doi.org/10.1529/biophysj.107.120154

4. Jaakkola P, Mole DR, Tian YM, Wilson MI, Gielbert J, Gaskell SJ, von Kriegsheim A, Hebestreit HF, Mukherji M, Schofield CJ, Maxwell PH, Pugh CW, Ratcliffe PJ (2001) Targeting of HIF-alpha to the von Hippel-Lindau ubiquitylation complex by O2-regulated prolyl hydroxylation. Science 292(5516):468–472. https://doi.org/10.1126/science.1059796

5. Wang GL, Jiang BH, Rue EA, Semenza GL (1995) Hypoxia-inducible factor 1 is a basic-helix-loop-helix-PAS heterodimer regulated by cellular O2 tension. Proc Natl Acad Sci U S A 92(12):5510–5514

6. Shaner NC, Steinbach PA, Tsien RY (2005) A guide to choosing fluorescent proteins. Nat Methods 2(12):905–909. https://doi.org/10.1038/nmeth819

7. Walther KA, Papke B, Sinn MB, Michel K, Kinkhabwala A (2011) Precise measurement of protein interacting fractions with fluorescence lifetime imaging microscopy. Mol Biosyst 7(2):322–336. https://doi.org/10.1039/c0mb00132e

8. Verveer PJ (2014) Advanced fluorescence microscopy: methods and protocols. Springer, New York

9. Ettinger A, Wittmann T (2014) Fluorescence live cell imaging. Methods Cell Biol 123:77–94. https://doi.org/10.1016/b978-0-12-420138-5.00005-7

10. Verveer PJ, Squire A, Bastiaens PI (2000) Global analysis of fluorescence lifetime imaging microscopy data. Biophys J 78(4):2127–2137. https://doi.org/10.1016/S0006-3495(00)76759-2

11. Clayton AH, Hanley QS, Verveer PJ (2004) Graphical representation and multicomponent analysis of single-frequency fluorescence lifetime imaging microscopy data. J Microsc 213(Pt 1):1–5

12. Grecco HE, Roda-Navarro P, Verveer PJ (2009) Global analysis of time correlated single photon counting FRET-FLIM data. Opt Express 17(8):6493–6508

13. Prost-Fingerle K, Hoffmann MD, Schutzhold V, Cantore M, Fandrey J (2017) Optical analysis of cellular oxygen sensing. Exp Cell Res 356(2):122–127. https://doi.org/10.1016/j.yexcr.2017.03.009

Transcriptional Profiling Using RNA-Seq to Study Hypoxia-Mediated Gene Regulation

Surendra K. Shukla, Ryan J. King, and Pankaj K. Singh

Abstract

Exposing cells to a hypoxic environment leads to significant physiological and molecular alterations. Most of the hypoxic responses are regulated by the transcription factors known as hypoxia-inducible factors (HIFs). HIF1, a heterodimer of hypoxia-stabilized subunit HIF-1alpha and a constitutively expressed subunit HIF-1beta, serves as a key transcription factor that regulates gene expressions which are involved in cell growth, metabolism, and proliferation. The global expression patterns can be analyzed by utilizing RNA-Seq to understand the cellular alterations in hypoxia. This technique enables us to understand the comprehensive regulation of gene expression by specific factors or environmental stimuli. Here, we describe the complete process of studying hypoxia-mediated gene expression by using RNA-Seq, including the hypoxic treatment of cells, RNA isolation, RNA quality check, cDNA library preparation, and library quality check.

Key words Hypoxia, RNA-Seq, HIF, Gene expression, High-throughput sequencing

1 Introduction

Hypoxia, defined as a deficiency of oxygen in the cellular environment, leads to several molecular and physiological adaptations. It activates a variety of complex pathways that help in establishing oxygen homeostasis [1]. This is achieved through the stabilization of hypoxia-inducible factors (HIFs), which function as key transcription factors to regulate the expression of multiple genes involved in cellular growth, proliferation, and metabolism [2]. Following hypoxic induction, HIFs selectively bind to hypoxia response elements (HRE) in promoters and regulate the transcriptional activity [3]. Due to the considerable range of transcriptional reprograming by HIFs under hypoxia, studying the wide range of expression alterations under hypoxic conditions can yield insight into the cellular and molecular components of the cell is altering to adapt to the stress.

L. Eric Huang (ed.), *Hypoxia: Methods and Protocols*, Methods in Molecular Biology, vol. 1742,
https://doi.org/10.1007/978-1-4939-7665-2_6, © Springer Science+Business Media, LLC 2018

The revolutionary advancements of next-generation sequencing (NGS) have fostered research to new heights and rapidly advanced the understanding of the transcriptional and genetic basis of diseases. Whole-transcriptome analysis coupled with cost reductions has allowed RNA sequencing (RNA-Seq) to spread and flourish across multiple fields. Transcriptome analysis can also be performed at the level of single cells owing to the advancements NGS has made in the last decade. RNA sequencing is a high-throughput transcriptomic analysis technique, which provides broader dynamic range and increased specificity over other methods of global gene expression analysis [4]. RNA-Seq provides discovery as well as well quantification of novel transcripts which makes it a powerful tool to study gene regulation [5]. HIF-1alpha is a key transcription factor that regulates transcription under hypoxic conditions and regulates overall metabolic reprogramming by regulating a number of metabolic genes [6–9]. RNAseq has been utilized to identifying such transcriptional changes mediated by HIF-1alpha [7]. To begin the process of RNA-Seq, with oligo-dT primers, a pool of RNA is converted to cDNA library, which is further fragmented by enzymatic digestion and subsequently ligated to adaptors. Further amplification of adapter-ligated fragments is performed before being assessed for quality and subjected to high-throughput sequencing.

In the current method below, we have described the detailed process of generating genome-wide gene expression analysis in response to hypoxia by utilizing RNA sequencing. An overview of the processes has been presented in Fig. 1.

2   Materials

1. Hypoxia chamber or hypoxia incubator.

2. TRIzol reagent (Invitrogen).

3. RNaseZap (Invitrogen).

4. RNase-free tubes.

5. RNase-free molecular grade water.

6. RNeasy Qiagen RNA isolation kit (Qiagen).

7. TruSeq RNA Library Prep Kit v2 (Illumina).

8. 96-well 0.3 mL PCR plate.

9. Adhesive seals.

10. SuperScript II reverse transcriptase (Invitrogen).

11. Ethanol (Molecular Biology Grade).

12. AMPure XP beads (Beckman Coulter).

13. NanoDrop (Thermo Scientific).

14. 2100 Bioanalyzer (Agilent).

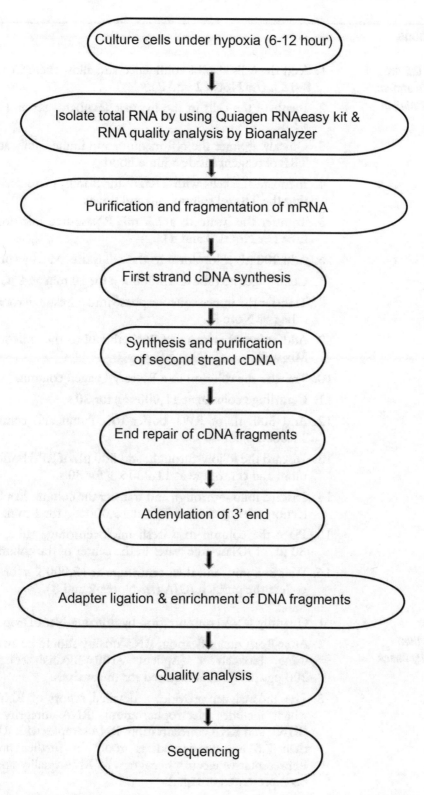

Fig. 1 An overview of the RNA-Seq method. Flowchart of sequential steps involved in RNA-Seq methods

3 Methods

3.1 Cell Culture, Hypoxia Treatment, and RNA Isolation

1. Seed the cells at 60% confluence and allow them to attach for 8–12 h (*see* **Notes 1** and **2**).

2. Incubate the cells in the hypoxic incubator set at 1% O_2 for 12 h.

3. Quickly aspirate the cell medium and immediately add 1 mL TRIzol reagent inside a fume hood.

4. Incubate the cells with TRIzol for 5 min, periodically pipetting the TRIzol solution up and down.

5. Transfer the lysate to a 1.5 mL RNase-free microcentrifuge tube (*see* **Notes 3** and **4**).

6. Add 200 μL chloroform to the cell lysate. Mix by vortexing.

7. Centrifuge the tube at 15,000 × *g* for 10 min at 4 °C.

8. Transfer the upper aqueous phase into a fresh microcentrifuge tube (*see* **Note 5**).

9. Add an equal amount of 70% ethanol to the aqueous phase. Mix well.

10. Transfer the solution to a RNeasy Qiagen column.

11. Centrifuge column at 11,000 × *g* for 30 s.

12. Add 500 μL of RW1 buffer to column and centrifuge at 11,000 × *g* for 30 s.

13. Discard the follow-through, add 500 μL of RPE buffer to column, and centrifuge at 11,000 × *g* for 30 s.

14. Discard follow-through and transfer the column to a fresh collection tube, and centrifuge at 15,000 × *g* for 1 min.

15. Place the column in a fresh microcentrifuge tube, and add 30 μL of RNase-free water in the center of the column.

16. Wait 3–5 min, and then centrifuge at 15,000 × *g* for 1 min to collect the purified RNA (*see* **Notes 7** and **8**).

3.2 RNA Quantification and Quality Check

1. Quantify RNA concentration by utilizing NanoDrop.

2. After RNA quantification, RNA quality should be analyzed by using Bioanalyzer (Agilent 2100 Bioanalyzer). Around 200 pg/μL RNA is required for the analysis.

3. The Bioanalyzer provides a detailed report of RNA quality which includes electropherogram, RNA integrity number (RIN), and RNA concentration. RNA samples with RIN above than 7.5 are considered as good for further processing. Representative electropherogram of RNA quality analysis has been presented in Fig. 2.

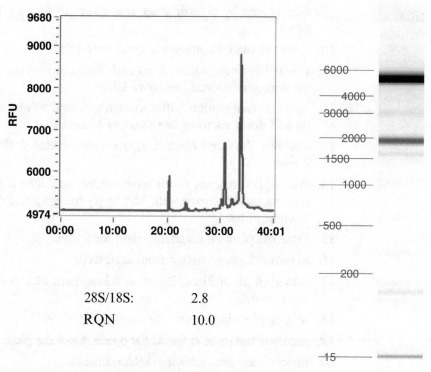

28S/18S: 2.8
RQN   10.0

Fig. 2 Representative electropherogram of RNA quality analysis. Total RNA was isolated from cultured mammalian cells and RNA quality analysis was performed using Bioanalyzer

Further steps like cDNA library preparation and adopter ligation and amplification will be performed by using Illumina TruSeq® RNA kit. Procedures are described below.

3.3 Purification and Fragmentation of mRNA

1. Dilute RNA samples in nuclease-free water (100 ng/µL) to final volume of 50 µL in a new 96-well 0.3 mL PCR plate.

2. Add 50 µL of RNA purification beads to each well of the plate to bind the polyA RNA to the oligo-dT beads. Mix it by pipetting up and down (*see* **Note 9**).

3. Seal the plate and facilitate the binding of mRNA to beads by incubating at 65 °C for 5 min in a thermocycler to denature the mRNA (*see* **Note 10**).

4. After denaturation, incubate the plate at room temperature to bind the beads to mRNA.

5. Remove the adhesive seal from the plate.

6. Place the plate on magnetic stand for 5 min to separate the polyA RNA bound magnetic beads (*see* **Note 11**).

7. Discard the supernatant from each well of the plate, and remove the plate from magnetic stand.

8. Add 200 µL of bead washing buffer to each well (*see* **Note 12**).

9. Gently mix by pipetting up and down around six to eight times.

10. Place the plate on magnetic stand for 5 min.

11. Discard the supernatant from each well and remove the plate from magnetic stand (*see* **Note 13**).

12. Add 50 μL of elution buffer to each well, and mix by pipetting up and down six to eight times (*see* **Note 14**).

13. Seal the plate and place it into thermal cycler at 80 °C for 2 min.

14. When plate reaches room temperature, add 50 μL of bead-binding buffer to each well. Mix by pipetting up and down six to eight times.

15. Place the plate on magnetic stand for 5 min.

16. Remove the supernatant from each well.

17. Add 19.5 μL of Elute, Prime, and Fragment Mix to each well of the plate.

18. Mix by pipetting up and down about six times.

19. Incubate the plate at 94 °C for 8 min. Cool the plate to 4 °C.

20. Immediately proceed for cDNA synthesis.

3.4 Synthesis of First-Strand cDNA

1. Place the plate on magnetic stand.

2. Take 17 μL of the supernatant (fragmented and primed mRNA) in 96-well 0.3 mL PCR plate.

3. Mix the First-Strand Master Mix with SuperScript II in a 9:1 ratio.

4. Mix thoroughly and pulse centrifuge.

5. Add 8 μL of mix to each tube. Mix by gently pipetting up and down six to eight times.

6. Seal the plate with adhesive tape.

7. Place the sealed plate to thermal cycler and run the following program:

   (a) Preheat lid (100 °C)

   (b) 25 °C for 10 min

   (c) 42 °C for 50 min

   (d) 70 °C for 15 min

   (e) Hold at 4 °C

   (*see* **Note 15**).

8. When the reaction is complete, remove the plate and proceed for second-strand cDNA synthesis.

3.5 Synthesis of Second-Strand cDNA

1. Add 25 µL of Second-Strand Master Mix to each well. Gently mix by pipetting up and down.
2. Seal the plate with adhesive seal.
3. Place the plate on the thermal cycler and incubate at 16 °C for 1 h.
4. When the reaction is complete, place the plate at room temperature.

3.6 Purification of cDNA

1. Add 90 µL of AMPure XP beads to each ds cDNA sample. Mix gently.
2. Incubate the plate at room temperature for 15 min.
3. Place the cDNA plate to the magnetic stand for 5 min. Make sure that all the beads are bound to the side of the well.
4. Discard the supernatant from each well.
5. Add freshly prepared 80% ethanol to each tube without disturbing the beads.
6. Incubate the plate at room temperature for 1 min, and then remove the supernatant.
7. Repeat **steps 5** and **6** one more time.
8. Dry the plate at room temperature for 15–20 min, and then remove it from magnetic stand (*see* **Note 16**).
9. Add 52.5 µL of resuspension buffer to each sample. Mix gently by pipetting up and down eight to ten times.
10. Incubate the plate at room temperature for 2 min.
11. Place the plate on the magnetic stand for 5 min at room temperature.
12. Transfer the 50 µL of supernatant to a fresh PCR plate.

(Procedure can be stopped at this point. Plate can be stored at −20 °C).

3.7 End Repair of cDNA Fragments

1. Add 10 µL of resuspension buffer to each cDNA sample.
2. Add 40 µL of End-Repair Mix to each well.
3. Mix gently by pipetting up and down around ten times.
4. Seal the plate with adhesive seal.
5. Incubate the plate at 30 °C in thermal cycler for 30° min.
6. Remove the plate from the thermal cycler and place it at room temperature.
7. Add 160 µL of AMPure XP beads to each well in the plate. Mix gently with pipette.
8. Incubate the plate at room temperature for 5 min.

9. Place the plate in the magnetic stand at room temperature for 5 min or more until the liquid becomes clear.

10. Discard the supernatant from each sample.

11. While the plate is on the magnetic stand, add 200 µL of freshly prepared 80% ethanol to each sample without disturbing the beads.

12. Incubate the plate at room temperature for 1 min. Remove the supernatant.

13. Repeat **steps 11** and **12**.

14. Let the plate dry at room temperature for 15 min and then remove the plate from the magnetic stand.

15. Add 17.5 µL of suspension buffer to each well of the plate. Mix gently and incubate at room temperature for 2 min.

16. Place the plate in the magnetic stand for 5 min at room temperature.

17. Transfer 15 µL of supernatant to a fresh PCR plate.

3.8 Adenylation of 3′ End

1. Add 12.5 µL of A-Tailing Mix to each sample. Gently mix by pipetting up and down.

2. Seal the plate with an adhesive seal.

3. Incubate the plate in the thermal cycler with following program:

   (a) Preheat lid at 100 °C

   (b) 37 °C for 30 min

   (c) 70 °C for 5 min

   (d) Hold at 4 °C

4. When thermal cycler's temperature reaches 4 °C, remove the plate.

3.9 Adapter Ligation

1. Add 2.5 µL of ligation mix to each sample (*see* **Note 17**).

2. Add 2.5 µL of RNA adapter index to each well. Gently mix by pipetting up and down (*see* **Note 17**).

3. Seal the plate with an adhesive seal.

4. Incubate the plate at 30 °C for 10 min in the thermal cycler.

5. Remove the plate from the thermal cycler, remove the adhesive seal, and add 5 µL of stop ligation buffer to each sample. Gently mix the contents by pipetting up and down.

6. Add 42 µL of AMPure XP beads to each sample and mix gently.

7. Incubate the plate at room temperature for 15 min.

8. Place the plate at room temperature for 5 min or until the liquid becomes clear.

9. Remove the supernatant from each well.

10. Add 200 μL of 80% ethanol to each sample without disturbing the beads.

11. Leave the plate at room temperature for 1 min.

12. Remove the supernatant and repeat **steps 10** and **11**.

13. Air-dry the samples at room temperature for 15 min.

14. Remove the plate from magnetic stand, and add 52.5 μL of suspension buffer to each sample. Gently mix by pipetting up and down 10–12 times.

15. Incubate the plate at room temperature for 2 min.

16. Place the plate on the magnetic stand for 5 min at room temperature.

17. Transfer 50 μL of supernatant to a fresh PCR plate without disturbing the beads.

18. Add 50 μL of AMPure XP beads to each well for a sec cleanup. Gently mix the beads and the sample by pipetting up and down.

19. Incubate the plate at room temperature for 15 min.

20. Place the plate on the magnetic stand and incubate at room temperature for 5 min.

21. Remove 95 μL of supernatant from each sample without disturbing the beads.

22. Add 200 μL of freshly prepared 80% ethanol to each sample without disturbing the beads.

23. Incubate the plate at room temperature for 1 min.

24. Remove all the supernatant and repeat **steps 22** and **23**.

25. Remove the supernatant from the plate, and air-dry the samples at room temperature for 15 min.

26. Remove the plate from magnetic stand, and add 22.5 μL of resuspension buffer to each well of the CAP plate. Gently mix by pipetting up and down.

27. Place the plate on the magnetic stand at room temperature for 5 min.

28. Transfer the clear supernatant to a fresh PCR plate.

3.10 Enrichment of DNA Fragments

1. Add 5 μL of PCR Primer Cocktail and 25 μL of PCR Master Mix to each well in the plate, mix gently, seal the plate with an adhesive seal, and place in the thermocycler with the following program:

(a) Preheat lid at 100 °C

(b) 98 °C for 30 s

(c) 15 cycles of

(d) 98 °C for 10 s

(e) 60 °C for 30 s

(f) 72 °C for 30 s, followed by

(g) 72 °C for 5 min (after 15 cycles)

(h) Hold at 10 °C

2. Add 50 μL of AMPure XP beads to each sample in the PCR plate with 50 μL of PCR-amplified library. Gently mix by pipetting up and down eight to ten times.

3. Incubate plate at room temperature for 15 min.

4. Place the plate on the magnetic stand and incubate at room temperature for 5 min.

5. Remove supernatant from each well of the plate.

6. Add 200 μL of freshly prepared 80% ethanol to each sample without disturbing the beads.

7. Incubate the plate at room temperature for 1 min and then discard all the supernatant.

8. Repeat **steps 6** and **7** one more time.

9. Air-dry the samples at room temperature for 15 min, and then remove the plate from the magnetic stand.

10. Resuspend the dried pellet in 32.5 μL of resuspension buffer. Gently mix the contents by pipetting up and down eight to ten times.

11. Incubate the PCR plate at room temperature for 2 min.

12. Place the plate in the magnetic stand and incubate for 5 min at room temperature.

13. Transfer the 30 μL of supernatant into a fresh PCR plate.

3.11 Quality Analysis

1. Add 1 μL of purified product to the Bioanalyzer.

2. Check the size and purity of purified DNA sample (*see* **Note 18**).

3.12 Sequencing

1. Prepared samples will be utilized on a high-throughput sequencing by using HiSeq 2000 (Illumina).

4   Notes

1. Before putting cells in a hypoxia chamber/incubator, ensure that the cells are properly attached. If they are not attached properly, they may die.

2. Add TRIzol immediately after taking out cells from the hypoxia incubator. If you want to isolate RNA later, incubate the cells in TRIzol for 5 min and store at −80 °C.

3. Use RNaseZap (Invitrogen) to clean pipette sets and workbench before starting RNA isolation. Always wear gloves and try to avoid any kind of contact with tips, pipette, etc. Always use disposable RNase-free plastic wares. All the reagents should be prepared in RNase-free components.

4. Always use RNase-free tubes and tips.

5. Avoid touching the interface between TRIzol and the upper aqueous phase.

6. Generally, 1–10 μg of RNA should be used as starting material.

7. It is good to have two set of RNA samples, one for quality analysis and one for sequencing procedures.

8. Avoid repeated freezing and thawing of RNA samples. It's better to aliquot the samples and store at −80 °C.

9. Thaw all the Illumina sequencing kit components on ice.

10. PCR plate should be properly sealed with an adhesive seal.

11. Always allow the beads to come to room temperature before use. Mix them thoroughly to make a homogenous solution. Provide sufficient time for the binding of all beads to the magnetic stand.

12. Beads should be thoroughly suspended in suspension buffer/binding buffer.

13. Removing the supernatant should be performed without disturbing the beads.

14. During the mixing process, avoid generating aerosol to avoid cross contamination.

15. Set up the thermal cycler before starting the reaction.

16. Samples should be not over dried.

17. Check the liquid volume very carefully during pipetting. If liquid is viscous, then it needs more precautions.

18. Size should be around 260 bp.

References

1. Kenneth NS, Rocha S (2008) Regulation of gene expression by hypoxia. Biochem J 414(1):19–29. https://doi.org/10.1042/BJ20081055

2. Bruick RK (2003) Oxygen sensing in the hypoxic response pathway: regulation of the hypoxia-inducible transcription factor. Genes Dev 17(21):2614–2623. https://doi.org/10.1101/gad.1145503

3. Semenza GL (2014) Oxygen sensing, hypoxia-inducible factors, and disease pathophysiology. Annu Rev Pathol 9:47–71. https://doi.org/10.1146/annurev-pathol-012513-104720

4. Wang Z, Gerstein M, Snyder M (2009) RNA-Seq: a revolutionary tool for transcriptomics. Nat Rev Genet 10(1):57–63. https://doi.org/10.1038/nrg2484

5. Conesa A, Madrigal P, Tarazona S, Gomez-Cabrero D, Cervera A, McPherson A, Szczesniak MW, Gaffney DJ, Elo LL, Zhang X, Mortazavi A (2016) A survey of best practices for RNA-seq data analysis. Genome Biol 17:13. https://doi.org/10.1186/s13059-016-0881-8

6. Surendra K. Shukla, Vinee Purohit, Kamiya Mehla, Venugopal Gunda, Nina V. Chaika, Enza Vernucci, Ryan J. King, Jaime Abrego, Gennifer D. Goode, Aneesha Dasgupta, Alysha L. Illies, Teklab Gebregiworgis, Bingbing Dai, Jithesh J. Augustine, Divya Murthy, Kuldeep S. Attri, Oksana Mashadova, Paul M. Grandgenett, Robert Powers, Quan P. Ly, Audrey J. Lazenby, Jean L. Grem, Fang Yu, José M. Matés, John M. Asara, Jung-whan Kim, Jordan H. Hankins, Colin Weekes, Michael A. Hollingsworth, Natalie J. Serkova, Aaron R. Sasson, Jason B. Fleming, Jennifer M. Oliveto, Costas A. Lyssiotis, Lewis C. Cantley, Lyudmyla Berim, Pankaj K. Singh (2017) MUC1 and HIF-1alpha Signaling Crosstalk Induces Anabolic Glucose Metabolism to Impart Gemcitabine Resistance to Pancreatic Cancer. Cancer Cell 32(1):71–87.e7

7. Kamiya Mehla, Pankaj K. Singh (2014) MUC1: A novel metabolic master regulator. Biochimica et Biophysica Acta (BBA) - Reviews on Cancer 1845(2):126–135

8. N. V. Chaika, T. Gebregiworgis, M. E. Lewallen, V. Purohit, P. Radhakrishnan, X. Liu, B. Zhang, K. Mehla, R. B. Brown, T. Caffrey, F. Yu, K. R. Johnson, R. Powers, M. A. Hollingsworth, P. K. Singh (2012) MUC1 mucin stabilizes and activates hypoxia-inducible factor 1 alpha to regulate metabolism in pancreatic cancer. Proceedings of the National Academy of Sciences 109(34):13787–13792

9. Nina V. Chaika, Fang Yu, Vinee Purohit, Kamiya Mehla, Audrey J. Lazenby, Dominick DiMaio, Judy M. Anderson, Jen Jen Yeh, Keith R. Johnson, Michael A. Hollingsworth, Pankaj K. Singh (2012) Differential Expression of Metabolic Genes in Tumor and Stromal Components of Primary and Metastatic Loci in Pancreatic Adenocarcinoma. PLoS ONE 7(3):e32996

Chapter 7

Chromatin Immunoprecipitation of HIF-α in Breast Tumor Cells Using Wild Type and Loss of Function Models

Danielle L. Brooks and Tiffany N. Seagroves

Abstract

Chromatin immunoprecipitation (ChIP) is a powerful method to determine whether a protein of interest binds to specific regulatory elements of the genome. Herein, we outline protocols optimized to detect binding of Hypoxia-Inducible Factor (HIF)-1α or HIF-2α to putative hypoxia response elements (HREs) within HIF target genes expressed in breast tumor epithelial cells.

Key words Hypoxia, Hypoxia response element (HRE), Hypoxia-Inducible Factor (HIF), Chromatin, Cross-linking, Immunoprecipitation, Input DNA, Gene deletion, Breast cancer, Polyomavirus middle T (PyMT)

1 Introduction

Identification of the cohort of genes directly regulated by a transcription factor network is key to understanding the biological activity of individual transcription factor complexes and downstream control of cellular processes. Chromatin immunoprecipitation (ChIP) was developed in the 1980s by David Gilmour while in the laboratory of John T. Lis to study interactions of RNA polymerase with genes in *Drosophila melanogaster* [1–3]. ChIP is now a routine protocol to understand the interactions of proteins with DNA. In cross-linking ChIP protocols (X-ChIP), proteins are first cross-linked to DNA using formaldehyde, which produces reversible cross-links. The DNA is then isolated and sheared by sonication into 200–1000 bp fragments. Sheared chromatin is then subjected to immunoprecipitation with an antibody of interest, and the cross-links are reversed. DNA fragments that were bound to the antibody are purified and are next subjected to quantitative, real-time PCR (qRT-PCR) using primers flanking known or putative binding elements. The data are then quantitated to calculate enrichment of protein binding to that sequence relative to binding observed for input DNA (sheared but not immunoprecipitated).

L. Eric Huang (ed.), *Hypoxia: Methods and Protocols*, Methods in Molecular Biology, vol. 1742,
https://doi.org/10.1007/978-1-4939-7665-2_7, © Springer Science+Business Media, LLC 2018

ChIP assays are a gold standard in the molecular biologist's toolkit in verifying whether a gene of interest, typically identified through a screen for differentially expressed genes (by microarray or RNA-Seq), is a bona fide transcription factor target or whether putative binding sites identified through a bioinformatic approach (scanning for consensus transcription factor binding sites) are functional. ChIP is informative to calculate the qualitative enrichment of a protein, or specific protein modifications, such as histone methylation, in chromatin at specific DNA sites within a composite gene regulatory element. ChIP can also be used to query how occupancy at each individual binding site fine-tunes individual gene expression levels and/or promoter activity. When followed by next generation sequencing, ChIP-Seq, ChIP is also useful to determine the relative abundance of a protein or specific protein modifications across the genome.

It is well known that tumor hypoxia rewires cellular metabolism, promotes invasion and metastasis, and contributes to therapeutic resistance through activation of a hypoxic response orchestrated by the oxygen-sensitive Hypoxia-Inducible Factor (HIF) transcription factors [4]. HIF stability is primarily regulated at the posttranslational level. In response to decreasing oxygen tensions, the alpha subunits of HIF-1 and HIF-2, HIF-1α and HIF-2α, are stabilized through inactivation of a family of prolyl hydroxylases (PHDs) that at normal oxygen tensions function to hydroxylate two key proline residues in the C-terminal oxygen-dependent degradation (ODD) domain. These modifications are required to mediate HIF-α interactions with von Hippel-Lindau (VHL) protein, which ubiquitinates the alpha subunits, targeting their destruction through the proteasome [5]. Unhydroxylated HIF-α subunits can dimerize with aryl hydrocarbon nuclear receptor translocator (ARNT), forming the functional HIF-1 or HIF-2 transcription factors, which bind to hypoxia response elements (HREs). The canonical HRE consensus sequence is 5′-RCGTG 3′ [6], which was refined by analysis of global ChIP-Seq data generated in MCF-7 breast cancer cells that demonstrated a sequence preference for A over G at the R position [7].

More than 100 target genes or microRNAs have been identified as regulated by the HIFs; several are implicated in metastasis [8]. The majority of HIF targets in breast cancer cells may be regulated by either HIF-1 or HIF-2, with some unique target genes identified [9, 10]. Although the majority of HIF-dependent genes are upregulated by hypoxia, a subset of genes are also repressed. The relative contribution of HIF-1α vs. HIF-2α in tumorigenesis is also highly tissue-type context dependent. For example, in breast cancer, HIF-1α has been shown to be required for efficient primary tumor growth and for lung metastasis [11, 12], whereas in renal clear cell carcinomas, HIF-2α is thought to play a more prominent role in tumor progression and aggressiveness than

HIF-1α [13]. In most solid tumors, tumors that express higher levels of HIF-1α, or have an enriched hypoxia-responsive gene signature, are associated with poorer overall survival [14, 15].

We recently employed the mouse mammary tumor virus (MMTV)-driven polyomavirus middle T oncoprotein (PyMT) transgenic mouse model [16] to identify HIF-dependent genes that regulate metastasis. In particular, we sought to confirm that differentially expressed genes of interest, identified by microarray profiling of HIF-1 wild type (WT) and knockout (KO) murine mammary tumor cell lines generated in [12], were direct HIF-1 targets. ChIP protocols described herein were first developed for murine breast cancer cells (PyMT) but were also applied to chromatin isolated from HIF-1 WT and shRNA-mediated *HIF1A* knockdown (KD) human breast cancer cell lines, such as MDA-MB-231 cells, as in [17]. Finally, we outline how the use of genetic loss of function models in ChIP assays assists in ruling out non-specific antibody interactions, as an alternative or a supplement to the use of IgG-only antibodies in the IP reaction.

2 Materials

2.1 Stock Solutions

Prepare all stock buffers using ultrapure deionized water (dH$_2$O). Sterile filter each solution through a 0.22 μm filter unit for long-term storage. All solutions can be stored at room temperature unless otherwise noted.

1. 1× PBS (1000 mL:): Add the following to 800 mL of dH$_2$O—8 g of NaCl, 0.2 g of KCl, 1.44 g of Na$_2$HPO$_4$, 0.24 g of KH$_2$PO$_4$. Adjust the pH to 7.4 with HCl. Bring to final volume of 1000 mL with dH$_2$O.

2. 125 mM glycine (250 mL): Add 4.7 g glycine to 200 mL dH$_2$O. Stir to dissolve. Bring to final volume to 250 mL with dH$_2$O.

3. 1 M Tris–HCl, pH 8.1 (500 mL): Add 60.57 g Tris base to 400 mL dH$_2$O. Adjust pH to 8.1 with 10 N HCl. Bring to final volume to 500 mL with dH$_2$O.

4. 1 M KCl (500 mL): Add 37.28 g KCl to 400 mL dH$_2$O. Stir to dissolve. Bring to final volume of 500 mL with dH$_2$O.

5. 0.5 M ethylenediaminetetraacetic acid (EDTA, 500 mL): Add 84.05 g EDTA to 400 mL dH$_2$O. Stir to dissolve. Adjust pH to 7.0 using HCl. Bring to final volume to 500 mL with dH$_2$O.

6. 2.5 M NaCl (500 mL): Add 73.05 g NaCl to 250 mL dH$_2$O. Dissolve on a stir plate on low heat. Once completely dissolved, bring final volume to 500 mL with dH$_2$O.

7. 2.5 M LiCl (250 mL): Add 26.49 g LiCl to 200 mL dH$_2$O. Dissolve on a stir plate. Once completely dissolved, bring final volume to 250 mL with dH$_2$O.

8. 1 M NaHCO$_3$ (250 mL): Add 21 g NaHCO$_3$ to 200 mL dH$_2$O. Dissolve on a stir plate. Once completely dissolved, bring final volume to 250 mL with dH$_2$O.

9. Formaldehyde stock, 10%: Open a fresh ampule of methanol-free, 16% formaldehyde, and dilute fresh with PBS for fixation of cells.

10. Proteinase K: 10 mg/mL in dH$_2$0, aliquot and freeze at −20°C.

11. RNase I: 1 mg/mL stock in dH$_2$O, aliquot and freeze at −20°C.

12. Phenylmethylsulfonyl fluoride, PMSF: 100 mM in 100% ethanol, aliquot and freeze at −20°C.

2.2 Lysis and Wash Buffers

Prepare all working stocks of lysis buffers fresh for every DNA harvest using the stock solutions listed in Subheading 2.1. Use dH$_2$O for all buffers. Recipes shown indicate the final concentrations of each component.

1. Hypotonic lysis buffer: 10 mM Tris–HCl, pH 8.1, 10 mM KCl, 2 mM MgCl$_2$, 2.5 mM sodium pyrophosphate, 1 mM beta-glycerophosphate, 1× protease inhibitor cocktail, 2 mM N-ethylmaleimide (NEM) (*see* **Note 1**), 2 mM sodium orthovanadate, 2 mM sodium fluoride, 1 mM dithiothreitol (DTT).

2. Nuclear lysis buffer: 50 mM Tris–HCl, pH 8.1, 10 mM EDTA, 1% sodium dodecyl sulfate (SDS), 1× protease inhibitor cocktail, 2 mM NEM, 2 mM sodium orthovanadate, 2 mM sodium fluoride.

3. ChIP dilution buffer: 10 mM EDTA, 1% Triton X-100, 150 mM NaCl, 20 mM Tris–HCl, pH 8.1, 1× protease inhibitor cocktail, 2 mM NEM, 2 mM sodium orthovanadate, 2 mM sodium fluoride.

4. Low-salt immune complex wash buffer: 0.1% sodium dodecyl sulfate (SDS), 1.0% Triton X-100, 2 mM EDTA, 20 mM Tris–HCl, pH 8.1, 150 mM NaCl.

5. High-salt immune complex wash buffer: 0.1% sodium dodecyl sulfate (SDS), 1.0% Triton X-100, 2 mM EDTA to 2 mM, 20 mM Tris–HCl, pH 8.1, 500 mM NaCl.

6. Lithium chloride (LiCl) immune complex wash buffer: 0.25 M LiCl, 1% IGEPAL CA630 (NP-40), 1% sodium deoxycholate, 1 mM EDTA, 10 mM Tris–HCl, pH 8.1.

7. Tris–EDTA wash buffer: 10 mM Tris–HCl, pH 8.1, and 1 mM EDTA.

2.3 Other Reagents

1. 15 cm cell culture dishes.

2. 1.5 mL microcentrifuge tubes.

3. 15 mL conical tubes.

4. Cell scrapers.

5. DNA cleanup kit (e.g., Zymo Research ChIP DNA Clean & Concentrator™).

6. Quantitative real-time PCR (qRT-PCR) multi-well plates.

7. qRT-PCR reagents (desalted primers, SYBR Green-based master mix, nuclease-free water).

8. Antibodies:

    (a) Anti-HIF-1α (Santa Cruz, sc-10790× or Abcam, ab2185).

    (b) Anti-HIF-2α (Novus, 100-122).

    (c) Anti-histone H3 (Abcam, ab1791).

    (d) Anti-IgG control for rabbit polyclonal antibodies.

    (e) Protein A agarose beads.

3 Methods

Carry out all procedures at room temperature unless otherwise specified.

3.1 Plating Cells for DNA Harvest

1. Use three P15 dishes per cell line and/or each HIF genotype per each hypoxia exposure time point (*see* **Note 2**).

    (a) Example: For HIF-1 wild type (WT) cells and HIF-1 knockout (KO) cells cultured at 0, 6, and 24 h hypoxia, you will need 9 plates per genotype.

2. When cells grow to ~80% confluence, change medium on all plates and incubate plates for the longest hypoxic time point in a hypoxia chamber, for example, 24 h (*see* **Note 3**).

3. Six hours prior to cell harvest, transfer cells for the 6-h hypoxic time point to the hypoxia chamber.

4. Collect all time points for DNA harvest by quickly removing cell plates from the incubators and transferring plates to a chemical fume hood.

    (a) Harvest cells in small batches. (e.g., if harvesting both the HIF-1 WT and KO cells at 24 h, this will mean you are working with six P15 dishes at a time.)

3.2 Cross-Linking Cultured Cells and DNA Extraction

1. Add 10% formaldehyde into the culture medium of each dish such that the final concentration is 1%. For a 15 cm plate containing 15 mL medium, this would be 1.5 mL per plate.

2. Recover plates and gently rock on shaker for 12 min.

3. Discard medium into chemical waste, and then wash cells once with 5 mL of 1× PBS.

4. Add 10 mL of 125 mM glycine solution to each dish in order to stop fixation, and return dishes to a shaker for 5 min.

5. Discard glycine and wash cells once with ice-cold 1× PBS.

6. Add 3 mL of 1× PBS containing freshly diluted PMSF (phenyl-methylsulfonyl fluoride; prepare by adding 10 μL PMSF stock/mL of PBS no sooner than 30 min prior to use) into each 15 cm dish. Scrape cells and pool cells from all three dishes into a single 15 mL tube. Centrifuge at 2000 × g for 3 min at 4°C in a swinging bucket centrifuge.

7. Nuclei preparation:

   (a) Add 1 mL of hypotonic lysis buffer to each cell pellet in the 15 mL tube, and incubate on ice for 10 min (*see* **Note 4**).

   (b) Add 55 μL of 10% NP-40 solution to the 1 mL of lysis buffer.

   (c) Vortex lysate for 5 s, then place on ice for 1 min, and then vortex again for 5 s.

8. Centrifuge lysate at 2000 × g for 5 min at 4°C in a swinging bucket centrifuge.

9. Discard supernatant and resuspend the isolated nuclei into 1 mL of nuclear lysis buffer per 15 mL tube (*see* **Note 5**).

10. At this step, the protocol can be stopped and the lysate stored at −80°C.

3.3 Sonication to Shear DNA

This protocol was optimized for use with a Bioruptor ultrasonicator; however, any cycling sonicator may be used. Alternatively, enzymatic shearing can be used. Optimal shearing conditions will likely need to be empirically determined for each cell type.

1. Cool sonicator by filling with a mixture of ice-H_2O 10–15 min before use.

2. Check that sonicator is set for a 30 s on/30 s off cycle and is set on high. Turn on sonicator for 12 min for one 15 mL tube (in no more than 1 mL volume, *see* **Note 6**).

3. Transfer sonicated sample to a clean 1.5 mL tube, and centrifuge at 15,000 × g for 10 min at 4 °C in a microcentrifuge.

4. Transfer supernatant into a clean 1.5 mL tube.

5. Check for DNA fragment length by analyzing 10 μL of supernatant on a 1% agarose gel at 100 V for approximately 20–30 min.

   (a) Fragmented DNA should be ~500–1000 bp, no shorter than 300 bp and no longer than 1500 bp.

6. Set aside 20% of the total volume (~200 μL) to use as input control and store at −80°C until proceeding to steps outlined in Subheading 3.6.

7. Dilute remaining sample tenfold into ChIP dilution buffer (*see* **Note 7**). Store all samples at −80°C until proceeding to steps outlined in Subheading 3.4.

3.4 Immuno-precipitation (IP), Day 1

1. Prepare agarose beads (*see* **Notes 8** and **9**) by washing with ChIP dilution buffer. Typically, 25–40 μL of beads per IP reaction is sufficient. Wash beads three times in 500 μL. Pellet beads at a low-speed centrifugation speed ($2000 \times g \times 2$ min) in a microcentrifuge between each wash, and aspirate the supernatant. After the final wash, resuspend beads as a 50% slurry in ChIP dilution buffer (*see* **Note 10**).

2. Prepare immunoprecipitation complexes as follows, one tube for each antibody and for each control (*see* **Note 11**):

Agarose beads	40 μL
Sonicated lysate	25 μg (~1 mL)
ChIP Grade Antibody	1–10 μg, depending on antibody
Histone H3	4 μg
HIF-1α	8 μg
HIF-2α	8 μg
Rabbit IgG	2 μg

3. Incubate overnight with 360° rotation at 4 °C.

3.5 Immuno-precipitation, Day 2

1. Centrifuge immunocomplexes at $2000 \times g$ for 2 min in a microcentrifuge to pellet beads. Carefully aspirate supernatant without disturbing the beads.

2. Wash the immunocomplexes one time each with the following buffers a–d. Incubate with 360° rotation at 4 °C for 5 min each. Pellet beads and aspirate buffer between each wash. After final wash, remove as much buffer as possible without drying beads:

   (a) Low-salt wash buffer, 500 μL/reaction.

   (b) High-salt wash buffer, 500 μL/reaction.

   (c) LiCl immune wash buffer, 500 μL/reaction.

   (d) TE buffer, 500 μL/reaction.

2. To elute the complexes from the beads, prepare elution buffer fresh, right before use. Elution buffer is prepared by adding 50 μL of 1 M $NaHCO_3$ to 1 mL 1% SDS.

3. Resuspend beads in 100 μL elution buffer. Rock for 15 min at room temperature.

4. Centrifuge to clear beads $2000 \times g$ for 2 min in a microcentrifuge. Transfer eluted DNA complexes to a clean 1.5 mL tube.

5. Samples can be stored at −20°C at this point, or the protocol can be continued.

3.6 Reverse Cross-Links and Treat with Proteinase K and RNase (Include Input Samples Reserved in Subheading 3.3, Step 6)

1. Add 8 μL of 2.5 M NaCl per 100 μL of sample (final concentration will be 0.2 M), and reverse cross-links by heating at 65°C for 2.5 h.

2. Add 1 μL of 1 mg/mL of RNase per 100 μL sample, and continue to reverse cross-links by heating at 65°C for an additional 2.5 h.

3. Add 2 μL of 0.5 M EDTA, 4 μL 1 M Tris–HCl, pH 6.5, and 2 μL of 10 mg/mL Proteinase K per 100 μL sample, and incubate for 1 h at 37°C.

3.7 DNA Cleanup, Concentration, and ChIP Quantification

1. Clean up reverse cross-linked DNA with a commercially available cleanup kit (*see* **Note 12**), and elute DNA in a final volume of 50 μL.

2. Using predesigned primers to the areas of interest, and a SYBR Green-based master mix, perform qRT-PCR (*see* **Notes 13** and **14**) using 2 μL of cleaned DNA per 10 μL total volume PCR reaction. An example PCR template setup sheet is shown in Fig. 1.

3. Analyze the input DNA in parallel with each immunoprecipitated ChIP sample (*see* **Note 15**, Fig. 1).

4. To determine the binding enrichment of HIF-1α, HIF-2α, or histone H3 to the DNA region of interest, first calculate the input-normalized ChIP fraction crossing point (Cp) values (ΔCp) per sample ($\Delta Cp = Cp_{IP}-Cp_{Input-DF}$, *see* **Note 15**). Next, adjust for non-specific background using the normalized IgG-only IP fraction Cp value ($\Delta\Delta Cp$, $\Delta\Delta Cp = \Delta Cp_{IP}-\Delta Cp_{IgG}$). Finally, the ChIP assay site fold enrichment above the sample-specific background can be calculated as $2^{(-\Delta\Delta Cp)}$ (Fig. 2a, b, c).

5. When comparing two different genotypes, such as HIF WT and KO cells, the fold change in site occupancy can be calculated as follows. First determine the normalized IP difference as $\Delta\Delta Cp = \Delta Cp_{IP:WT}-\Delta Cp_{IP:KO}$. Then, calculate the fold change as $2^{(-\Delta\Delta Cp)}$ (Fig. 2a, d).

4 Notes

1. NEM irreversibly inhibits cysteine peptidases and is used as an inhibitor of deubiquitinases to further enhance HIF-α protein stability.

2. Plate the cells at a density such that they are between 60 and 70% confluence the next day.

3. It is essential to ensure when using rapidly dividing cells (such as murine PyMT or human MDA-MB-231 breast cancer cells) that the cells are not >80% confluent before exposure at hypoxia for 24 h. Otherwise, media may become very acidic,

	1	2	3	4	5	6	7	8	9	10	11	12	13	14	15
A	HIF1A WT INPUT	HIF1A WT INPUT	HIF1A WT INPUT	HIF1A WT w/ HIST H3	HIF1A WT w/ HIST H3	HIF1A WT w/ HIST H3	HIF1A WT w/ HIF1	HIF1A WT w/ HIF1	HIF1A WT w/ HIF1	HIF1A WT w/ HIF2A	HIF1A WT w/ HIF2A	HIF1A WT w/ HIF2A	HIF1A WT w/ IgG	HIF1A WT w/ IgG	HIF1A WT w/ IgG
B	HIF1A KO INPUT	HIF1A KO INPUT	HIF1A KO INPUT	HIF1A KO w/ HIST H3	HIF1A KO w/ HIST H3	HIF1A KO w/ HIST H3	HIF1A KO w/ HIF1	HIF1A KO w/ HIF1	HIF1A KO w/ HIF1	HIF1A KO w/ HIF2A	HIF1A KO w/ HIF2A	HIF1A KO w/ HIF2A	HIF1A KO w/ IgG	HIF1A KO w/ IgG	HIF1A KO w/ IgG
C	HIF1A WT INPUT	HIF1A WT INPUT	HIF1A WT INPUT	HIF1A WT w/ HIST H3	HIF1A WT w/ HIST H3	HIF1A WT w/ HIST H3	HIF1A WT w/ HIF1	HIF1A WT w/ HIF1	HIF1A WT w/ HIF1	HIF1A WT w/ HIF2A	HIF1A WT w/ HIF2A	HIF1A WT w/ HIF2A	HIF1A WT w/ IgG	HIF1A WT w/ IgG	HIF1A WT w/ IgG
D	HIF1A KO INPUT	HIF1A KO INPUT	HIF1A KO INPUT	HIF1A KO w/ HIST H3	HIF1A KO w/ HIST H3	HIF1A KO w/ HIST H3	HIF1A KO w/ HIF1	HIF1A KO w/ HIF1	HIF1A KO w/ HIF1	HIF1A KO w/ HIF2A	HIF1A KO w/ HIF2A	HIF1A KO w/ HIF2A	HIF1A KO w/ IgG	HIF1A KO w/ IgG	HIF1A KO w/ IgG
E	HIF1A WT INPUT	HIF1A WT INPUT	HIF1A WT INPUT	HIF1A WT w/ HIST H3	HIF1A WT w/ HIST H3	HIF1A WT w/ HIST H3	HIF1A WT w/ HIF1	HIF1A WT w/ HIF1	HIF1A WT w/ HIF1	HIF1A WT w/ HIF2A	HIF1A WT w/ HIF2A	HIF1A WT w/ HIF2A	HIF1A WT w/ IgG	HIF1A WT w/ IgG	HIF1A WT w/ IgG
F	HIF1A KO INPUT	HIF1A KO INPUT	HIF1A KO INPUT	HIF1A KO w/ HIST H3	HIF1A KO w/ HIST H3	HIF1A KO w/ HIST H3	HIF1A KO w/ HIF1	HIF1A KO w/ HIF1	HIF1A KO w/ HIF1	HIF1A KO w/ HIF2A	HIF1A KO w/ HIF2A	HIF1A KO w/ HIF2A	HIF1A KO w/ IgG	HIF1A KO w/ IgG	HIF1A KO w/ IgG
G	HIF1A WT INPUT	HIF1A WT INPUT	HIF1A WT INPUT	HIF1A WT w/ HIST H3	HIF1A WT w/ HIST H3	HIF1A WT w/ HIST H3	HIF1A WT w/ HIF1	HIF1A WT w/ HIF1	HIF1A WT w/ HIF1	HIF1A WT w/ HIF2A	HIF1A WT w/ HIF2A	HIF1A WT w/ HIF2A	HIF1A WT w/ IgG	HIF1A WT w/ IgG	HIF1A WT w/ IgG
H	HIF1A KO INPUT	HIF1A KO INPUT	HIF1A KO INPUT	HIF1A KO w/ HIST H3	HIF1A KO w/ HIST H3	HIF1A KO w/ HIST H3	HIF1A KO w/ HIF1	HIF1A KO w/ HIF1	HIF1A KO w/ HIF1	HIF1A KO w/ HIF2A	HIF1A KO w/ HIF2A	HIF1A KO w/ HIF2A	HIF1A KO w/ IgG	HIF1A KO w/ IgG	HIF1A KO w/ IgG

Primer set: HRE Site 1   Primer set: HRE Site 2   Primer set: Non-HRE Site   Primer set: Positive Control HRE Site (e.g. EPO, Vegf)

Fig. 1 Example plate setup template for ChIP assay qRT-PCR reactions. For multiple samples and antibodies, it is best to use a 384-well plate. In this example assay, a template is shown for HIF-1α WT and KO PyMT chromatin that has been isolated from cells cultured overnight at hypoxia (1 % oxygen) and then immunoprecipitated with one of four antibodies: HIF-1α, HIF-2α, histone H3 (Hist H3), and IgG control (w/IgG). Each individual chromatin prep (WT or KO) also includes the input control (INPUT). A SYBR Green-based master mix is used in the PCR reaction. All DNA samples are first plated in triplicate into the 384-well plate, then the master mix is added, and the plate is analyzed on a Roche LightCycler 480 PCR instrument using standard amplification conditions for a total of 45 cycles. Each set of WT or KO input and antibody immunoprecipitated samples is profiled using four separate primer sets: two putative HRE sites ("HRE site 1," black font, and "HRE site 2," red font), a non-HRE site (blue font; no binding should be observed at this site) and a positive control HRE site (green font; primers will span a locus known to bind HIF-α proteins, such as the human *EPO* 3′ UTR, or the mouse *Vegf* proximal promoter). Individual crossing point (Cp) values are calculated using the absolute quantification method of the LightCycler software package. The average of the triplicate well crossing point (Cp) values is calculated and is used to quantitate the antibody binding enrichment at each genomic locus as described in the methods

and the cells may detach from the plate. To buffer cells that approach confluence, the cell culture growth medium may be supplemented with HEPES to a final concentration of 15–25 mM to facilitate prolonged exposures to hypoxia.

4. 1 mL is a general volume to use during lysis. If the cell pellet seems larger or smaller than expected, then adjust lysis buffer volume accordingly.

5. As with the hypotonic lysis buffer, adjust the volume of the nuclear lysis buffer according to the original size of cell pellet.

6. If samples require multiple rounds of sonication, let the sonicator cool 30 min between usages and fill with fresh ice-H₂O before proceeding to the next sample. It is important not to overheat the samples (which may cause degradation) or the sonicator.

7. Alternatively, do not dilute the sample, but use tenfold less volume as input chromatin for the immunoprecipitation reactions in order to decrease the total volume of the IP reactions.

8. As an alternative to agarose beads, magnetic beads may also be used; however, this protocol was optimized using protein A agarose beads.

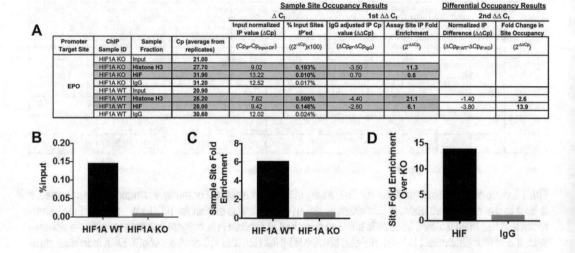

Promoter Target Site	ChIP Sample ID	Sample Fraction	Cp (average from replicates)	Input normalized IP value (ΔCp) $(Cp_{IP}-Cp_{Input\text{-}DF})$	% Input Sites IP'ed $((2^{-\Delta Cp})\times 100)$	IgG adjusted IP Cp value (ΔΔCp) $(\Delta Cp_{IP}-\Delta Cp_{IgG})$	Assay Site IP Fold Enrichment $(2^{-\Delta\Delta Cp})$	Normalized IP Difference (ΔΔCp) $(\Delta Cp_{P:WT}-\Delta Cp_{P:KO})$	Fold Change in Site Occupancy $(2^{-\Delta\Delta Cp})$
	HIF1A KO	Input	21.00						
	HIF1A KO	Histone H3	27.70	9.02	0.193%	-3.50	11.3		
	HIF1A KO	HIF	31.90	13.22	0.010%	0.70	0.6		
EPO	HIF1A KO	IgG	31.20	12.52	0.017%				
	HIF1A WT	Input	20.90						
	HIF1A WT	Histone H3	26.20	7.62	0.508%	-4.40	21.1	-1.40	2.6
	HIF1A WT	HIF	28.00	9.42	0.146%	-2.60	6.1	-3.80	13.9
	HIF1A WT	IgG	30.60	12.02	0.024%				

Table header spans: **Sample Site Occupancy Results** (ΔC_t, 1st ΔΔ C_t) | **Differential Occupancy Results** (2nd ΔΔ C_t)

Fig. 2 Example template for ChIP analysis calculations and example bar graphs of analyzed data. Table summary generated from qRT-PCR results after immunoprecipitation of chromatin from HIF-1 WT and KO PyMT cells using histone H3 or HIF-1α antibodies for the HRE site present in the human *EPO* 3′ UTR (**a**). The table shows the average Cp values of triplicate wells. Sample site occupancy (at *EPO* HRE) for each genotype of cells was calculated first by normalizing the input fraction ($\Delta Cp = Cp_{IP}-Cp_{Input\text{-}DF}$). Percent input was calculated as %input = $(2^{(-\Delta Cp)} \times 100)$ and presented in graph form in (**b**). Non-specific background was adjusted using the data from IgG-only samples ($\Delta\Delta Cp = \Delta Cp_{IP}-\Delta Cp_{IgG}$), and then the sample HRE site fold enrichment above background was calculated as $2^{(-\Delta\Delta Cp)}$ and presented in graph form in (**c**). Differential occupancy at the EPO HRE site in WT versus KO cells ("site fold enrichment over KO") was determined by first calculating the normalized IP difference $\Delta\Delta Cp = \Delta Cp_{IP:WT}-DCp_{IP:KO}$. Then, the fold change in WT over KO binding efficiency was calculated as $2^{(-\Delta\Delta Cp)}$ for the HIF-1α and the IgG precipitation reactions and the data presented in graph form in (**d**)

9. Agarose beads are available as bound to protein A, protein G, or a mix of A/G. The beads chosen should be optimized for the host of the antibodies used in the IP reaction. For this protocol, we selected rabbit antibodies to the HIF-α subunits, which have a higher binding affinity to protein A agarose beads. Make sure to check the antibody affinity prior to choosing the agarose beads.

10. Occasionally, high background and non-specific binding to the agarose beads can occur. If this happens, when analyzing the qRT-PCR results, it will be observed that the crossing point (Cp) value from your IgG-only sample is close to the Cp value of your sample with the antibody of interest. Ideally, the IgG Cp values should be at least ≥2 Cp values higher (indicating lower abundance). If this happens, it is recommended to pre-clear the beads by incubating the beads alone with the lysed samples for 1 h prior to adding the antibody to the IP reaction.

11. There are many controls needed for ChIP experiments. First, there are controls for the IP steps. First is a negative control that consists of lysate, agarose beads, and an IgG antibody

from the same species as your IP antibody of interest. qRT-PCR data for these samples will report the background binding within the assay (non-specific interactions of protein with agarose beads). Second, a positive control is needed, such as histone H3, which is highly enriched in actively transcribed regions. Including this control helps to ensure that each step of the ChIP protocol is working, even if an individual binding site does not show enrichment of your protein of interest. Third, there are controls for the PCR amplification steps, such as a no template control for the qRT-PCR reaction to check for contamination of PCR reagents. Finally, the input DNA (non-IP) samples should produce positive signals for all primers sets tested and should always be included (*see* **Note 15**).

12. We have optimized this protocol to use the ChIP DNA Clean & Concentrator™ kit available from Zymo Research. However, other commercially available DNA cleanup kits, such as those offered by Qiagen, work, although we found in our hands that the components of the Zymo kit were more stable over time than those purchased from Qiagen, and that the cleanup yield was higher with the Zymo cleanup kit than with the Qiagen kit.

13. When designing primers, select primers with a Tm of 58–60 °C. Additionally, keep the GC content between 30 and 80%, with 50% being ideal. Do not choose primers with long runs of the same nucleotide (>3 bp), and for the 3′ end of the primer sequence, do not include more than two G and/or C bases. Finally, to ensure that the primers you have designed are unique, search them through National Center for Biotechnology Information (NCBI) nucleotide BLAST program.

14. When designing primers, make sure to include all binding site loci controls. First, an assay positive control is needed, which is a HRE sequence known to bind HIF-1 or HIF-2, such as the HRE site within the human erythropoietin (EPO) 3′ UTR enhancer [18] or the HRE site within the murine *Vegf* proximal promoter [19]. Second, a negative control is required, which is a sequence to which HIF-α proteins should not bind. This sequence may be a random site in the same regulatory region of interest, but does not match the HRE consensus sequence, which we refer to as the assay "non-HRE site." This latter control will report the extent of non-specific binding to DNA.

hEPO primers, forward, 5′-GCTGGCCTCTGGCTCTCATGG-3′, and reverse, 5′-CAGGGTTGGCAGCTGCCTTACTG-3′.

mVEGF primers, forward, 5′-CTGGCTTCAGTTCCCTGGCAACATCTCT-3′, and reverse, 5′-CCTGGGGTGAATGGGATCCTCTGG-3′.

15. Input DNA is all of the sheared genomic DNA not subjected to the IP steps, and is a required control to include in the qRT-PCR assay step for every independent ChIP assay in order to estimate enrichment of binding of HIF-1 or HIF-2 to the HRE binding site. PCR amplification products should be detectable in all primer sets tested using the input DNA. Enrichment is calculated from the qRT-PCR values observed for the IP fraction relative to the values observed for the input DNA sample using the same primer set. We typically setup PCR reactions to contain 2 μL of input chromatin per 10 μL of total volume (or 20% of the original input DNA concentration), which is a dilution factor (DF) of 5. To adjust the input Cp value, herein referred to as the "$Cp_{INPUT-DF}$," take the $\log_2$ of 5, or 2.32, and subtract from the raw Cp_{INPUT} value (refer to Fig. 2 for example calculations). Then, subtract the $Cp_{INPUT-DF}$ from all the IP fraction Cp values (Cp_{IP}) to determine the input-normalized Cp, or ΔCp. If you prefer to present the data as % input, enter the ΔCp value for each individual IP sample in the following equation: %input = $((2^{-\Delta Cp}) \times 100)$ (Figs. 2a, b).

Acknowledgments

This work was supported by NIH grant CA138488 and the Department of Defense Breast Cancer Research Program award BC083846 (to TNS). The HIF-1 WT and KO PyMT cells were originally created by Dr. Luciana P. Schwab, and the HIF shRNA knockdown MDA-MB-231 cell line models were provided by Dr. Roland Wenger of the University of Zurich. We also thank Dr. Meiyun Fan at the University of Tennessee Health Science Center for providing us with technical advice and for assisting us with troubleshooting these protocols during their optimization.

References

1. Gilmour DS, Lis JT (1985) In vivo interactions of RNA polymerase II with genes of Drosophila melanogaster. Mol Cell Biol 5(8):2009–2018

2. Gilmour DS, Pflugfelder G, Wang JC, Lis JT (1986) Topoisomerase I interacts with transcribed regions in Drosophila cells. Cell 44(3):401–407

3. Gilmour DS, Rougvie AE, Lis JT (1991) Protein-DNA cross-linking as a means to determine the distribution of proteins on DNA in vivo. Methods Cell Biol 35:369–381

4. Semenza GL (2010) Defining the role of hypoxia-inducible factor 1 in cancer biology and therapeutics. Oncogene 29(5):625–634. https://doi.org/10.1038/onc.2009.441

5. Jaakkola P, Mole DR, Tian YM, Wilson MI, Gielbert J, Gaskell SJ, Kriegsheim A, Hebestreit HF, Mukherji M, Schofield CJ, Maxwell PH, Pugh CW, Ratcliffe PJ (2001) Targeting of HIF-alpha to the von Hippel-Lindau ubiquitylation complex by O2-regulated prolyl hydroxylation. Science 292(5516):468–472

6. Wenger RH, Stiehl DP, Camenisch G (2005) Integration of oxygen signaling at the consensus HRE. Sci STKE 306:re12. https://doi.org/10.1126/stke.3062005re12

7. Schodel J, Oikonomopoulos S, Ragoussis J, Pugh CW, Ratcliffe PJ, Mole DR (2011) High-resolution genome-wide mapping of HIF-binding sites by ChIP-seq. Blood 117(23):e207–e217. https://doi.org/10.1182/blood-2010-10-314427

8. Tsai YP, Wu KJ (2012) Hypoxia-regulated target genes implicated in tumor metastasis. J Biomed Sci 19:102. https://doi.org/10.1186/1423-0127-19-102

9. Hu CJ, Wang LY, Chodosh LA, Keith B, Simon MC (2003) Differential roles of hypoxia-inducible factor 1alpha (HIF-1alpha) and HIF-2alpha in hypoxic gene regulation. Mol Cell Biol 23(24):9361–9374

10. Stiehl DP, Bordoli MR, Abreu-Rodriguez I, Wollenick K, Schraml P, Gradin K, Poellinger L, Kristiansen G, Wenger RH (2011) Non-canonical HIF-2alpha function drives autonomous breast cancer cell growth via an AREG-EGFR/ErbB4 autocrine loop. Oncogene 31(18):2283–2297. https://doi.org/10.1038/onc.2011.417

11. Liao D, Corle C, Seagroves TN, Johnson RS (2007) Hypoxia-inducible factor-1alpha is a key regulator of metastasis in a transgenic model of cancer initiation and progression. Cancer Res 67(2):563–572. https://doi.org/10.1158/0008-5472.CAN-06-2701

12. Schwab LP, Peacock DL, Majumdar D, Ingels JF, Jensen LC, Smith KD, Cushing RC, Seagroves TN (2012) Hypoxia-inducible factor 1alpha promotes primary tumor growth and tumor-initiating cell activity in breast cancer. Breast Cancer Res 14(1):R6. https://doi.org/10.1186/bcr3087

13. Raval RR, Lau KW, Tran MG, Sowter HM, Mandriota SJ, Li JL, Pugh CW, Maxwell PH, Harris AL, Ratcliffe PJ (2005) Contrasting properties of hypoxia-inducible factor 1 (HIF-1) and HIF-2 in von Hippel-Lindau-associated renal cell carcinoma. Mol Cell Biol 25(13):5675–5686. https://doi.org/10.1128/MCB.25.13.5675-5686.2005

14. Chi JT, Wang Z, Nuyten DS, Rodriguez EH, Schaner ME, Salim A, Wang Y, Kristensen GB, Helland A, Borresen-Dale AL, Giaccia A, Longaker MT, Hastie T, Yang GP, van de Vijver MJ, Brown PO (2006) Gene expression programs in response to hypoxia: cell type specificity and prognostic significance in human cancers. PLoS Med 3(3):e47

15. Semenza GL (2002) HIF-1 and tumor progression: pathophysiology and therapeutics. Trends Mol Med 8(4 Suppl):S62–S67

16. Lin EY, Jones JG, Li P, Zhu L, Whitney KD, Muller WJ, Pollard JW (2003) Progression to malignancy in the polyoma middle T oncoprotein mouse breast cancer model provides a reliable model for human diseases. Am J Pathol 163(5):2113–2126. https://doi.org/10.1016/S0002-9440(10)63568-7

17. Brooks DL, Schwab LP, Krutilina R, Parke DN, Sethuraman A, Hoogewijs D, Schorg A, Gotwald L, Fan M, Wenger RH, Seagroves TN (2016) ITGA6 is directly regulated by hypoxia-inducible factors and enriches for cancer stem cell activity and invasion in metastatic breast cancer models. Mol Cancer 15:26. https://doi.org/10.1186/s12943-016-0510-x

18. Wang GL, Semenza GL (1995) Purification and characterization of hypoxia-inducible factor 1. J Biol Chem 270(2):1230–1237

19. Oosthuyse B, Moons L, Storkebaum E, Beck H, Nuyens D, Brusselmans K, Van Dorpe J, Hellings P, Gorselink M, Heymans S, Theilmeier G, Dewerchin M, Laudenbach V, Vermylen P, Raat H, Acker T, Vleminckx V, Van Den Bosch L, Cashman N, Fujisawa H, Drost MR, Sciot R, Bruyninckx F, Hicklin DJ, Ince C, Gressens P, Lupu F, Plate KH, Robberecht W, Herbert JM, Collen D, Carmeliet P (2001) Deletion of the hypoxia-response element in the vascular endothelial growth factor promoter causes motor neuron degeneration. Nat Genet 28(2):131–138. https://doi.org/10.1038/88842

Chapter 8

Evaluating the Metabolic Impact of Hypoxia on Pancreatic Cancer Cells

Divya Murthy, Enza Vernucci, Gennifer Goode, Jaime Abrego, and Pankaj K. Singh

Abstract

Hypoxia is frequently observed in human cancers and induces global metabolic reprogramming that includes an increase in glucose uptake and glycolysis, alterations in $NAD(P)H/NAD(P)^+$ and intracellular ATP levels, and increased utilization of glutamine as the major precursor for fatty acid synthesis. In this chapter, we describe in detail various physiological assays that have been adopted to study the metabolic shift propagated by exposure to hypoxic conditions in pancreatic cell culture model that includes glucose uptake, glutamine uptake, and lactate release by pancreatic cancer cell lines. We have also elaborated the assays to evaluate the ratio of $NAD(P)H/NAD(P)^+$ and intracellular ATP estimation using the commercially available kit to assess the metabolic state of cancer cells.

Key words Hypoxia, HIF, Pancreatic cancer, Cancer metabolism, Glucose uptake, Glutamine uptake, Lactate release, ATP estimation, NADPH/NADP+ ratio

1 Introduction

Hypoxia and altered energy metabolism are key features of cancer cells that capture the major phenotypic characteristics of pancreatic cancer progression [1]. The increased hypoxia is manifested by the significant desmoplasia exhibited by pancreatic tumors [2]. Such induction of hypoxia in the tumor microenvironment has been demonstrated to stabilize a class of transcription factors known as hypoxia-inducible factors (HIFs) [3, 4]. HIFs, in turn, modulate the biology of a hypoxic cancer cell by rewiring metabolic pathways that facilitate cancer cell progression and survival [5]. HIF1α has been shown to regulate the transcriptional expression of genes that encode glucose transporters and glycolytic enzymes while limiting the carbon flux through the tricarboxylic acid cycle [6–8]. To assess such enhanced metabolic activity and glucose uptake in cancer cells, the radiolabeled glucose uptake assay is utilized routinely [9–12]. Thus, the enhanced glycolytic flux allows

L. Eric Huang (ed.), *Hypoxia: Methods and Protocols*, Methods in Molecular Biology, vol. 1742,
https://doi.org/10.1007/978-1-4939-7665-2_8, © Springer Science+Business Media, LLC 2018

channeling of glycolytic intermediates into various biosynthetic pathways resulting in increased biomass and cellular assembly. In addition to increased glucose uptake, HIF-dependent changes in glutamine metabolism have also been studied [13].

The intensified glycolysis is closely intertwined with the increased lactate secretion in the tumor microenvironment resulting in extracellular acidification. In tumor cells, HIF stabilization in hypoxia directly regulates the expression of LDHA and MCT4. These alterations upregulate the secretion of lactate from cells in a hypoxic microenvironment [14]. The abnormally high level of lactate in the tumor extracellular environment has been quantified using various colorimetric and fluorimetric reaction-based kits [15, 16]. Increased activity of LDHA in tumor cells implies an increase of NADH relative to NAD^+. Such an increase in the ratio of $NADH/NAD^+$ has previously been reported in breast cancer [14]. Owing to the high glycolytic rate in cancer cells, there is a significant routing of the glycolytic intermediates to the pentose phosphate pathway (PPP) resulting in augmented NADPH production that counteracts oxidative stress.

The ratio of $NAD(P)H$ to $NAD(P)^+$ is closely associated with another prototypical metabolite of cancer cells, adenosine triphosphate (ATP). An increased synthesis of ATP within the cancer cells ensues to recompense for the augmented energy demand required for cancer cell growth. In cancer cells, ATP plays a fundamental role not just in maintaining cellular energetics but also in sustaining proliferative cellular signals and inhibiting growth suppressors [17]. Metabolic changes in response to hypoxia are elicited through a reduction in ATP generation by oxidative phosphorylation while increasing the rate of glycolysis even in the presence of sufficient oxygen that can support mitochondrial metabolism, a phenomenon known as Warburg effect in cancer cells [18]. Clearly, induction of hypoxia promotes a major metabolic shift in cancer cells, and it is pertinent to choose appropriate in vitro assays to monitor the hallmarks of cancer. The combination of assays would enable an in-depth data collection providing several biological parameters to study the metabolic impact of hypoxia on the physiology of cancer cells.

2 Materials

2.1 Cell Culture Reagents

1. Dulbecco's Modified Eagle's Medium (DMEM) (13.4 g/L) (with phenol red).

2. Dulbecco's Modified Eagle's Medium (without phenol red, glucose, glutamine, and pyruvate).

3. L-sodium bicarbonate (3.7 g/L).

4. Sodium pyruvate (0.110 g/L).

5. Fetal bovine serum (FBS).

6. Phosphate buffered saline—Add 8 g of NaCl, 0.2 g of KCl, 1.44 g of Na_2HPO_4, 0.24 g of KH_2PO_4 to 800 mL Milli-Q H_2O. Adjust the pH to 7.4 with hydrochloric acid. Add Milli-Q H_2O to a total volume of 1 L.

7. Trypsin (0.25%).

8. Penicillin-streptomycin solution (100×) (penicillin 10,000 units/mL; streptomycin 10,000 μg/mL).

2.2 Glucose Uptake Assay

1. [^3H] 2-Deoxy-D-glucose (^3H-2-DG).

2. Glucose stock 50× (1250 mM).

3. 1% sodium dodecyl sulfate (SDS): Add 1 g of SDS to 100 mL of Milli-Q H_2O.

4. Scintillation vials.

5. Scintillation fluid: Ecoscint H.

2.3 Glutamine Uptake Assay

1. L-[3,4-^3H(N)]-glutamine.

2. 1% SDS 1%.

3. Scintillation vials.

4. Scintillation fluid: Ecoscint H.

2.4 Lactate Release Assay

1. L-lactate Assay Kit I, Eton Bioscience Inc. Catalog#120001100A.

2. L-lactate assay standards (3 mM).

3. L-lactate assay solution contains lactate dehydrogenase (LDH), DMSO, Tris–HCl/pH 7.5, and tetrazolium INT salt (light-sensitive reagents).

2.5 NADP+/NADPH Estimation

1. NADP/NADPH Quantification Colorimetric kit components, Catalog # K347–100.

2. NADP Cycling Buffer, Catalog # K347–100-2.

3. NADP Cycling Enzyme Mix, Catalog # K347–100-3.

4. NADP/NADPH Extraction Buffer, Catalog # K347–100-1.

5. NADPH Developer, Catalog # K347–100-4.

6. Stop solution, Catalog # K347–100-5.

7. NADPH Standard, Catalog # K347–100-6.

2.6 ATP Estimation

ATP Bioluminescence Assay Kit HS II, Catalog # 11699709001.

1. Luciferase reagent, lyophilized.

2. ATP standard, lyophilized.

3. Cell lysis reagent.

4. Dilution buffer.

2.7 Equipment	1. 12-well plate.
	2. 24-well plate.
	3. 96-well plate.
	4. Cell scrapers.
	5. O_2/CO_2 incubator.
	6. Microfuge tubes.
	7. Beckman Coulter Multipurpose Scintillation Counter LS6500
	8. Cytation 3 Imaging Reader by BioTek.
	9. Gilson pipettes.
	10. Pipette tips: 0.1–20 μL, 200 μL, 1000 μL.
	11. Refrigerated microfuge.
	12. Orbi-Shaker™ XL, Benchmark Scientific.
	13. Automated cell counter.

2.8 Software	1. GEN5 3.0 Microplate Reader and Imager Software (for Cytation 3).

3 Methods

3.1 Cell Culture	1. Seed approximately 0.2×10^6 cells in a 100 mm cell culture dish with DMEM (+ phenol red) containing 10% FBS.
	2. Incubate cells at 37 °C in an incubator containing 5% CO_2 and 95% humidity. Maintain oxygen levels at 20%.
	3. After 24–36 h of seeding the cells, or when the cells reach 80–90% confluence, aspirate and discard medium. Wash the cells twice with 1× PBS and detach cells by incubating with trypsin at 37 °C.
	4. After detachment, add DMEM with 5% FBS to neutralize trypsin. Transfer the detached cell suspension to a 15 mL centrifuge tube and centrifuge at $135 \times g$ for 2 min.
	5. Discard the supernatant and resuspend the pellet in fresh DMEM.
	6. From the cell suspension, aliquot 20 μL in a 1.5 mL centrifuge tube, and add 20 μL of trypan blue. Mix the solution thoroughly and incubate at room temperature for 5 min (min).
	7. Load 10 μL of the cell suspension to a counting slide and insert it into an automated cell counter.
	8. Seed cells as recommended for each assay protocol (as described in individual method section).

3.2 Glucose Uptake Assay

3.2.1 Hypoxic Culture Conditions

1. Seed cells at a confluence of 40% (*see* **Note 1**) in serum-containing medium (10% FBS) in a 24-well plate and incubate in 5% CO_2 incubator at 37 °C.

2. Once the cells have attached, exchange DMEM with fresh phenol red-free DMEM medium and incubate cells in normoxia or hypoxia incubators for 24 h (*see* **Note 2**).

3.2.2 Glucose Uptake Assay

1. Rinse cells with PBS 1× twice.

2. Starve the cells using 500 μL (per well) of medium without glucose and FBS for 2 h (*see* **Note 3**).

3. After 2 h, add 20 μL from the glucose 50× stock only in the wells that will be used for measuring the background for 20 min (*see* **Note 4**).

4. Meanwhile, count the number of cells for each condition and every cell type (*see* Subheading 3.1) (*see* **Note 5**).

5. Add 1 μL of [³H]-2-DG in all the wells including the background wells and incubate the plate at 37 °C for 30 min in the incubator (*see* **Note 6**).

6. Add 5 mL of scintillation fluid to the scintillation vials.

7. After 30 min, rinse each well with 1 mL of 1× PBS.

8. Add 500 μL of 1% SDS for each well.

9. Pipette up and down vigorously to lyse the cells. Transfer the lysate in the scintillation vials (*see* **Note 7**).

10. Read the incorporated [³H] signal using a liquid scintillation counter.

3.2.3 Data Analysis

1. Calculate the average CPM values of the background controls and subtract it from all the samples (*see* **Note 8**).

2. Divide each value by the respective cell count number to normalize the data.

3. Divide the individual normalized value of the hypoxic cells by the mean of values obtained from cells in normoxic conditions.

3.3 Glutamine Uptake Assay

See Subheading 3.1, **step** 1 in *Glucose Uptake Assay*.

3.3.1 Hypoxic Culture Conditions

3.3.2 Glutamine Uptake Assay

1. Add 3 μL of L-[3,4-³H(N)]-glutamine in all the wells except for the background wells.

2. After 3 min, wash each well with 1× PBS (*see* **Note 9**).

3. Add 500 μL of 1% SDS for each well.

4. Pipette up and down vigorously to lyse the cells. Transfer the lysate to the scintillation vials.

5. Read the incorporated [³H] signal using a liquid scintillation counter.

3.3.3 Data Analysis

The analysis is performed similarly using the method described in Subheading 3.2.3 in Glucose Uptake Assay.

3.4 Lactate Release Assay

3.4.1 Hypoxic Culture Conditions

1. Seed cells at approximately 0.3×10^6 cells per well in a 12-well dish. Include 4–8 replicates per dish. Calculate seeding volume at 500 μL per well.

2. After the cells have attached (overnight), exchange DMEM with fresh phenol red-free DMEM medium and incubate cells in normoxia or hypoxia incubators for 24 h.

3.4.2 Preparation of Samples

1. After incubation period, use a pipette to collect the medium from all wells in 1.5 mL microcentrifuge tubes and centrifuge at $135 \times g$ for 2 min.

2. Transfer the supernatant into a new microcentrifuge tube. Keep samples on ice or store at −80 °C for future use.

3. Wash cells in wells with 1× PBS twice and count the number of cells for each condition and every cell type (*see* Subheading 3.1).

3.4.3 Lactate Measurement in Culture Medium

1. Use a 96-well plate for lactate assay.

2. Standard curve preparation: aliquot 50 μL, 40 μL, 30 μL, 20 μL, 10 μL, 5 μL, 2.5 μL, and 0 of the L-lactate supplied in the kit. Bring up the volume of each well to 50 μL using Milli-Q H_2O. Prepare standard curve in duplicates.

3. Aliquot 50 μL of cell medium per sample in the 96-well plate.

4. Add 50 μL of L-lactate assay solution for a total volume of 100 μL per well. Incubate samples at 37 °C for 30 min. Do not use a CO_2 incubator (*see* Notes 10 and 11).

5. After incubation, take the plate out of 37 °C and add 50 μL of 0.5 M acetic acid to stop the reaction. Perform this step without exposing plate to light. Shake the contents gently and remove any bubbles from the wells (*see* Note 12).

6. Set the microplate reader to measure absorbance of samples at OD490 nm.

3.4.4 Data Analysis

1. To generate the standard curve, first subtract the absorbance of the 0 L-lactate (blank) well from all other wells that contain the sample or L-lactate standard.

2. Prepare a standard curve by plotting the reading of standards at 490 nm as a function of L-lactate concentration.

3. Use equation from the linear regression of the standard curve to determine L-lactate concentration in samples. Equation— the L-lactate (μM): [(corrected absorbance)-(Y-intercept)]/ slope

4. To determine the concentration of secreted L-lactate in cells in hypoxia versus normoxia, first, divide the sample L-lactate concentration of all samples by the number of cells counted with the cells in the corresponding location in the 12-well plate. Second, determine the average of the L-lactate release of samples in normoxia and divide all samples (both normoxia and hypoxia) by this average to determine the fold change.

3.5 NADP⁺/NADPH Estimation

3.5.1 Hypoxic Culture Conditions

1. Seed cells at approximately 0.5×10^6 cells per well in a 6 cm dish.

2. After the cells have attached (overnight), exchange DMEM with fresh phenol red-free DMEM medium and incubate cells in normoxia or hypoxia incubators for 24 h.

3.5.2 Assay Preparation

1. Carry out all procedures at room temperature and protected from light unless otherwise specified.

2. NADPH Developer: Spin vial to ensure no loss of contents. Add 1.2 mL of Milli-Q H_2O and pipet several times until powder is completely dissolved.

3. NADPH Standard: Spin vial to ensure no loss of contents. To make 1 nmol/μL standard stock solution, add 0.2 mL of DMSO and pipet several times until powder is completely dissolved.

4. NADP cycling mix: Add 98 μL of NADP cycling buffer with 2 μL of NADP cycling enzyme mix per sample.

3.5.3 Sample Collection

1. After the incubation period, wash the cells twice with 1× PBS. Trypsinize the cells, neutralize with complete medium, and then collect by centrifugation at $135 \times g$ for 2 min.

2. Wash cell pellet with cold 1× PBS and collect in a microcentrifuge tube at a concentration of 4×10^6 cells/reaction.

3. Centrifuge cells at $376 \times g$ for 5 min.

4. Decant PBS and lyse the cell pellet with 800 μL of NADP/ NADPH extraction buffer. Keep the sample on ice for 10 min then, centrifuge at $10,000 \times g$ for 10 min, and transfer supernatant to a fresh tube ensuring that the pellet is not disturbed.

3.5.4 Standard Curve Preparation

1. To make standard NADPH 10 pmol/μL, dilute 10 μL of 1 nmol/μL with 990 μL NADP/NADPH extraction buffer.

2. Ensure standard is used within 4 h and discard unused portions.

3. Create a standard curve in duplicate by adding diluted NADPH (0, 2, 4, 6, 8, and 10 μL) to create 0, 20, 40, 60, 80, and 100 pmoles per well in a 96-well plate.

4. Bring the total volume of each well to 50 μL by adding NADP/NADPH extraction buffer.

3.5.5 Sample Preparation

1. To detect total NADP⁺/NADPH in the sample, transfer 50 and 25 μL of each sample in duplicate to the 96-well plate.

2. Bring the total volume of each well to 50 μL by adding NADP⁺/NADPH extraction buffer.

3. To detect NADPH only, aliquot 200 μL samples into 1.5 mL microcentrifuge tubes and heat the samples at 60 °C for 30 min in a heating block or a water bath. Heating the samples results in decomposition of all NADP⁺ whereas the NADPH remains intact. Cool samples on ice and spin samples if precipitate is observed. Transfer 50 and 25 μL of each sample in duplicate to the 96-well plate.

3.5.6 Assay

1. Add 100 μL of NADP cycling mix to well and mix by pipetting gently not to create bubbles.

2. Incubate the 96-well plate at room temperature for 5 min to convert NADP⁺ to NADPH.

3. To initiate the reaction, add 10 μL of NADPH Developer to each well.

4. For kinetics, place the plate in the plate reader and read absorbance at 450 nm every 10 min up to 4 h while the color develops or until the signal is saturated.

3.5.7 Data Analysis

1. The standard curve created in this assay is linear over the range of 0–100 pmol (*see* **Note 13**).

2. Subtract the OD for 0 pmoles standard from all the readings.

3. Determine the total NADP ($NADP_t$) and NADPH concentrations by applying the sample OD at 450 nm to the standard curve.

4. Calculate the amount of NADP⁺ by subtracting concentration of NADPH from total NADP.

5. The ratio of NADP⁺/NADPH may be calculated as

$$NADP + /NADPH = (NADP_t - NADPH)/NADPH$$

3.6 **ATP Estimation**

3.6.1 Hypoxic Culture Conditions

1. Seed approximately 5×10^5 cells per well in a 12-well tissue culture plate and culture the cells in DMEM containing 10% FBS. Set up the experiment in quadruplet keeping one well for cell counting for normalization.

2. After 16–18 h, aspirate the culture medium and replace with 1 mL fresh medium. Incubate the cells under 20% O_2 for nor-

moxic exposure and 1% O_2 for hypoxic exposure for multiple time durations.

3. After 24 h, remove the cells from the incubators and immediately aspirate the medium.

4. Place the cells on ice and wash the cells twice with 1× PBS (*see* **Note 14**).

5. Count the number of cells only in the wells set for cell counting (*see* Subheading 3.1).

3.6.2 ATP Estimation Assay Reagent Preparation and Storage

1. To prepare the luciferase reagent, the whole content of vial 1 from the kit must be carefully dissolved in 10 mL of dilution buffer and incubate for 5 min at 0–4 °C without stirring or shaking. To mix the contents, carefully rotate the bottle. Shaking must be avoided to prepare the homogeneous solution (*see* **Note 15**).

2. To prepare the ATP standard, dissolve the content of vial 2 by the addition of 1.65 mL of dilution buffer to generate 10 mM stock solution.

3.6.3 Cell Lysis and Sample Preparation

1. Add 100 µL of cell lysis reagent to the cells while placing the plate on ice (*see* **Note 16**).

2. Shake the plate on ice on an orbital shaker for 10 min.

3. Scrape the cells using a cell scraper and collect the lysed cell fractions in a pre-chilled microfuge tube.

4. Centrifuge the cells at $16,000 \times g$ at 4 °C for 10 min.

5. Collect the supernatant in a fresh microfuge tube and place on ice (*see* **Note 17**).

3.6.4 ATP Standard Curve Generation

1. The ATP standard curve is generated by serial dilutions of the 10 mM ATP stock solution. Dilute ATP standard with dilution buffer by serial dilution in the range of 10^{-6} to 10^{-12} M as shown in Fig. 1 (*see* **Note 18**).

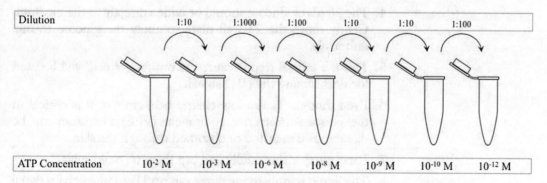

Fig. 1 Serial dilution of the ATP standard to generate the ATP standard curve. The ATP standard is diluted using the dilution buffer provided in the kit by the manufacturer

2. Add 50 μL of the ATP standard or the sample (in duplicates) in each well of a multiwell plate (*see* **Note 19**).

3. To the ATP standard or the sample, add 50 μL of the dilution buffer. Incubate the mix for approximately 5 min at room temperature.

4. Add 50 μL of the luciferase reagent to the standard and sample wells by automated injection. Do not forget to keep background control samples without luciferase reagent. Instead of the luciferase reagent, you can add 50 μL of dilution buffer (*see* **Note 20**).

5. Measure the luminescence using a luminometer (Cytation 3 cell imaging multi-mode reader).

6. Start the measurement after a 1-s delay and integrate the bioluminescence measurements for 1–10 s.

3.6.5 Data Analysis

1. Correct the background by subtracting the blank from the raw data.

2. Calculate the concentration of ATP in cells cultured under normoxia and hypoxia by analyzing the log-log plot generated from the ATP standard curve data.

3. Normalize the concentration of ATP per cell using the cell count.

4 Notes

1. Seeding the cells for the assay is very critical, and we must avoid generation of cell aggregates. For this purpose, cells should not be left in trypsin for an extended time (Fig. 2).

2. Duration of exposure to hypoxia may change in accordance to the experimental design.

3. Cell washing should be performed carefully to avoid sloughing off the cells from the plate.

4. The 20 μL of glucose should be added directly to the medium. Gently move the 24-well plate to allow the glucose to mix uniformly.

5. This is a crucial step because the number of cells will be used for normalizing the [^{3}H] signal.

6. Even though ^{3}H is a low-energy beta-emitter, it is critical to use personal protective equipment (PPE) as tritium can be dangerous if ingested or absorbed through the skin.

7. Label every vial carefully and cap them to avoid spills.

8. The liquid scintillator analyzer can read the radioactivity signal as CPM (count per minute) or DPM (disintegration per min-

Normoxia					
B	GUN				
B	GUN				
B	GUN				
	CN				

Hypoxia					
B	GUH				
B	GUH				
B	GUH				
	CH				

Fig. 2 A represented template. Above is a template for seeding cells for normoxia and hypoxia experiment before glucose uptake. The acronym in each well indicates for which purpose those cells will be used

ute); both are equally accepted. For radioactivity disposal, the institutional guidelines must be followed.

9. Glutamine uptake must be performed very quickly as it can be rapidly imported and exported from the cells.

10. At higher L-lactate concentration, the colorimetric reaction oversaturates fairly quickly. 30 min is the maximum incubation time to determine changes in L-lactate concentration. Cells must be in constant observation at 20–30 min of incubation as to not oversaturate readings of the absorbance generated from adducts of samples and high concentration standards.

11. For optimal readout using the Eton Bioscience Lactate Assay Kit I, it is recommended to dilute the samples. To 10 μL of sample, add 40 μL of Milli-Q H_2O with an incubation period of 30 min. The data should be adjusted to calculate concentration correctly.

12. If the reaction oversaturates, the L-lactate assay solution can be diluted by addition of 25 μL of Milli-Q H_2O to 25 μL of the sample. However, the time of the reaction must be extended to 45–60 min.

13. Always create a new standard curve for each experiment to account for variations in experimental conditions.

14. As ATP is known to be highly labile, using fresh samples for the assay is recommended. Snap-freeze the samples to be assayed at a later date using liquid N_2 or dry ice.

15. The luciferase activity is highly sensitive to freeze/thaw cycle. Therefore, it is suggested to avoid repeated freezing and thawing.

16. It is recommended to use autoclaved or heat-sterilized labware to avoid any ATP or microbial contamination.

17. As the pH optima of the luciferase reaction is in the range of 7.75, it is highly advisable to correct the pH of samples with extreme pH values to a value between 7.6 and 8.0.

18. For unknown samples, a pilot experiment must be performed using various sample dilutions to ensure that the readings are within the standard curve range.

19. The ATP estimation must be performed in black or white microtiter plates.

20. Many biological samples have the inherent problem of possessing sample background. Therefore, parallel well(s) must be prepared to contain the same amount of sample as in the test well.

References

1. Chaika NV, Gebregiworgis T, Lewallen ME, Purohit V, Radhakrishnan P, Liu X, Zhang B, Mehla K, Brown RB, Caffrey T, Yu F, Johnson KR, Powers R, Hollingsworth MA, Singh PK (2012) MUC1 mucin stabilizes and activates hypoxia-inducible factor 1 alpha to regulate metabolism in pancreatic cancer. Proc Natl Acad Sci U S A 109(34):13787–13792. https://doi.org/10.1073/pnas.1203339109

2. Erickson LA, Highsmith WE Jr, Fei P, Zhang J (2015) Targeting the hypoxia pathway to treat pancreatic cancer. Drug Des Devel Ther 9:2029–2031. https://doi.org/10.2147/DDDT.S80888

3. Bertout JA, Patel SA, Simon MC (2008) The impact of O2 availability on human cancer. Nat Rev Cancer 8(12):967–975. https://doi.org/10.1038/nrc2540

4. Kamiya Mehla, Pankaj K. Singh, (2014) MUC1: A novel metabolic master regulator. Biochimica et Biophysica Acta (BBA) - Reviews on Cancer 1845(2):126–135

5. Surendra K. Shukla, Vinee Purohit, Kamiya Mehla, Venugopal Gunda, Nina V. Chaika, Enza Vernucci, Ryan J. King, Jaime Abrego, Gennifer D. Goode, Aneesha Dasgupta, Alysha L. Illies, Teklab Gebregiworgis, Bingbing Dai, Jithesh J. Augustine, Divya Murthy, Kuldeep S. Attri, Oksana Mashadova, Paul M. Grandgenett, Robert Powers, Quan P. Ly, Audrey J. Lazenby, Jean L. Grem, Fang Yu, José M. Matés, John M. Asara, Jung-whan Kim, Jordan H. Hankins, Colin Weekes, Michael A. Hollingsworth, Natalie J. Serkova, Aaron R. Sasson, Jason B. Fleming, Jennifer M. Oliveto, Costas A. Lyssiotis, Lewis C. Cantley, Lyudmyla Berim, Pankaj K. Singh, (2017) MUC1 and HIF-1alpha Signaling Crosstalk Induces Anabolic Glucose Metabolism to Impart Gemcitabine Resistance to Pancreatic Cancer. Cancer Cell 32(1):71–87.e7

6. Denko NC (2008) Hypoxia, HIF1 and glucose metabolism in the solid tumour. Nat Rev Cancer 8(9):705–713. https://doi.org/10.1038/nrc2468

7. Iyer NV, Kotch LE, Agani F, Leung SW, Laughner E, Wenger RH, Gassmann M, Gearhart JD, Lawler AM, Yu AY, Semenza GL (1998) Cellular and developmental control of O2 homeostasis by hypoxia-inducible factor 1 alpha. Genes Dev 12(2):149–162

8. Chen C, Pore N, Behrooz A, Ismail-Beigi F, Maity A (2001) Regulation of glut1 mRNA by hypoxia-inducible factor-1. Interaction between H-ras and hypoxia. J Biol Chem 276(12):9519–9525. https://doi.org/10.1074/jbc.M010144200

9. Hansen PA, Gulve EA, Holloszy JO (1994) Suitability of 2-deoxyglucose for in vitro mea-

surement of glucose transport activity in skeletal muscle. J Appl Physiol 76(2):979–985

10. Abrego J, Gunda V, Vernucci E, Shukla SK, King RJ, Dasgupta A, Goode G, Murthy D, Yu F, Singh PK (2017) GOT1-mediated anaplerotic glutamine metabolism regulates chronic acidosis stress in pancreatic cancer cells. Cancer Lett 400:37–46. https://doi.org/10.1016/j.canlet.2017.04.029

11. Goode G, Gunda V, Chaika NV, Purohit V, Yu F, Singh PK (2017) MUC1 facilitates metabolomic reprogramming in triple-negative breast cancer. PloS one 12(5):e0176820. doi:10.1371/journal.pone.0176820

12. Surendra K. Shukla, Aneesha Dasgupta, Kamiya Mehla, Venugopal Gunda, Enza Vernucci, Joshua Souchek, Gennifer Goode, Ryan King, Anusha Mishra, Ibha Rai, Sangeetha Nagarajan, Nina V. Chaika, Fang Yu, Pankaj K. Singh, (2015) Silibinin-mediated metabolic reprogramming attenuates pancreatic cancer-induced cachexia and tumor growth. Oncotarget 6(38):41146–41161

13. Sun RC, Denko NC (2014) Hypoxic regulation of glutamine metabolism through HIF1 and SIAH2 supports lipid synthesis that is necessary for tumor growth. Cell Metab 19(2):285–292. https://doi.org/10.1016/j.cmet.2013.11.022

14. Allison SJ, Knight JR, Granchi C, Rani R, Minutolo F, Milner J, Phillips RM (2014) Identification of LDH-A as a therapeutic target for cancer cell killing via (i) p53/NAD(H)-dependent and (ii) p53-independent pathways. Oncogene 3:e102. https://doi.org/10.1038/oncsis.2014.16

15. Henry RJ, Chiamori N, Golub OJ, Berkman S (1960) Revised spectrophotometric methods for the determination of glutamic-oxalacetic transaminase, glutamic-pyruvic transaminase, and lactic acid dehydrogenase. Am J Clin Pathol 34:381–398

16. Lloyd B, Burrin J, Smythe P, Alberti KG (1978) Enzymic fluorometric continuous-flow assays for blood glucose, lactate, pyruvate, alanine, glycerol, and 3-hydroxybutyrate. Clin Chem 24(10):1724–1729

17. Parks SK, Mazure NM, Counillon L, Pouyssegur J (2013) Hypoxia promotes tumor cell survival in acidic conditions by preserving ATP levels. J Cell Physiol 228(9):1854–1862. https://doi.org/10.1002/jcp.24346

18. Fan J, Kamphorst JJ, Mathew R, Chung MK, White E, Shlomi T, Rabinowitz JD (2013) Glutamine-driven oxidative phosphorylation is a major ATP source in transformed mammalian cells in both normoxia and hypoxia. Mol Syst Biol 9:712. https://doi.org/10.1038/msb.2013.65

Chapter 9

Hypoxia-Induced Metabolomic Alterations in Pancreatic Cancer Cells

Venugopal Gunda, Sushil Kumar, Aneesha Dasgupta, and Pankaj K. Singh

Abstract

Hypoxic conditions in the pancreatic tumor microenvironment lead to the stabilization of hypoxia-inducible factor-1 alpha (HIF-1α), which acts as the master regulator of cancer cell metabolism. HIF-1α-mediated metabolic reprogramming results in large-scale metabolite perturbations. Characterization of the metabolic intermediates and the corresponding metabolic pathways altered by HIF-1α would facilitate the identification of therapeutic targets for hypoxic microenvironments prevalent in pancreatic ductal adenocarcinoma and other solid tumors. Targeted metabolomic approaches are versatile in quantifying multiple metabolite levels in a single platform and, thus, enable the characterization of multiple metabolite alterations regulated by HIF-1α. In this chapter, we describe a detailed metabolomic approach for characterizing the hypoxia-induced metabolomic alterations using pancreatic cancer cell lines cultured in normoxic and hypoxic conditions. We elaborate the methodology of cell culture, hypoxic exposure, metabolite extraction, and relative quantification of polar metabolites from normoxia- and hypoxia-exposed cell extracts, using a liquid chromatography-coupled tandem mass spectrometry approach. Herein, using our metabolomic data, we also present the methods for metabolomic data representation.

Key words Hypoxia, HIF, Pancreatic Cancer, Metabolomics, Tandem Mass Spectrometry, Relative Quantification, Metabolic Pathways, Cancer Metabolism

1 Introduction

Metabolites constitute low-molecular-weight polar and nonpolar compounds of biological origin. Alteration in cellular metabolite levels can be indicators of the pathological state of the cells. Diseases associated with metabolic alterations, such as pancreatic cancer, reflect an overall derailment in cellular metabolism. Pancreatic tumors exhibit desmoplasia that results in hypoxic microenvironment [1]. Hypoxia-inducible factor-1α (HIF-1α) is one of the key proteins stabilized under hypoxic conditions that further regulates the expression of metabolic genes especially those involved in glucose uptake and glycolysis [2, 3]. This enables the pancreatic cancer cells to increase glucose flux through glycolysis and decrease its flux into the tricarboxylic acid cycle. Thus, hypoxic

L. Eric Huang (ed.), *Hypoxia: Methods and Protocols*, Methods in Molecular Biology, vol. 1742,
https://doi.org/10.1007/978-1-4939-7665-2_9, © Springer Science+Business Media, LLC 2018

metabolic alterations occurring in pancreatic cancer cells show an overall metabolic shift due to alterations in the multiple interconnected metabolic pathways [3].

The enumeration of metabolic shifts in cultured cancer cells in vitro under hypoxic conditions is possible either by physiological assays of cells exposed to hypoxia-mimicking agents or by comparing the metabolite levels in cells exposed to normoxic and hypoxic conditions [4, 5]. The physiological assays indicate the overall metabolic shift; however, they cannot directly indicate the alterations in individual metabolite levels. Complementary experimental techniques for measuring the perturbations in individual metabolite levels such as oxygen consumption rates; direct colorimetric assays for individual metabolites; enzyme-linked colorimetric, either fluorometric or luminescence, assays; and radioactive metabolite uptake or release assays are commonly employed to ascertain the alterations at individual metabolite levels [3, 6–8]. Although these approaches for evaluating alterations in individual metabolite levels exposed to hypoxia or hypoxia-mimicking agents can serve as an indirect readout for the altered metabolite levels, it cannot reflect a total shift in metabolic pathways. Thus, techniques for analyzing relative levels of multiple metabolites from individual and interconnected pathways are required for establishing the metabolic shift occurring in cells exposed to metabolic regulatory microenvironments such as hypoxia [9].

Metabolomics is a rapidly emerging technology with amenities required for analyzing multiple metabolite levels within a single platform [10]. Liquid chromatography-coupled tandem mass spectrometry (LC-MS/MS) platform can be utilized to quantify multiple metabolites of interest by chromatographic separation of the metabolites, followed by their quantification based on their characteristic fragmentation patterns obtained through single reaction monitoring (SRM) mode of data acquisition [10]. The availability of triple quadrupole mass spectrometry instruments with high-speed data acquisition capabilities has simplified the metabolomic approaches, thus enabling the wide applicability of mass spectrometer-based metabolomic techniques. Recently, we demonstrated the role of MUC1-mediated metabolic alterations in imparting gemcitabine-resistance and radiation-resistance to pancreatic cancer cells by utilizing a targeted metabolomics platform [11, 12]. Especially, the role of HIF-1α-mediated metabolic reprogramming in gemcitabine-resistance was unraveled through metabolomic analysis of cultured cells in vitro [11]. This chapter describes a detailed methodology of metabolomic analysis using pancreatic cancer cell line, S2–013 exposed to normoxic and hypoxic conditions. We describe this method in a series of steps starting from cell culture in hypoxic conditions, polar metabolite extraction, setting up the LC-MS/MS, and analysis of the generated data along with the inferences derived from our hypoxia-metabolomic experiment.

2 Materials

2.1 Cell Culture Reagents

1. DMEM/HIGH with L-glutamine, without sodium pyruvate, powdered medium (HyClone). Reconstitute as per the manufacturer's instruction along with the addition of L-sodium bicarbonate (3.7 g/L) and sodium pyruvate (0.110 g/L).

2. Fetal bovine serum (FBS).

3. Phosphate-buffered saline (PBS).

4. Trypsin (0.25%).

5. Penicillin-streptomycin solution (100×) (penicillin 10,000 units/mL; streptomycin 10,000 μg/mL).

2.2 LC-MS/MS Reagents

1. Water, mass spectrometry grade.

2. Acetonitrile, mass spectrometry grade.

3. Methanol, HPLC grade.

4. Ammonium hydroxide, 25% (w/v), certified ACS.

5. Ammonium acetate.

2.3 Equipment

1. Column: XBridge BEH Amide column (4.6 mm × 100 mm, 3.5 μm, 130 Å, Waters).

2. Guard column: ACQUITY UPLC BEH Amide VanGuard pre-column (2.1 mm × 5 mm, 130 Å, 1.7 μm, Waters).

3. UPLC: ACQUITY UPLC H-Class system (Waters).

4. Mass spectrometer: Xevo TQ-S triple quadrupole mass spectrometer (Waters). This instrument has polarity switch time 20 ms between positive and negative ion modes and a scan speed of 10,000 Da/S.

5. Bottles for preparation and storage of mobile phase: reusable Pyrex bottles.

6. Cell culture incubator: O_2/CO_2 incubator, equipped with gas-tight split inner doors for each rack.

7. Tissue culture dishes: TC dish 100 mm standard or TC dish 60 mm standard.

8. Dry ice, 15 mL polypropylene tubes, and glass pipettes.

9. Centrifugal evaporator: SpeedVac (SVC 200H concentrator, Savant)

10. Lyophilizer.

11. Centrifuge: Sorvall Legend RT refrigerated centrifuge.

12. Tissue homogenizer: Ultrasonic converter connected to Sonic Dismembrator (Fisher Scientific Series 60, Model F60).

13. Fisher Scientific mini vortex mixer.

14. Ultrasonic bath: Elmasonic S100H (Elma).

15. Vials: Autosampler vials: Verex™ certified glass vials with 300 μL insert (usable volume: 200 μL) and preSlit PTFE/Silicone Cap (Phenomenex).

16. Softwares: MassLynx (V4.1): Used for configuration and operation of UPLC and mass spectrometer and raw data acquisition. TargetLynx application manager was used for data processing and reporting.

2.4 LC-MS and Software Requirements

1. **UPLC-MS setup**: UPLC has a high resolution, which reduces the peak width at half maxima. This requires a mass spectrometer having high scan speed, shorter time to switch between positive and negative polarity mode, and high MRM acquisition rate, to obtain sufficient data points per peak. Waters TQ-S mass spectrometer has a scan speed of 20 scans/s for mass range 50–1000 Da, polarity switch time of 20 ms, and MRM acquisition rate of 250 data points per second, thus reducing the duty cycle time and providing a sufficient number of scans per metabolite. However, UPLC columns with sub-2 μ particle size offer less flexibility with mobile phase flow rates and aqueous phase composition due to high backpressure (*see* **Notes 1** and **2**).

2. **Data processing and analysis:** The software used for metabolite data processing should be compatible with acceptable input format of commonly used software such as MetaboAnalyst (http://www.metaboanalyst.ca/MetaboAnalyst). This allows direct uploading of the processed data without any manual conversion.

3 Methods

3.1 Mobile Phase Preparation

1. Methanol (80% (v/v)): Mix 80 mL of methanol with 20 mL of LC-MS grade water in a glass bottle. Store in −20 °C freezer for further use (*see* **Note 3**).

2. Mobile phase A: (95% v/v 20 mM ammonium acetate: 5% v/v acetonitrile; pH 9.0). Transfer accurately 1.5416 g of ammonium acetate to a 1 L HPLC mobile phase bottle. Add 950 mL of LC-MS grade water, cap the bottle, and mix. Add 2.8 mL of 25% (w/v) ammonium hydroxide and 50 mL of LC-MS grade acetonitrile. Cap the bottle and mix. Ensure that the pH of mobile phase A is 9.0. If not, adjust the pH to 9.0 with ammonium hydroxide or acetic acid. Mobile phase A can be stored for up to 2 weeks at room temperature (20–25 °C) (*see* **Notes 4** and **5**).

3. Mobile phase B: Use 100% LC-MS grade acetonitrile as mobile phase B.

3.2 Cell Culture and Hypoxia Exposure

1. Seed approximately 2×10^6 cells in 10 cm cell culture dishes, and culture them in DMEM containing 10% FBS.

2. Incubate cells at 37 °C in an incubator containing 5% CO_2 and 95% humidity. Maintain oxygen composition at 20% for normoxia conditions.

3. Replace the medium with fresh DMEM containing 10% FBS the next day, after the cells are attached. Use DMEM containing 10% FBS for all the subsequent cultures.

4. After 3 days, discard the medium. Wash the cells thrice with PBS and detach cells by incubating cells at 37 °C with trypsin (*see* **Note 6**).

5. After detachment, neutralize trypsin with culture medium (DMEM containing 5% FBS). Transfer the detached cell suspension to 15 mL centrifuge tube and centrifuge at $310 \times g$ for 2 min.

6. Discard the supernatant and resuspend the pellet in fresh medium. Distribute appropriate volume of cell suspension in culture dishes, and make up the volume in each dish with fresh medium. The total volume of medium should be 10 mL for 10 cm culture dish, respectively.

7. Expand the cells into culture dishes with minimum five replicates for each condition. Let the cells reach 80% confluence.

8. Exchange with fresh medium before and incubate cells for 24 h in normoxia and hypoxia incubators (*see* **Note 7**). Hypoxia treatments are maintained at 1% oxygen.

3.3 Metabolite Extraction

1. After 24-h incubation, aspirate the medium, and rinse the cells with mass spectrometry grade water (*see* **Note 8**).

2. Place the culture dishes on dry ice.

3. Add 4 mL of 80% methanol to the plate on dry ice and transfer it to −80 °C freezers. Follow the rinsing and freezing steps for other replicates (*see* **Note 9**).

4. After 10 min incubation at −80 °C, scrape the cells and collect the lysate into a 15 mL polypropylene, conical bottom tube.

5. Incubate at −80 °C for 10 min.

6. Centrifuge the tube at $3000 \times g$ for 10 min (4 °C).

7. Transfer the supernatants into a new 15 mL centrifuge tube.

8. Evaporate the samples to dryness using SpeedVac without heating to remove the methanol.

9. Freeze the aqueous fractions at −80 °C.

10. Remove the caps of the tubes and wrap the open ends with parafilm. Punch holes in the parafilm and lyophilize the samples to remove the water until the sample becomes dry (*see* **Note 10**).

3.4 Liquid Chromatography-Coupled Tandem Mass Spectrometry

1. Reconstitute the samples in 100 µL of 50% acetonitrile (v/v). Vortex for 15 s and centrifuge at $16,000 \times g$ for 10 min at (4–8 °C). Transfer 75 µL of supernatant to auto-injector vials for subsequent analysis using LC-MS/MS.

2. Perform liquid chromatography using the following UPLC parameter gradient method at a constant flow rate of 0.280 mL/min.

Time (min)	0	3	7	12	15	17	19	22	26	30
% A	85	84	65	60	55	50	50	70	85	85
% B	15	16	35	40	45	50	50	30	15	15

3. MS parameters: Source temperature should be 150 °C, desolvation temperature 600 °C, cone gas flow 149 L/h, desolvation gas flow 1191 L/h, and collision gas flow 0.24 mL/min, and switch time between positive and negative polarity should be 50 ms.

3.5 Data Acquisition

Acquire data for each metabolite of interest using:

1. Single Q1/Q3, selected reaction monitoring (SRM) transition.
2. Cone voltage (V).
3. Collision energy (eV).
4. Optimal dwell time (ms) (*see* **Note 11**).

3.6 Data Processing

Open the raw file in TargetLynx software and follow the steps for integrating individual peaks as per the vendor's instructions.

1. Export the final data containing peak areas, retention times, sample, and individual metabolite labels, and organize the data by grouping the samples by organizing the replicates of cells treated in normoxic and hypoxic conditions.

3.7 Selecting the Metabolite Peaks for Comparative Analysis

1. Check the variance among the individual metabolite peak areas within the biological replicates of a group by using statistical variance analysis.
2. Metabolites with a fold variance value >2 between the replicates should be excluded from the analysis.
3. Metabolites with low abundance (peaks with low signal/noise, S/N) are identified with their very low peak area values and excluded from analysis (*see* **Note 12**).

3.8 Data Formatting for Fold Change Analyses

1. Normalize the data by using the factor obtained from the ratios of protein content obtained from normoxic and hypoxic group samples. **Formula for normalization**: peak area from

hypoxic cells × (protein concentration from cells cultured in normoxia/protein concentration from cells cultured in hypoxia) (*see* **Note 13**).

2. Obtain statistical mean for individual metabolites in one group (e.g., normoxia).

3. Divide each metabolite peak area value in experimental replicate samples from normoxia group with the mean area value obtained for the same metabolite in normoxia.

4. Divide each normalized peak area of the experimental replicate samples from the hypoxia group with the corresponding mean value obtained for the same metabolite from the normoxia group (*see* **Note 14**).

5. Use the fold change values obtained in **steps 3** and **4** for the graphical data analysis (using GraphPad Prism), and data can be represented as relative values as shown in Fig. 1.

3.9 Data Analysis Using MetaboAnalyst and Data Interpretation

1. Organize the normalized peak area data according to the instructions provided on the MetaboAnalyst website (http://www.metaboanalyst.ca/MetaboAnalyst).

2. Unsupervised hierarchical clustering analysis of the data sets show clustering of the individual samples based on their relatedness as shown in Fig. 2

3. Apply partial least square-discriminant analysis (PLS-DA) from MetaboAnalyst tools to obtain the clustering patterns among the replicates and the groups under study as shown in Fig. 3.

4. The close clustering of the biological replicates in the heatmaps and the PLS-DA analysis is an indication of low variance among the biological replicates of the samples under each condition.

5. Apply pathway impact analysis available in MetaboAnalyst tools to identify the individual metabolic pathways altered under hypoxic conditions as shown in Fig. 4.

4 Notes

1. Optimize the flow rate and composition of the buffers, if using sub-2 μ particle-sized columns to maintain the column backpressure below 12,000 psi. Avoid use of the columns having higher particle size (>3.5 μ) which lower the chromatographic resolution.

2. Maintain constant temperature in autosampler using refrigerated autosampler, and the column temperature must be constant throughout the analysis.

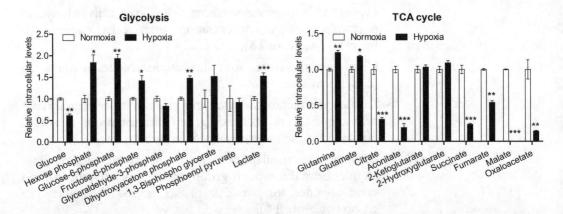

Fig. 1 Relative quantification of glycolysis and TCA cycle metabolites. The graphical data representation of fold change in glycolysis and TCA cycle metabolite levels from S2–013 cells cultured under normoxia and hypoxia. Peak areas were normalized in comparison to normoxia values, and fold changes obtained from the normalized values are used for the graphical presentation using Prism software. Statistical analysis presented in the figure represents mean ± SEM obtained for four biological replicates obtained by Student's t-test. Data represented in the bar plot indicate an increase in glycolysis and an overall decrease in TCA cycle metabolites under hypoxia in S2–013 cells

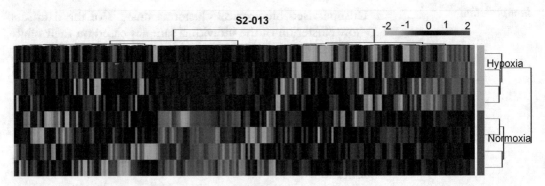

Fig. 2 Heatmap showing unsupervised hierarchical clustering analysis. The heatmap shows the output of relative metabolite analysis performed using polar metabolite data obtained from the S2–013 cells subjected to 24-h normoxia and hypoxia treatments (four replicates for each condition). Each row in the heatmap represents a replicate and each column represents a metabolite. The red, black, and green colors indicate the relative abundance intensity of each metabolite within a sample. Scale bar for color intensity is in the top right corner of the figure. Dendrogram on the right side of the figure indicates that the S2–013 cells cultured under normoxia and hypoxia possess different metabolic profiles, and the replicates are tightly clustered together

3. Internal standard (IS) of choice can be added to 80% methanol v/v. Internal standards can be used either for evaluating the consistency in the detection method or evaluating the stability of metabolite during processing and analysis.

4. Buffer concentration > 25 mM is not recommended, and the pH of the mobile phase should be within the tolerance range of the column.

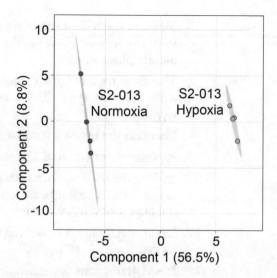

Fig. 3 Partial least square-discriminant analysis plot. PLS-DA plot showing the separation S2–013-normoxia (red) and S2–013-hypoxia (green) groups based on their polar metabolite contents. Each colored circle indicates one replicate of the group. Component 1 indicates the degree of variation between the normoxia and hypoxia groups based on their total polar metabolite content, and component 2 indicates the differences within the groups. It can be noted that component 1 value is sevenfold greater than component 2, indicating that the difference between the groups is greater than within the group

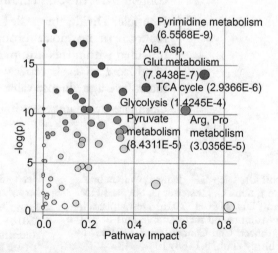

Fig. 4 Pathway impact analysis. Pathways significantly altered in S2–013 cells upon hypoxic treatment. Sizes of the circles correlate with the number of metabolites matching between the input data and the databases utilized by the pathway analysis tool. In general, human database is applicable for this kind of analysis. The color gradient used in filling the circles indicates the statistical significance values. Dark red circles indicate pathways highly, significantly altered, and the significance decreases, as the color becomes light yellow. Consider pathways represented by larger circles and having higher significant values for further analysis (*see* **Note 15**)

5. Measure the pH of the buffers in aliquots in a different tube from the mobile phase bottle. Do not dip electrodes into the mobile phase bottle.

6. The time required for detachment of cells by trypsin varies with the cell type. Thus, avoid overexposure of cells to trypsin.

7. Maintain the hypoxia incubator at 1% O_2 using nitrogen gas.

8. Aspirate water within 5 s to avoid osmotic lysis of the cells.

9. Adding the methanol for each plate after transferring to dry ice makes it easier to scrape the cells rather than adding the methanol after aspirating water from the plates.

10. Dried metabolite samples can be stored at −80 °C for several weeks.

11. MRM transitions for individual metabolites can be obtained either from published reports or mass bank databases.

12. Exclude peaks with areas <10,000 from analysis. However, if the metabolite peak area is high (>10,000) among all the replicates in one condition (e.g., in normoxia) but low in other condition, such data can be retained for further comparative analyses.

13. Normalization can also be performed using cell count or DNA content from the cells cultured in normoxia and hypoxia.

14. Example: Divide the peak areas of glucose values from each replicate of normoxia group with the mean/average value obtained within this group. Divide the normalized peak areas of glucose in each replicate from hypoxia group with the mean/average glucose value obtained from normoxia group.

15. Compare two data sets using pathway impact analysis.

References

1. Whatcoff CJ, Diep CH, Jiang P, Watanabe A, LoBello J, Sima C, Hostetter G, Shepard HM, Von Hoff DD, Han H (2015) Desmoplasia in primary tumors and metastatic lesions of pancreatic cancer. Clin Cancer Res 21(15):3561–3568. http://doi:10.1158/1078-0432

2. Mehla K, Singh PK (2014) MUC1: a novel metabolic master regulator. Biochim Biophys Acta 1845(2):126–135. http://doi:10.1016/j.bbcan.2014.01.001

3. Chaika NV, Gebregiworgis T, Lewallen ME, Purohit V, Radhakrishnan P, Liu X, Zhang B, Mehla K, Brown RB, Caffrey T, Yu F, Johnson KR, Powers R, Hollingsworth MA, Singh PK. (2012) MUC1 mucin stabilizes and activates hypoxia-inducible factor 1 alpha to regulate metabolism in pancreatic cancer. Proc Natl Acad Sci U S A 109(34):13787–13792. http://doi:10.1073/pnas.1203339109

4. Byrne MB, Leslie MT, Gaskins HR, Kenia PJA (2014) Methods to study the tumor microenvironment under controlled oxygen conditions. Trends Biotechnol 32(11):556–563. http://:doi:10.1016/j.tibtech.2014.09.006

5. Hsu SH, Chen CT, Wei YH (2013) Inhibitory effects of hypoxia on metabolic switch and osteogenic differentiation of human mesenchymal stem cells. Stem Cells 31(12):2779–2788. http://doi: 10.1002/stem.1441

6. Solaini G, Baracca A, Lenaz G, Sgarbi G (2010) Hypoxia and mitochondrial oxidative metabolism. Biochim Biophys Acta 1797(6–7):1171–1177. http://doi: 10.1016/j.bbabio.2010.02.011

7. Jiang J, Auchinvole C, Fisher K, Campbell CJ (2014) Quantitative measurement of redox potential in hypoxic cells using SERS nanosensors. Nanoscale 6(20):12104–12110. http://doi: 10.1039/c4nr01263a

8. Winer PLS, Wu M (2014) Rapid analysis of glycolytic and oxidative substrate flux of cancer cells in a microplate. PLoS One 9(10):e109916. http://doi: 10.1371/journal.pone.0109916

9. Feala JD, Coquin L, Zhou D, Haddad GG, Paternostro G, McCulloch AD (2009) Metabolism as means for hypoxia adaptation: metabolic profiling and flux balance analysis. BMC Syst Biol 3:91. http:// doi: 10.1186/1752-0509-3-91

10. Gunda V, Yu F, Singh PK (2016) Validation of metabolic alterations in microscale cell culture lysates using hydrophilic interaction liquid chromatography (HILIC)-tandem mass spectrometry-based metabolomics. PLoS One 11(4):e0154416. https://doi.org/10.1371/journal.pone.0154416

11. Shukla SK, Purohit V, Mehla K, Gunda V, Chaika NV, Vernucci E, King RJ, Abrego J, Goode GD, Dasgupta A, Illies AL, Gebregiworgis T, Dai B, Augustine JJ, Murthy D, Attri KS, Mashadova O, Grandgenett PM, Powers R, Ly QP, Lazenby AJ, Grem JL, Yu F, Matés JM, Asara JM, Kim J, Hankins JH, Weekes C, Hollingsworth MA, Serkova NJ, Sasson AR, Fleming JB, Oliveto JM, Lyssiotis CA, Cantley LC, Berim l, Singh PK (2017) MUC1 and HIF-1alpha Signaling Crosstalk Induces Anabolic Glucose Metabolism to Impart Gemcitabine Resistance to Pancreatic Cancer. Cancer Cell 32(1):71–87.e7. http://doi: 10.1016/j.ccell.2017.06.004

12. Gunda V, Souchek J, Abrego J, Shukla SK, Goode GD, Vernucci E, Dasgupta A, Chaika NV, King RJ, Li S, Wang S, Yu F, Bessho T, Lin C, Singh PK (2017) MUC1-Mediated Metabolic Alterations Regulate Response to Radiotherapy in Pancreatic Cancer. Clinical Cancer Research 23(19):5881–5891. http://doi: 10.1158/1078-0432.CCR-17-1151

Chapter 10

Hypoxia-Mediated In Vivo Tumor Glucose Uptake Measurement and Analysis

Surendra K. Shukla, Scott E. Mulder, and Pankaj K. Singh

Abstract

Most solid tumors are hypoxic in nature due to the limited supply of oxygen to internal tissues. Hypoxia plays an important role in metabolic adaptations of tumors that contribute significantly to cancer pathogenesis. Among the several metabolic alterations induced by hypoxia, hypoxia-mediated increased glucose uptake serves as the hallmark of metabolic reprogramming. Hypoxia-mediated stabilization of hypoxia-inducible factor-1 alpha (HIF-1α) transcription factor leads to altered expression of several glycolytic genes and glucose transporters, which results in increased glucose uptake by tumor cells. Here we describe an easy and simple way of measuring the hypoxia-mediated tumor glucose uptake in vivo. The method is based on fluorescent imaging probe, RediJect 2-DG, which is a nonradioactive fluorescent-tagged glucose molecule. We have discussed orthotopic tumor implantation of HIF-1α knockdown and control pancreatic cancer cells and glucose uptake measurement in vivo by using IVIS imaging system along with reagent preparations.

Key words 2-DG, Glucose uptake in mice tumors, In vivo glucose uptake, Hypoxia, HIF-1α, Cancer metabolism

1 Introduction

Cancer cells exhibit several metabolic alterations in order to fulfill their energy and biomass requirements to support uncontrolled growth and proliferation [1]. In most cancer cells glucose serves as a major source of energy that gets utilized primarily through aerobic glycolysis, a phenomenon known as Warburg effect [2]. Tumor cells not only utilize glucose for energy and biomass requirements but also as a regulator of several signaling pathways, which drive cellular proliferation and death [3]. Since several studies have deciphered a close relationship between metabolic reprogramming and cancer pathogenesis, metabolic imaging has gained considerable interest in the field of tumor identification [4]. ^{18}F–FDG-PET, utilizing a radiolabeled glucose analog, is routinely employed in imaging several solid tumors [5]. Oncogenic signaling and molecular pathways are closely connected with metabolic alterations in

L. Eric Huang (ed.), *Hypoxia: Methods and Protocols*, Methods in Molecular Biology, vol. 1742,
https://doi.org/10.1007/978-1-4939-7665-2_10, © Springer Science+Business Media, LLC 2018

tumor cells. It has clearly been shown that several oncogenes and tumor suppressors regulate the glycolytic nature of cancer cells [6]. Since oncogenic signals play a critical role in metabolic reprogramming of cancer cells, metabolic imaging can be considered as readout of their activity.

Most of the solid tumors are poorly oxygenated and, hence, are hypoxic [7]. Hypoxia regulates several aspects of cancer cells and their response toward chemotherapy and radiotherapy [8]. Hypoxia leads to increased glucose uptake by regulating expression of several glycolytic genes [9, 10]. The hypoxic environment in the cells leads to stabilization of hypoxia-inducible factor-1a (HIF-1α), which is a key transcriptional regulator of several glycolytic genes [11–14]. Here we describe the method of glucose uptake measurement in response to knockdown of HIF-1α gene in cancer cells. We also describe the orthotopic implantation of control and HIF-1α knockdown pancreatic cancer cells in the pancreas of athymic nude mice, in vivo glucose uptake measurement in the tumor using IVIS Spectrum in vivo imaging system, and data analysis.

2   Materials

Cell culture medium (DMEM).

Fetal bovine serum.

Cancer cell lines (HIF-1α knockdown and control).

Athymic nude mice.

RediJect 2-DG fluorescent probe (PerkinElmer).

Normal saline (NaCl, 0.85% w/v).

Syringes with needle (27G).

Animal surgical instruments.

Ethilon nylon sutures.

Chromic gut sutures.

  Betadine™ Surgical Scrub (Fisher Scientific).

  Betadine Solution Swab (Moore Medicals).

  Cotton swabs.

  Anesthesia chamber with oxygen supply.

Alcohol wipes.

Heating pads.

IVIS Spectrum in vivo imaging system (PerkinElmer).

3 Methods

3.1 Orthotopic Implantation of Pancreatic Cancer Cells into the Pancreas of Athymic Nude Mice

All procedures should be performed according to the IACUC guidelines of the institute.

1. Athymic nude mice (6–8 weeks) should be utilized for orthotopic implantation of human pancreatic cancer cell lines into the mice pancreas.

2. Prepare the desired volume of cells with the required concentration in PBS. Place the cells on ice (see **Note 1**).

3. Before surgery, prepare all the required instruments and accessories. Autoclave the instruments properly.

4. Arrange the anesthesia chamber, nose cones attached to the anesthesia chamber, surgery instruments, and accessories inside a laminar hood (see **Note 2**).

5. Anesthetize the mice by using an isoflurane and oxygen mixture in the anesthesia chamber. When the mouse is completely anesthetized, take it out and place laterally on a surgery stand. Place the nose of mouse in a nose cone connected with an anesthesia machine (see **Note 3**).

6. To begin with, prepare the site of surgery by wiping the skin thrice with Betadine followed by cleaning the area thrice with ethanol wipes.

7. First, open the upper skin (dermis) of the mouse with the help of a sharp scissors, and then make a small cut in the peritoneum just above the spleen area.

8. Take the desired volume of cell suspension into a syringe.

9. Gently take the spleen out and hold it with the help of forceps and inject the cell suspension into the pancreas (see **Note 4**).

10. Gently push the pancreas and spleen inside the peritoneum.

11. Suture the peritoneum with chromic gut sutures (Ethicon).

12. Suture the skin with nylon sutures (Ethilon).

13. Place the mice in a fresh, sterile mouse cage on a heating pad (see **Note 5**).

14. When mice are completely awake, gently transfer them to a fresh cage with food and water.

15. For the next 2 days, inject analgesic as per the institutional guidelines.

3.2 In Vivo Glucose Uptake Imaging Using IVIS Spectrum

When the tumor becomes palpable, glucose uptake can be measured by using IVIS Spectrum.

1. Bring the RediJect 2-DG probe to room temperature before injection (see **Note 6**).

2. Wipe the injection surface of mice twice with alcohol wipes.

3. Inject 100 µL of the probe in each mouse by intraperitoneal injections.

4. Images can be acquired at time points such as 3, 12, and 24 h post injection.

5. Bring the mice cages to the imaging room.

6. Switch on the IVIS system (if it's not already on, generally we don't switch off IVIS system). Click on *Living Image* software icon on the desktop.

7. Initialize the software. The computer may take several minutes for checking the system and cooling down the camera to −90°C. Camera temperature can be checked by clicking the red bar at the bottom of the software window. Once the system is initialized, the red bar at bottom will turn to green.

8. Set up the nose cones in the IVIS machine (*see* **Note 7**).

9. When the machine is ready, start the anesthesia chamber. First, turn on the oxygen supply, and then turn on the pump.

10. Place mice in the induction chamber. Turn on the isoflurane knob to 2–2.5. Turn the induction chamber on (*see* **Note 8**).

11. Once the mice are anesthetized, switch ON the IVIS flow and place the mice inside IVIS machine laterally.

12. Set up the field view, it should be E for five mice.

13. Set up the exposure time which is generally 10 s for fluorescence imaging. The ideal spectrum setup for imaging is 745 nm excitation and 820 nm emission.

14. Click the acquire button and wait till the image acquisition is completed (*see* **Note 9**).

15. Save the image by activating the autosave mode.

16. After imaging is completed, turn the isoflurane dial to zero.

17. Close the oxygen supply and turn the pump and anesthesia machine off.

18. Clean the induction chamber with Clidox-S.

19. Clean the imaging platform of IVIS machine with Clidox-S (*see* **Note 10**).

20. Refill the isoflurane in the anesthesia machine vaporizer, quit the *Living Image* software, and log off the computer.

3.3 Analysis of Glucose Uptake

1. Open the image by utilizing *Living Image* software (represented in Fig. 1).

2. On the right side in tool icon, select ROI measurements.

3. According to the analysis, you can select a single or multiple ROI.

4. Set up photons or counts on the top left side of the image.

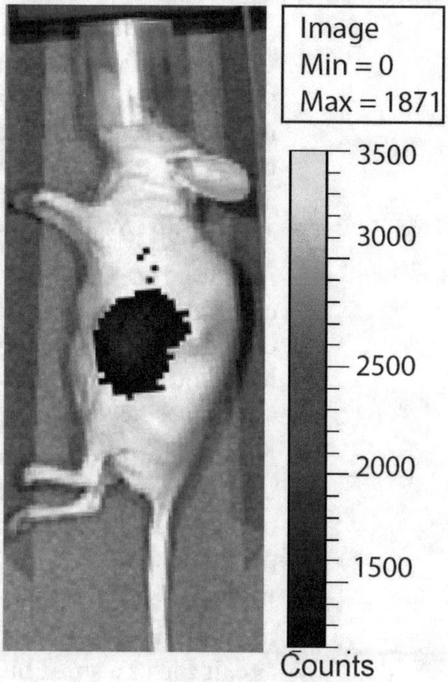

Fig. 1 Representative image of in vivo glucose uptake. A tumor-bearing mouse was injected with 100 μL RediJect 2-DG probe, and after 12 h of injection, fluorescent images were captured by using IVIS Spectrum in vivo imaging system (PerkinElmer). The images were processed by using **Living Image software**

5. Set up the minimum and maximum counts/photon density on image adjustment scale bar on the right side of the image. It should be constant for all the images that are being compared.

6. Click the measurement icon. A new window will pop up that will show total photon/counts along with the standard deviation. Note down the total photon numbers/counts.

7. Determine the tumor volume of corresponding mice by palpation and measuring with a caliper.

8. Normalize the photons/counts per unit volume the tumor.

9. Represent the tumor glucose uptake relative to the control group.

10. Minimum 6–8 mice should be measured in each group.

11. Images of individual mice can be exported in JPEG/png format.

12. Representative images can be shown along with a bar chart with intensity scale.

4   Notes

1. Cells should not be stored more than 3 h as they lose viability significantly. Cell number per mL should be adjusted according to the requirements of cells that need to be implanted per mice. We generally use cell number between 100,000 and 500,000 cells/mouse. Defined number of cells should be present in a 50 μL of volume.

2. All the procedures on the animals should be performed in a completely sterile environment. Avoid touching the surroundings during the surgery procedure. Frequently sterilize your hand and working surfaces with 70% ethanol and change gloves as often as needed.

3. Nose cone regulator in the anesthesia chamber should be in the "off" position while not using the nose cone. It will avoid unnecessary inhalation of anesthesia by the working personnel.

4. Be very careful at this stage; avoid the leakage of cancer cell through the pancreas. If you inject cells properly without leakage, it forms a small bubble.

5. Heating pads should be set up a half an hour before the surgery. Keep heating pad's temperature low/medium; it should never be on a high.

6. Solution can be diluted in PBS for more economic use. We generally use 1:10 diluted solution for imaging after 24 h of injection.

7. To avoid cross contamination or infection from other mice, use your own set of nose cones on a shared equipment.

8. Ear tag the mice for further analysis.

9. If the signal is saturated, you can reduce the exposure time.

10. Never use bleach or alcohol for any part of the IVIS machine. Spray Clidox-S on a paper towel to wipe the machine platform. Never spray anything inside the machine.

References

1. Galluzzi L, Kepp O, Vander Heiden MG, Kroemer G (2013) Metabolic targets for cancer therapy. Nat Rev Drug Discov 12(11):829–846. https://doi.org/10.1038/nrd4145

2. Warburg O (1956) On the origin of cancer cells. Science 123(3191):309–314.

3. Zhao Y, Wieman HL, Jacobs SR, Rathmell JC (2008) Mechanisms and methods in glucose metabolism and cell death. Methods Enzymol 442:439–457. https://doi.org/10.1016/S0076-6879(08)01422-5

4. Plathow C, Weber WA (2008) Tumor cell metabolism imaging. J Nucl Med 49(Suppl 2):43S–63S. https://doi.org/10.2967/jnumed.107.045930

5. Zhu A, Lee D, Shim H (2011) Metabolic positron emission tomography imaging in cancer detection and therapy response. Semin Oncol

38(1):55–69.   https://doi.org/10.1053/j.
seminoncol.2010.11.012

6. Bui T, Thompson CB (2006) Cancer's sweet
tooth. Cancer Cell 9(6):419–420. https://doi.
org/10.1016/j.ccr.2006.05.012

7. Brown JM (2007) Tumor hypoxia in cancer
therapy. Methods Enzymol 435:297–321.
h t t p s : / / d o i . o r g / 1 0 . 1 0 1 6 /
S0076-6879(07)35015-5

8. Wilson WR, Hay MP (2011) Targeting hypoxia
in cancer therapy. Nat Rev Cancer 11(6):393–
410. https://doi.org/10.1038/nrc3064

9. Chaika NV, Gebregiworgis T, Lewallen ME,
Purohit V, Radhakrishnan P, Liu X, Zhang B,
Mehla K, Brown RB, Caffrey T, Yu F, Johnson
KR, Powers R, Hollingsworth MA, Singh PK
(2012) MUC1 mucin stabilizes and activates
hypoxia-inducible factor 1 alpha to regulate
metabolism in pancreatic cancer. Proc Natl
Acad Sci U S A 109(34):13787–13792.
https://doi.org/10.1073/pnas.1203339109

10. Mehla K, Singh PK (2014) MUC1: a novel
metabolic master regulator. Biochim Biophys
Acta   1845(2):126–135.   https://doi.
org/10.1016/j.bbcan.2014.01.001

11. Eales KL, Hollinshead KE, Tennant DA (2016)
Hypoxia and metabolic adaptation of cancer

cells.   Oncogene   5:e190.   https://doi.
org/10.1038/oncsis.2015.50

12. Shukla SK, Gebregiworgis T, Purohit V, Chaika
NV, Gunda V, Radhakrishnan P, Mehla K,
Pipinos II, Powers R, Yu F, Singh PK (2014)
Metabolic reprogramming induced by ketone
bodies diminishes pancreatic cancer cachexia.
Cancer Metab 2(1):18.

13. Abrego J, Gunda V, Vernucci E, Shukla SK,
King RJ, Dasgupta A, Goode G, Murthy D, Yu
F, Singh PK (2017) GOT1 -mediated anaple-
rotic glutamine metabolism regulates chronic
acidosis stress in pancreatic cancer cells. Cancer
Lett 400:37–46.

14. Shukla SK, Purohit V, Mehla K, Gunda V,
Chaika NV, Vernucci E, King RJ, Abrego J,
Goode GD, Dasgupta A, Illies AL,
Gebregiworgis T, Dai B, Augustine JJ, Murthy
D, Attri KS, Mashadova O, Grandgenett PM,
Powers R, Ly QP, Lazenby AJ, Grem JL, Yu F,
Matés JM, Asara JM, Kim J-W, Hankins JH,
Weekes C, Hollingsworth MA, Serkova NJ,
Sasson AR, Fleming JB, Oliveto JM, Lyssiotis
CA, Cantley LC, Berim L, Singh PK (2017)
MUC1 and HIF-1alpha signaling crosstalk
induces anabolic glucose metabolism to impart
gemcitabine resistance to pancreatic cancer.
Cancer Cell 32(1):71–87.e7.

Chapter 11

Measurement of Sensory Nerve Activity from the Carotid Body

Ying-Jie Peng and Nanduri R. Prabhakar

Abstract

Carotid bodies are sensory organs for monitoring chemical composition of the arterial blood, especially the O_2 levels. Carotid bodies are located bilaterally at the bifurcation of the common carotid artery. Hypoxia increases sensory nerve activity of the carotid body, which is transmitted to the brainstem neurons triggering reflex stimulation of breathing and blood pressure. Measurement of action potentials from the carotid sinus nerve is a widely used approach for understanding the mechanisms of hypoxic sensing by the carotid body. Here, we describe the detailed methodology for recording action potential signals from the carotid sinus nerve from in vivo and ex vivo carotid bodies from rats.

Key words Action potential, Oxygen sensing, Chemoreflex, Blood pressure, Single unit

1 Introduction

The carotid body is a major sensory organ for detecting changes in arterial blood O_2 levels. Anatomically, carotid bodies are located bilaterally at the bifurcations of the common carotid arteries. The carotid sinus nerve, which is a branch of the glossopharyngeal nerve, provides sensory innervation to the carotid body. The chemoreceptor tissue is composed of two cell types: type I (glomus) and type II (sustentacular) cells. A substantial body of evidence suggests that glomus cells are the primary site of sensory transduction and they work in concert with the nearby afferent nerve ending as a "sensory unit" [1, 2]. Sensory nerve activity of the carotid body is low under normoxia (arterial pO_2 ~100 mmHg), which increases in a curvilinear manner in response to progressive hypoxia. The increased sensory nerve activity is conveyed to the brainstem neurons, triggering reflex activation of breathing and blood pressure. Much of the information on O_2 sensing by the carotid body came from the studies on cats [1]. In recent years, many investigators employ rats for studying carotid body function, because of the cost effectiveness [3, 4]. However, unlike cats, carotid bodies from

L. Eric Huang (ed.), *Hypoxia: Methods and Protocols*, Methods in Molecular Biology, vol. 1742,
https://doi.org/10.1007/978-1-4939-7665-2_11, © Springer Science+Business Media, LLC 2018

rats are physically small. On average, a rat carotid body weighs ~60 µg [1, 2]. This chapter is intended to provide detailed methodologies for monitoring the sensory nerve activity from in vivo and ex vivo carotid bodies from rats (*see* **Note 1**).

2   Materials

2.1   Surgery

1. Urethane.
2. Saline.
3. Heparin.
4. Isopropyl alcohol (70%).
5. Cotton-tipped applicators.
6. Gauze sponges.
7. Venous catheter (PE 50 tubing).
8. Arterial catheter (PE 60 tubing).
9. Tracheal catheter (PE 240 tubing).
10. Silk threads (2-0, 4-0).
11. Syringes (1 cm³, 5 cm³).
12. Scotch tape.
13. Slings.
14. Small animal surgical board.
15. Mayo scissors.
16. Iris scissors.
17. Micro iris scissors.
18. Micro scissors (SC-MS-152, Braintree Scientific, Inc.).
19. Dressing forceps.
20. Vessel cannulation forceps.
21. Fine forceps (#5.5).
22. Alm minor surgery retractor (Braintree Scientific, Inc.).
23. Scalpel handle and blade.
24. Moria vascular clamp (Fine Science Tools).
25. Electric clipper for hair removal.

2.2   In Vivo Carotid Body

1. Rat head holder.
2. Heating pad (Gaymar).
3. Thermistor probe (Gaymar).
4. Mineral oil.
5. Pancuronium bromide (Sigma).

6. Surgical spears (Braintree Scientific, Inc.).

7. Unipolar platinum-iridium electrode (A-M Systems).

8. Electrode holder (A-M Systems).

9. Reference electrode (silver-silver chloride).

10. Gas tanks (100% O_2, 12% O_2-balanced N_2).

2.3 Ex Vivo Carotid Body

1. Syringes (60 cm^3).

2. Physiological saline: 125 mM NaCl, 5 mM KCl, 1.8 mM $CaCl_2$, 2 mM $MgSO_4$, 1.2 mM NaH_2PO_4, 25 mM $NaHCO_3$, 10 mM glucose, and 5 mM sucrose, bubbled with 5% CO_2-balanced 95% O_2, pH 7.38–7.4. The physiological saline should be prepared fresh before the experiment.

3. Ca^{2+}-/Mg^{2+}-free modified Tyrode solution: 140 mM NaCl, 5 mM KCl, 10 mM HEPES, 5 mM glucose, and pH 7.2 (adjusted by NaOH).

4. 0.1% collagenase in Ca^{2+}-/Mg^{2+}-free modified Tyrode solution.

5. Ice and ice bucket.

6. Transfer pipette.

7. Conical tube (15 mL).

8. Petri dish (40 mm, 90 mm).

9. Polymerized Sylgard.

10. Insect pins (Fine Science Tools).

11. Glass capillaries (1.2 mm × 0.68 mm, A-M Systems).

12. Suction electrode (A-M Systems).

13. Circulating water bath (Cole-Parmer).

14. Gas tanks (95% O_2 + 5% CO_2, 1% O_2 + 5% CO_2).

2.4 Equipment and Software

1. Weighing scale for rodents.

2. pH meter.

3. Binocular dissecting microscope (Wild Heerbrugg).

4. Thermal cautery unit (Geiger).

5. Blood pressure transducer (CWE, Inc.).

6. Transducer amplifier (TA-100, CWE, Inc.).

7. Animal respirator (Model 683, Harvard Apparatus).

8. Blood gas analyzer (ABL 80, Radiometer America).

9. Micromanipulator (WPI).

10. Pipette puller (PP-83, Narishige).

11. Tissue chamber (Cell MicroControls).

12. Temperature controller (Cell MicroControls).

13. AC preamplifier (P511K, Grass Instrument).

14. Oscilloscope (DL708E, Yokogawa).

15. Audio monitors (Model 3300, A-M Systems).

16. Window discriminator (Model 121, WPI).

17. Rate meter (RIC-830, CWE, Inc.).

18. Power Lab (Model 8/35, AD Instruments).

19. LabChart 7 Pro (AD Instruments).

20. Personal computer.

3 Methods

3.1 Recording Action Potentials from the Carotid Sinus Nerve in Vivo

1. Anesthetize rats with intraperitoneal administration of urethane (1.2 g/kg). If the animal exhibits limb withdrawal reflex in response to noxious pinching of the toe, administer supplemental doses of anesthesia (urethane, 15% of initial dose) (*see* **Note 2**).

2. Shave the surgical field (i.e., neck and right groin) with an electric clipper. Clean the area with 70% alcohol. Place the rat on its back on a surgical board covered with a heating pad. Mount the head of the rat to a head holder. Secure the limbs using slings and tapes. The rectal temperature of the rat is monitored by a thermistor probe and maintained between 37 and 38 °C with the thermostatically controlled heating pad.

3. Using the scalpel, make a 2-cm-long midline incision in the sub-hyoid region. Expose the trachea along with the esophagus. Pass a 2-0 silk thread under the trachea and esophagus and make a loose ligature around the tissue. Make a 2-mm-long transverse incision in the trachea cranial to the ligature using micro iris scissors. Insert a tracheal catheter (PE 240 tubing) into the trachea toward the lungs, and tighten the ligature to secure the catheter in the trachea.

4. *This step is done under a well-lit dissecting microscope.* Make a 3-cm-long incision in the right inguinal region (*see* **Note 3**). Expose the femoral artery and vein of about 1–2 cm. Make two loose ligatures about 1 cm apart around the artery with 4-0 silk threads. Tighten the posterior ligature to occlude blood flow. Clamp the artery cranial to the anterior ligature using a vascular clamp, and then make a small transverse incision on the artery between the two ligatures with micro iris scissors. Insert the arterial catheter into the artery and slowly push it toward the abdominal aorta over a length of 0.5 cm (*see* **Note 4**). Then tighten the ligature, and release the clamp. Perform a similar procedure to cannulate the femoral vein. Cover the incision with moisturized gauze.

5. Connect the femoral artery catheter to a blood pressure transducer connected to an amplifier (*see* **Note 5**). The blood pressure signals are collected using a data acquisition system (Power Lab, AD Instruments) and stored in a computer for later analysis.

6. Connect the tracheal cannula to a rodent respirator for mechanical ventilation. Set the tidal volume at 1.5 mL and the rate respiration at 70 strokes/min. Administer pancuronium bromide (2.5 mg/kg/h; i.v.) to paralyze the animal (*see* **Note 6**).

7. Ligate the trachea and esophagus cranial to the tracheal cannula with 2-0 silk thread. Transect all tissues between the ligature and the tracheal cannula, and retract the ligature along with the tissue in the rostral direction, and fix it with a piece of tape. Place an Alm retractor in the neck region to enlarge the surgical field. Make a loose ligature (2-0 silk thread) around the left external carotid artery, and retract it laterally (Fig. 1) (*see* **Note 7**).

8. *This step should be performed under well-lit dissecting microscope.* Locate the superior cervical sympathetic ganglion (SCG),

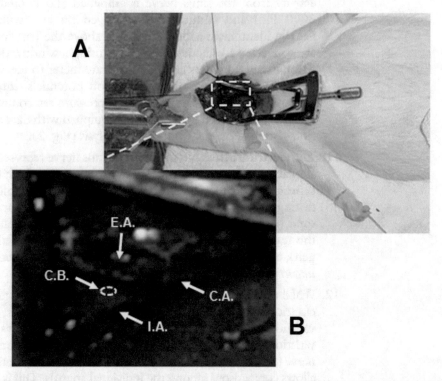

Fig. 1 (**a**) Rat placed in a head holder for exposing the carotid bifurcation. Note that the head holder is turned left slightly to provide good visualization of the left carotid bifurcation. The left external artery is gently retracted laterally to further improve the visualization. (**b**) Enlarged image of the carotid bifurcation enclosed in dashed rectangle in A. *EA* external carotid artery, *IA* internal carotid artery, *CA* common carotid artery, *CB* carotid body

where it joins the cervical sympathetic nerve. Ablate the SCG using micro iris scissors, and identify the carotid body, which lies underneath the SCG. By blunt dissection clean the carotid body from the surrounding connective tissue, and identify the carotid sinus nerve. Using fine forceps (#5.5), carefully isolate the nerve from the surrounding tissue. Transect the nerve where it joins the glossopharyngeal nerve. With fine forceps (#5.5), remove the sheath surrounding the carotid sinus nerve. To prevent drying, immerse the nerve in warm mineral oil.

9. Collect arterial blood samples (100 μL each) from the arterial catheter in heparinized syringes. Measure arterial pO_2, pCO_2, and pH using a blood gas analyzer (*see* **Note 8**).

10. A unipolar platinum-iridium electrode is used for recording action potentials from the carotid sinus nerve. Place the electrode in an electrode holder connected to micromanipulator. Adjusting the micromanipulator, lower the recording electrode close to the carotid sinus nerve. Gently place the nerve on the recording electrode. Cover the electrode and the nerve with warm mineral oil to prevent drying of the nerve. Place the reference electrode in the neck muscle (*see* **Note 9**). Electrical activity from the sinus nerve is amplified and filtered (100–3000 Hz) and continuously displayed on an oscilloscope. Clearly identifiable action potentials above the baseline noise are converted to standardized pulses using a window discriminator. Standard pulses are fed into a rate meter to measure the frequency of action potentials. Action potentials, integrated output of the rate meter, and blood pressure are continuously recorded and stored in a computer equipped with data acquisition system (Power Lab) for later analysis (Fig. 2).

11. The electrical activity in the carotid sinus nerve represents sensory activity from the carotid body and as well as from baroreceptors (*see* **Note 10**). The following criteria are employed to distinguish the carotid body sensory activity from baroreceptors: (a) action potentials should *increase* during ventilating the rat with a hypoxic gas mixture (12% O_2 balanced nitrogen), and (b) action potential frequency should *decrease* or be *absent* during ventilation with 100% O_2.

12. While recording from whole carotid sinus nerve, the quality of electrical signals varies from experiment to experiment, due to variations in the contact of the nerve to the electrode as well as variations in cleaning and removing the sheath that surrounds the nerve (*see* **Note 11**). Therefore, the data needs normalization that allows comparisons among the individual animals. This is accomplished by the following procedure. At the end of the experiment, turn off the respirator for about 2 min (asphyxia). This procedure will result in intense increase in the carotid sinus nerve electrical activity, which is to be taken as 100% (asphyxic response).

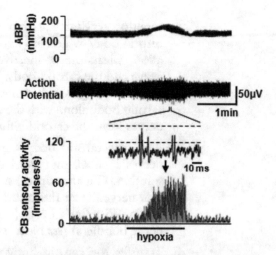

Fig. 2 Example illustrating simultaneous recording of arterial blood pressure (ABP, top panel) and action potentials from the carotid sinus nerve (middle panel) in an anesthetized rat challenged with hypoxia (12% O_2). The integrated sensory nerve activity is presented in the bottom panel. *Inset*: setting of the window discriminator for selection of action potentials above the baseline, on a different timescale

13. Chemoreceptor activity is analyzed during challenging the rats with different levels of inspired oxygen and expressed as impulses per second. The magnitude of the sensory response is normalized as percentage of asphyxic response and plotted against arterial pO_2 values, which is measured at the end of each gas challenge.

3.2 Recording of Action Potentials from Carotid Body Ex Vivo

1. After anesthetizing the rat with urethane (1.2 g/kg, i.p.), intubate the trachea, and cannulate the femoral vein as described in Subheading 3.1, **steps 3** and **4**. Administer heparin (150 U/kg in 0.3 mL saline; i.v.), and wait for 15 min to prevent blood clots in the carotid body.

2. Using the dissecting microscope, expose the carotid artery bifurcation as described in Subheading 3.1, **step 7**, and ligate the common, internal, and external carotid arteries with 4-0 silk thread. Dissect the carotid artery bifurcation free of connective tissue (*see* **Note 12**). Place the carotid bifurcation in ice-old physiological saline bubbled with 95% O_2/5% CO_2 gas mixture in a 15-mL conical tube.

3. Add 3-mL ice-cold physiological saline to a dissection chamber that is made of 90-mm Petri dish filled partially with polymerized Sylgard (*see* **Note 13**). Using the dissecting microscope, place the carotid bifurcation with the superior cervical ganglion facing up, and secure the common, internal, and external carotid arteries on the Sylgard with insect pins. Remove the

superior cervical ganglion with micro scissors. Identify the carotid body, which lies underneath the superior cervical ganglion. Clean the connective tissue around the carotid sinus nerve and the carotid body. Trace the carotid sinus nerve until it joins the glossopharyngeal nerve. Surgically separate the carotid body along with the carotid sinus and glossopharyngeal nerves from the carotid bifurcation.

4. Place the carotid body, carotid sinus, and glossopharyngeal nerves in a 40-mm Petri dish containing 0.1% collagenase solution. Cut the carotid sinus nerve from the glossopharyngeal nerve. Treat the carotid sinus nerve with collagenase for 5 min. With fine forceps (#5.5), tease the nerve to obtain fine nerve bundle(s) (*see* **Note 14**).

5. Transfer the carotid body along with the sinus nerve to a recording chamber (volume 200 μL; Cell MicroControls) with a glass bottom with a heating device connected to a temperature controller (Cell MicroControls). A series of 60-cm^3 syringes are placed ~ 12 in. above the recording chamber and are filled with physiological saline. The syringes are preheated to 37 °C using a circulating water bath. The physiological saline in each syringe is bubbled with an appropriate gas to obtain desired pO_2, pCO_2, and pH, which are monitored by a blood gas analyzer (ABL 80). The carotid body along with the sinus nerve in chamber is continuously irrigated with gravity-fed physiological saline at a rate of 2.5–3 mL/min.

6. Action potentials from the carotid sinus nerve are recorded with a suction electrode. Electrodes are pulled from a glass capillary tubing using a two-stage Narishige pipette puller. Cut the tip of the electrode to a diameter of ~20 μm. Place the electrode in a holder and lower the tip into the recording medium. Bring the electrode close to the sinus nerve using a micromanipulator. Apply a slight negative pressure to the electrode, and let the nerve bundle enter the electrode. A reference electrode made of Ag-AgCl is placed in the recording chamber close to the carotid body. Electrical activity from the nerve bundle is amplified, filtered, displayed, and stored as described in Subheading 3.1, **step 10**.

7. Carotid body sensory response to graded hypoxia can be accomplished by switching the superfusate to a medium with desired pO_2.

8. The module "Spike Histogram" in the software package LabChart 7 Pro is used to discriminate the "single" units from others on the basis of height and duration of individual action potentials. A rate meter is usually used to plot the frequency of a single chemoreceptor action potential (Fig. 3). "Spike Histogram" program also allows analysis of the interspike interval histogram, amplitude histogram, peri-stimulus time histogram, and autocorrelation histogram.

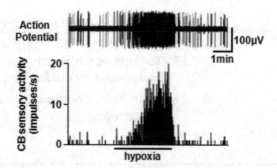

Fig. 3 Example illustrating action potentials recorded from the carotid sinus nerve of an ex vivo rat carotid body. Note the increased action potential frequency during irrigation of the carotid body with medium bubbled with hypoxic gas mixture (medium pO_2~ 40 mmHg). The action potential frequency of a "single unit" is analyzed using "Spike Histogram" of LabChart 7 Pro

4 Notes

1. All protocols involving the use of live animals need to be approved by the Institutional Animal Care and Use Committee (IACUC).

2. Six-week-old rats are usually chosen because they exhibit sufficient length of the carotid sinus nerve that allows to be placed on the electrode. At this age, the rats have little or no adipose tissue around the carotid body.

3. Prior to cannulation, the arterial and venous catheters must be fully filled with diluted heparin saline (20 units of heparin per milliliter). Catheters should devoid of air bubbles.

4. Placing forceps underneath the blood vessels will help for introducing the catheter into the vessels.

5. Prior to recording blood pressure, the blood pressure transducer needs to be calibrated.

6. Ventilating the rat with O_2-enriched gas (~50% O_2) helps to maintain blood pressure and acid-base status of the arterial blood.

7. To provide good visualization of the carotid body, turn the head holder toward left at a small angle.

8. Adjust the tidal volume and rate of the respirator to keep blood gas values within physiological range.

9. The reference electrode inserted in the deep neck muscle minimizes the electrical noise levels.

10. To eliminate baroreceptor activity, ablate the sinus nerve branches originating from the carotid sinus region.

11. To maintain the quality of electrical signals, use surgical spears to absorb any fluids around the nerve.

12. To avoid inadvertent cutting of the carotid sinus nerve, leave a portion of the glossopharyngeal nerve.

13. To improve visualization of the carotid sinus nerve, darken the background by placing black tape on top of Sylgard.

14. The whole procedure should be performed under dissecting microscope.

Acknowledgments

This work was supported by National Institutes of Health grant PO1-HL- 90554.

References

1. Kumar P, Prabhakar NR (2012) Peripheral chemoreceptors: function and plasticity of the carotid body. Compr Physiol 2:141–219

2. Prabhakar NR (2006) O2 sensing at the mammalian carotid body: why multiple O2 sensors and multiple transmitters? Exp Physiol 91(1):17–23. https://doi.org/10.1113/expphysiol.2005.031922

3. Peng Y-J, Nanduri J, Raghuraman G, Souvannakitti D, Gadalla MM, Kumar GK, Snyder SH, Prabhakar NR (2010) H2S mediates O2 sensing in the carotid body. Proc Natl Acad Sci U S A 107(23):10719–10724. https://doi.org/10.1073/pnas.1005866107

4. Conde SV, Obeso A, Rigual R, Monteiro EC, Gonzalez C (2006) Function of the rat carotid body chemoreceptors in ageing. J Neurochem 99(3):711–723. https://doi.org/10.1111/j.1471-4159.2006.04094.x

Chapter 12

Monitoring Functional Responses to Hypoxia in Single Carotid Body Cells

Ana María Muñoz-Cabello, Hortensia Torres-Torrelo, Ignacio Arias-Mayenco, Patricia Ortega-Sáenz, and José López-Barneo

Abstract

The carotid body is the main arterial chemoreceptor in mammals that mediates the cardiorespiratory reflexes activated by acute hypoxia. Here we describe the protocols followed in our laboratory to study responsiveness to hypoxia of single, enzymatically dispersed, glomus cells monitored by microfluorimetry and the patch-clamp technique.

Key words Acute oxygen sensing, Hypoxia, Glomus cells, Enzymatic cell dispersion, O_2-sensitive ion channels, Ionic currents, Patch clamp, Microfluorimetry, Cytosolic calcium

1 Introduction

Arterial chemoreceptors are sensory organs responsible for the detection of acute changes in oxygen tension (PO_2) in the blood, which in response to hypoxia trigger fast (in seconds) reflex compensatory hyperventilation and sympathetic activation. The carotid body (CB), the main arterial chemoreceptor, is strategically located in the bifurcation of the carotid artery and is composed of clusters of cells surrounded by a dense network of capillaries and afferent sensory fibers [1]. CB chemoreceptor cells (glomus cells) contain O_2-sensitive K^+ channels, which are inhibited in response to hypoxia, thereby leading to cell depolarization, opening of voltage-gated Ca^{2+} channels and Ca^{2+} influx. The rise of cytosolic Ca^{2+} triggers the release of neurotransmitters that stimulate afferent sensory fibers activating brain centers involved in respiration and autonomic regulation [2–6]. The CB has an essential role in body homeostasis; however, it can also be maladaptive and contribute to the exaggerated sympathetic outflow underlying refractory

L. Eric Huang (ed.), *Hypoxia: Methods and Protocols*, Methods in Molecular Biology, vol. 1742, https://doi.org/10.1007/978-1-4939-7665-2_12, © Springer Science+Business Media, LLC 2018

hypertension and other comorbidities highly frequent in the human population [7, 8].

Research on the functional responses to hypoxia of glomus cells has progressed steadily during the last decades due to the development of methodologies for enzymatic dispersion and culture of cells from the CB and other O_2-sensitive organs (*see* for review ref. [1]). In addition, single-cell studies have become available thanks to the development of techniques for monitoring cellular physiological parameters (e.g., cytosolic Ca^{2+} or pH) by microfluorimetry or to record the cell's electrophysiological properties (membrane potential, ionic currents, or action potential firing) with the patch-clamp technique. Herein, we summarize the main features of these methodologies, as adapted to our laboratory (*see* ref. [9]), which have been used to investigate responsiveness to hypoxia of single peripheral chemoreceptor cells.

2 Materials

2.1 Dispersion and Culture of Carotid Body Cells

1. The procedure followed to culture dispersed CB cells requires a standard culture room facility with a laminar flow cabinet, a benchtop centrifuge for conical tubes, a 37 °C incubator with a water-saturated atmosphere containing 20% O_2 and 5% CO_2, and an inverted microscope with phase contrast.

2. After dissection from the animals, CBs are placed on the filtered enzymatic solution (PBS pH 7.4 supplemented with 50 µM $CaCl_2$, 0.6 mg/mL collagenase II, 0.27 mg/mL trypsin, and 1.25 U/mL porcine elastase).

3. The culture medium contains DMEM/F-12 (no glutamine, no HEPES) supplemented with penicillin (100 U/mL)/streptomycin (10 µg/mL), 2 mM L-glutamine, 10% fetal bovine serum and 84 U/L insulin.

2.2 Patch-Clamp Setup and Electrophysiology

1. Micropipettes (2–4 MΩ) are pulled from capillary glass tubes with a horizontal pipette puller (Sutter instruments model P-1000) and fire polished with a microforge MF-830 (Narishige).

2. Ionic currents are recorded with an EPC-7 amplifier (HEKA Elektronik). The signal is filtered (3–10 kHz depending on the experiment), subsequently digitized with an analog/digital converter (ITC-16 Instrutech Corporation), and finally sent to the computer. Data acquisition and storage on a Macintosh computer is done using the Pulse/Pulsefit software (HEKA Elektronik) at a sampling interval ranging from 20 µs to 500 ms depending on the experiment. Data analysis is performed with the Igor Pro Carbon (Wavemetrics) and Pulse/Pulsefit (HEKA Elektronik) programs.

Table 1

Composition of external and internal solutions used in patch-clamp experiments performed with carotid body glomus cells

External solution			Internal solution				
			Whole-cell			Perforated-patch	
	I_{Na}, I_{Ca}	I_k		I_{Na}, I_{Ca}	I_k		I_k
NaCl	120	120	CsCl	110	–	K_2SO_4	70
BaCl$_2$	9	–	Kglutamate	–	80	KCl	30
CaCl$_2$	2.5	2.5	KCl	–	50	HEPES	10
KCl	2.5	4.5	CsF	30	–	EGTA	1
NaHCO$_3$	–	23	HEPES	10	10	MgCl$_2$	2
HEPES	10	–	MgCl$_2$	–	1	Amphotericin B (µg/mL)	240
MgCl$_2$	1	1	ATP-Mg	4	4		
Sucrose	5	5					
Glucose	5	5					

Concentrations are expressed in mM

3. Table 1 shows the composition of the external (bathing solution) and internal (inside the micropipette) solutions that we normally employ in experiments using two different patch-clamp configurations.

2.3 Microfluorimetry

1. Our microfluorimetric setup consists of an inverted microscope equipped with a 40×/0.60 NA objective and a filter wheel, a 150 W xenon lamp, a monochromator, a CCD camera, and a computer (see Fig. 3a *below*). The monochromator (Polychrome V, Till Photonics) allows the selection of the wavelength (λ) that, depending on the experiment, will excite the sample. The light beam travels through an optical fiber to the microscope where it is deflected by a 45° inclination dichroic mirror (which transmits and reflects the radiation depending on the wavelength). The beam of light reaches the sample, causing the excitation of the indicator and the emission of fluorescence. Then, the emitted fluorescence passes through a band-pass filter (variable according to the indicator) and is detected by a CCD camera (Orca Flash 4, Hamamatsu Photonics). Monochromator, CCD camera, and image acquisition

are controlled by Aquacosmos software (Hamamatsu Photonics).

2. Coverslips with CB glomus cells are transferred to the recording chamber (≈ 0.2 mL) with a continuous flow of solution. The composition of the bathing solution is 117 mM NaCl, 23 mM $NaHCO_3$, 5 mM Glucose, 5 mM Sucrose, 4.5 mM KCl, 2.5 mM $CaCl_2$ and 1mM $MgCl_2$. To perform experiments in normoxic conditions, the external solution is bubbled with a gas mixture of 5% CO_2, 20% O_2, and 75% N_2. To expose cells to hypoxia, the external solution is bubbled with 5% CO_2 and 95% N_2, to reach an O_2 tension in the chamber of ~10–15 mm Hg. Microfluorimetric recordings are performed at 30–33 °C.

3   Methods

3.1   Enzymatic Dispersion and Culture of Carotid Body Cells

1. Dispersion and culture of CB cells (most frequently from mice and rats) are normally used to monitor single-cell parameters by microfluorimetry and the patch-clamp techniques. Animals are euthanized with an overdose of sodium thiopental (intraperitoneal, 120–150 mg/kg). Then, the carotid bifurcations are carefully removed from the animals and transferred to ice-cold PBS. Once in PBS, CBs are dissected from the bifurcations under a stereomicroscope ($\times 10$ magnification) with the aid of Dumont #5 tweezers and surgical scissors (*see* **Note 1**).

2. A sterile 35 mm Ø petri dish is filled with enzymatic solution in the laminar flow cabinet (*see* **Note 2**). After dissection, CBs are transferred under the stereomicroscope to the enzymatic solution using fine tweezers and are maintained in the 37 °C incubator for 20 min (*see* **Note 3**).

3. The petri dish containing the CBs is placed under a phase-contrast microscope to perform a mechanical dissociation of the tissue by stretching it with the help of two 25G needles (*see* **Note 3**). The sample is then incubated at 37 °C for 5 min.

4. CBs are transferred with a P1000 pipette to a 15 mL conical tube containing ≈ 10 mL of cold culture medium to stop the enzymatic reaction. The sample is centrifuged at $250 \times g$ and 10 °C for 5 min, and the supernatant is discarded, leaving ≈ 100 µL of the solution to gently resuspend the pellet with a P200 pipette (*see* **Note 4**).

5. Cells are plated in small pieces of sterile coverslips coated with 1 mg/mL poly-L-lysine (*see* **Note 5**). They are kept at the 37 °C incubator for 2 h to allow the cells to settle. Afterward,

the petri dish is carefully filled with 2 mL of culture medium (*see* **Note 6**) and returned to the incubator (*see* **Note 7**).

3.2 Recordings of Ionic Currents with the Patch-Clamp Technique: General Aspects of the Technique

This section is not intended to provide a detailed description of the patch-clamp technique, which can be found elsewhere [10, 11]. Instead, we describe how the patch-clamp methodology can be applied to monitor electrophysiological responses to acute hypoxia from single CB glomus cells (*see* **Note 8**).

1. The patch-clamp setup is schematized in Fig. 1. Coverslips with isolated CB cells are transferred to a recording chamber ($\approx$0.2 mL) mounted on the stage of an inverted microscope equipped with a 40×/0.60 NA objective. The chamber is continuously perfused with the external solution at a constant rate (~2.5 mL/min) by a flow pump. The micropipette is placed in the pipette holder in contact with the chloride-silver electrode, which is connected to a head-stage preamplifier (*see* Subheading in Fig. 1).

2. To perform experiments in normoxic conditions, the external solution is bubbled with a gas mixture of 5% CO_2, 20% O_2, and 75% N_2 (O_2 tension ~145 mmHg). To expose cells to hypoxia, the external solution is bubbled with 5% CO_2 and 95% N_2 to reach an O_2 tension in the chamber of ~10–15 mmHg. Experiments are conducted at 30–33 °C.

3.3 Electro-physiological Methods: Whole-Cell Configuration

In this section, we describe the procedure to perform patch-clamp recordings of macroscopic currents from single CB cells (*see* **Note 9**). This is normally done using the whole-cell configuration in the voltage-clamp mode. In this configuration cells are dialyzed with

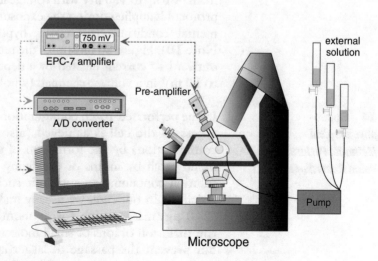

Fig. 1 Schematic representation of the equipment used in a patch-clamp setup. *See* text for details

the pipette solution, as in few minutes this solution replaces the solution inside the cytosol.

1. Most of the patch-clamp configurations require the initial formation of a high-resistance seal in the gigaohm (10^9 Ω) range (gigaseal) between the micropipette tip and the cell membrane to prevent leakage of ions between the pipette and the membrane, so that any ion movement takes place through ion channels in the isolated membrane patch. To form the gigaseal and reach the cell-attached configuration (Fig. 2a *left*), the pipette tip is located near the cell using a micromanipulator, and a small amount of negative pressure is applied. To obtain the whole-cell configuration (Fig. 2a *middle*), a brief pulse of negative pressure is applied to produce the rupture of the membrane patch inside the tip of the micropipette. Then intracellular dialysis with the micropipette solution starts. The whole-cell configuration is the most appropriate methodology used to record currents generated by the ensemble of ion channels existing in the entire cell membrane (normally called "macroscopic currents"). Using blockers of specific types of ion channels added to the internal or external solutions, it is possible to separate the currents mediated by K^+, Na^+, or Ca^{2+} ions. Representative macroscopic outward (carried by K^+ ions) and inward (carried by Na^+ and Ca^{2+} ions) currents recorded in mouse CB cells with a depolarizing pulse to 10 mV are illustrated in Fig. 2b, c, respectively.

2. To test the effect of changes in O_2 tension on the macroscopic ionic currents in CB glomus cells, we apply depolarizing steps of variable duration (normally between 10 and 200 ms) from a holding potential of –70 mV to different voltages ranging from −50 up to +30 mV with voltage increments of 10 mV. This protocol is applied during the exposure to the different experimental conditions (normoxia, hypoxia, and recovery) (*see* **Note 10**). Figure 2d illustrates the reduction by hypoxia of the outward K^+ current elicited by a depolarizing pulse from −70 to 10 mV in a voltage-clamped (whole-cell) mouse CB cell.

3.4 Electrophysiological Methods: Perforated-Patch Recordings

1. In the perforated-patch configuration (Fig. 2a *right*), electrical access to the cell is achieved (also from the cell-attached configuration) by the formation of small pores in the membrane patch by means of backfilling the micropipette with a solution containing an antibiotic such as nigericin or amphotericin B. In our case, we usually make use of amphotericin B (240 µg/mL), which forms membrane pores that allow the diffusion of ions or small molecules (MW < 200 daltons) but prevent the passage of larger cytosolic molecules. The perforated-patch technique is useful to study the function of

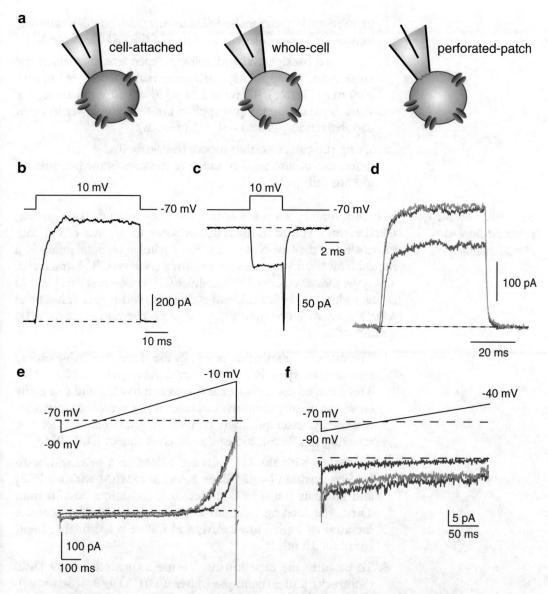

Fig. 2 Recording of ionic currents with the patch-clamp technique. (**a**) Schematic representation of the different patch-clamp configurations: cell-attached (*left*), whole-cell (*middle*), and perforated-patch (*right*). (**b–d**) Patch-clamp recordings of macroscopic currents in whole-cell configuration (voltage-clamp mode) in mouse CB cells. (**b**) *Top*, Depolarizing pulse protocol from −70 mV to 10 mV. *Bottom*, Macroscopic outward current recorded during the depolarizing pulse. (**c**) *Top*, Depolarizing pulse protocol from −70 mV to 10 mV. *Bottom*, Macroscopic inward current recorded during the depolarizing pulse. (**d**) Currents recorded during a depolarization pulse from −70 mV to 10 mV under normoxia (red), hypoxia (blue), and recovery in normoxia (gray) conditions. (**e–f**) Patch-clamp recordings in perforated-patch configuration (voltage-clamp mode) in mouse glomus cells. (**e**) *Top*, Voltage ramp protocol. *Bottom*, Currents recorded in normoxia (red), hypoxia (blue), and recovery in normoxia (gray) conditions. (**f**) *Top*, Voltage ramp protocol. *Bottom*, Holding currents recorded in normoxia, hypoxia, and recovery conditions. Recordings are the same as in (**e**). (**b–f**) Modified from Ref. 13

membrane ion channels, without the dialysis of potentially important intracellular components of the cell (*see* **Note 11**).

2. To record background and voltage-dependent currents at the same time, we apply depolarization ramps from −90 mV to −10 mV (Fig. 2e). To record background currents alone, the same depolarizing ramp is applied but reaching lower levels of depolarization (−90 to −40 mV; Fig. 2f).

3. Using the current-clamp mode, the perforated-patch configuration can be also used to measure the membrane potential of glomus cells [12].

3.5 Microfluorimetry: Cytosolic Calcium Measurements

Microfluorimetry is a valuable methodology to study the signaling mechanisms involved in the acute response to hypoxia of CB chemosensitive cells (*see* **Note 12**). Extracellular calcium influx is a requisite for the acute release of transmitters from CB glomus cells. Thus, the measurement of intracellular Ca^{2+} concentration ($[Ca^{2+}]_i$) by microfluorimetry is a relevant technique, and as such it has been widely used to investigate the process of chemotransduction by glomus cells.

1. The calcium indicator we normally use is the permeable variety Fura-2-acetoxymethyl ester (Fura-2 AM) [6] (*see* **Note 13**). This form of the indicator is insensitive to Ca^{2+} and can easily cross the plasma membrane. Once in the cytoplasm, the acetoxymethyl ester group is cleaved by endogenous esterases, releasing the Fura-2 molecule, which is able to bind Ca^{2+}.

2. Coverslips with the CB cells are placed in a petri dish with loading medium (4 µM Fura-2 AM in DMEM without FBS) and incubated at 37 °C under dark conditions for 30 min. Then, the loading medium is replaced by complete culture medium to wash Fura-2 AM, and CB cells are further incubated for 15 min.

3. To perform the experiments, we use a dichroic FF409-Di03 (Semrock) and a band-pass filter FF01-510/84 (Semrock). The acquisition protocol is established at the beginning of the experiment using Aquacosmos software. For Fura-2 AM recordings, we select a spatial resolution of 4 × 4 pixels, an excitation time of 20 ms, and an acquisition interval of 2 s. Camera information is monitored in real time, representing the intensities for each λ (340 nm and 380 nm) and the resulting ratio F340/F380 in arbitrary units (*see* **Note 14**). Finally, F340/F380 ratios from the selected regions of interest (ROIs) are monitored online in a graph. Figure 3b shows typical cytosolic calcium responses to hypoxia (and other stimuli) in glomus cells loaded with Fura-2 AM.

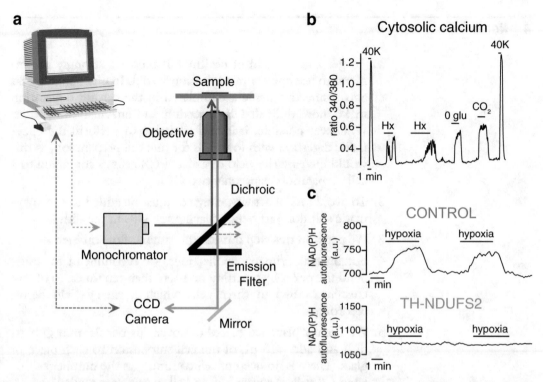

Fig. 3 Microfluorimetric recordings in mouse CB cells. (**a**) Scheme of the setup used for microfluorimetric recordings from single dispersed cells. (**b**) Increases in cytosolic [Ca^{2+}] elicited by the indicated stimuli (*40 K*, depolarization induced by incubation with a bathing solution containing 40 mM KCl; *Hx*, hypoxia; *0 glu*, bathing solution with no glucose; *CO$_2$*, hypercapnia; bathing solution is bubbled with 20% CO$_2$) in dispersed mouse glomus cells. (**c**) Monitoring of NAD(P)H autofluorescence changes in arbitrary units (a.u.) during exposure to hypoxia of glomus cells from control and TH-NDUFS2 mice. (**b, c**) Modified from ref. 13

3.6 Microfluorimetry: NAD(P)H Autofluorescence

1. Monitoring NAD(P)H autofluorescence by microfluorimetry is a useful procedure to test responsiveness of glomus cell mitochondria to changes in O$_2$ tension (*see* **Note 15**). To this end, coverslips with CB cells are transferred from the culture dish to the recording chamber. Our acquisition protocol has a spatial resolution of 4 × 4 pixels, an excitation time of 150 ms ($\lambda = 360$ nm), and an acquisition interval of 5 s. NAD(P)H emitted fluorescence is measured at 460 nm (*see* **Note 14**). For these experiments, we use a dichroic FF409-Di03 (Semrock) and a band-pass filter FF01-510/84 (Semrock).

2. The reversible increase in NAD(P)H levels during exposure to hypoxia in a typical glomus cell (control) is illustrated in Fig. 3c. Note that this signal completely disappears in glomus cells from TH-NDUFS2 animals, which lack the gene encoding the Ndufs2 MCI subunit necessary for ubiquinone binding. These animals also exhibit selective abolishment of the hyperventilatory response to hypoxia [13].

4 Notes

1. To get a good yield of healthy CB cells, we strongly recommend to use mice or rats 1–4 months old. In older animals, fat and connective tissue accumulated in the carotid bifurcation make more difficult CB dissection and enzymatic digestion. Whenever possible, it is also advisable to perform the enzymatic digestion with four CBs for one cell preparation, as this would increase the yield of isolated CB cells as compared to a cell preparation using only two CBs.

2. To avoid loss of activity, enzymes must be added to the enzymatic solution just before placing the CBs in the dish.

3. We perform this step outside the laminar flow cabinet.

4. For microfluorimetry applications, we recommend to resuspend the pellet by pipetting no more than ten times, to obtain clusters of two to three cells which normally yield better recordings.

5. We usually place ≈6 pieces of coverslips per 35 mm Ø petri dish and add ≈15 μL of the cell suspension to each piece of glass. There is no need of cell counting, as the number of isolated CB cells is always low and all of them are seeded.

6. The culture medium should be prepared in the cell culture cabinet and placed in the incubator for at least 1 h before use. This preincubation is necessary to achieve pH equilibration.

7. We recommend the use of cells within 16–48 h after dispersion, although CB cells are still responsive to hypoxia even 96 h after enzymatic dissociation.

8. The basis for the patch-clamp technique is to electrically isolate a patch of cell membrane from the external solution using a glass micropipette. Electrical isolation is achieved because the micropipette is tightly sealed onto the cell membrane. Ions moving through the membrane patch flow into the pipette, and the current is recorded by a chloride-silver wire electrode connected to the head-stage amplifier. Depending on the experimental objective, patch-clamp recordings are carried out using either voltage-clamp or current-clamp modes. In the voltage-clamp mode, ionic currents can be recorded at a constant voltage. The membrane potential is changed at any desired value to measure the current flowing through the ion channels open at that voltage. In current-clamp mode, current is kept constant to monitor changes in membrane potential. Depolarizing or hyperpolarizing current stimuli are applied at different intensities, and membrane potential variations are then recorded.

9. Glomus cells are excitable and contain numerous classes of ion channels, including voltage-gated Na^+, Ca^{2+}, and K^+ channels. Patch-clamp measurements have been decisive to establish the basis for the hypoxia chemotransduction process in the CB as recordings in single glomus cells showed that K^+ channel open probability is inhibited during exposure to hypoxia (*see* refs. [1–5]). CB glomus cells contain several classes of K^+ channels inhibited by hypoxia, including background K^+ channels [14–16], maxi-K^+ channels [17, 18], and voltage-dependent channels [2–4, 19].

10. We usually wait 3–4 min after changing to the new solution before performing the activation protocol.

11. As mentioned in Note 9, CB glomus cells have background K^+ channels that are inhibited by hypoxia. Background K^+ channels are open at the resting membrane potential, and their inhibition under hypoxic condition leads to an increase in input resistance. We use the perforated-patch configuration to measure resting membrane potential, input resistance, and inward holding current.

12. The microfluorimetry technique allows the detection of changes in the concentration of cellular components by monitoring the emission of fluorescence under a microscope. This is achieved by either loading the cells with fluorescent indicators (such as Fura-2) that emit fluorescence with different intensities depending on the concentration of the particular molecule to which they specifically bind or taking advantage of the autofluorescence of some biomolecules such as NADH, after being excited by a suitable energy source.

13. Among the different calcium indicators, Fura-2 is one of the most commonly used. It shows an absorption spectral shift, which allows performing ratiometric fluorescence measurements for the detection of $[Ca^{2+}]_i$. Fura-2 is a Ca^{2+} BAPTA chelator derivative that binds to Ca^{2+} with high affinity. It has a dissociation constant Kd ~ 135 nM (Mg^{2+}-free) and Kd ~ 224 nM (1 mM Mg^{2+}). These values are within the appropriate range to measure cytosolic $[Ca^{2+}]$ in living mammalian cells with the required sensitivity [20]. Fura-2 is excited at 340 nm and 380 nm and emits at 510 nm. When Fura-2 is excited at $\lambda = 340$ nm, the intensity of the emitted light is directly proportional to the increase in $[Ca^{2+}]_i$. On the other hand, when it is excited at $\lambda = 380$ nm, the opposite happens. This spectral shift is used to establish a F340/F380 ratio, which is obtained after alternately exciting the sample at 340 nm and 380 nm and monitoring the emission at 510 nm. One of the major advantages of using ratiometric indicators is the elimination of

artifacts. F340/F380 ratio values do not depend either on the dye loading or the intensity of illumination, and the effects of photobleaching in the records are minimized [21].

14. Background correction is performed by subtracting the fluorescent signal of a cell-free region from the intensity of the different regions of interest selected for the experiment.

15. During acute hypoxia, there is a reversible increase in NADH concentration in CB glomus cells [13, 22, 23], which can be explained by a decrease in the rate of NADH oxidation due to slowdown of the electron transport chain [13]. NAD(P)H levels can be monitored by microfluorimetry, as NADH and NADPH are autofluorescent molecules with an emission peak at 450 ± 15 nm when excited at 340–360 nm.

Acknowledgments

This work was supported by the Botín Foundation, the Spanish Ministry of Economy and Innovation (SAF2012-39343, SAF2016-74990-R), and the European Research Council (ERC Advanced Grant PRJ201502629).

References

1. López-Barneo J, González-Rodríguez P, Gao L, Fernández-Agüera MC, Pardal R, Ortega-Sáenz P (2016) Oxygen sensing by the carotid body: mechanisms and role in adaptation to hypoxia. Am J Physiol Cell Physiol 310:C629–C642

2. López-Barneo J, López-López JR, Ureña J, González C (1988) Chemotransduction in the carotid body: K+ current modulated by PO_2 in type I chemoreceptor cells. Science 241:580–582

3. Peers C (1990) Hypoxic suppression of K+ currents in type I carotid body cells: selective effect on the Ca^{2+}-activated K+ current. Neurosci Lett 119:253–256

4. Stea A, Nurse CA (1991) Whole-cell and perforated-patch recordings from O_2-sensitive rat carotid body cells grown in short- and long-term culture. Pflügers Arch 418:93–101

5. Buckler KJ, Vaughan-Jones RD (1994) Effects of hypoxia on membrane potential and intracellular calcium in rat neonatal carotid body type I cells. J Physiol 476:423–428

6. Ureña J, Fernández-Chacón R, Benot AR, Álvarez de Toledo GA, López-Barneo J (1994) Hypoxia induces voltage-dependent Ca^{2+} entry and quantal dopamine secretion in carotid body glomus cells. Proc Natl Acad Sci U S A 91:10208–10211

7. Paton FR, Sobotka A, Fudim M, Engelman ZJ, Hart EC, McBryde FD et al (2013) The carotid body as a therapeutic target for the treatment of sympathetically mediated diseases. Hypertension 61:5–13

8. Marcus NJ, Del Río R, Schultz EP, Xia XH, Schultz HD (2014) Carotid body denervation improves autonomic and cardiac function and attenuates disordered breathing in congestive heart failure. J Physiol 592:391–408

9. López-Barneo J, Pardal R, Montoro RJ, Smani T, García-Hirschfeld J, Ureña J (1999) K+ and Ca^{2+} channel activity and cytosolic $[Ca^{2+}]$ in oxygen-sensing tissues. Respir Physiol 115:215–227

10. Sakmann B, Neher E (1995) Single cannel recording. Plenum Press, New York

11. Molnar P, Hickman JJ (2007) Patch-clamp methods and protocols. Humana Press, New York

12. García-Fernández M, Ortega-Sáenz P, Castellano A, López-Barneo J (2007) Mechanisms of low-glucose sensitivity in carotid body glomus cells. Diabetes 56:2893–2900

13. Fernández-Agüera MC, Gao L, González-Rodríguez P, Pintado CO, Arias-Mayenco I, García-Flores P et al (2015) Oxygen sensing by arterial chemoreceptors depends on mitochondrial complex I signaling. Cell Metab 22:825–837

14. Delpiano MA, Hescheler J (1989) Evidence for a PO$_2$-sensitive K$^+$ channel in the type-I cell of the rabbit carotid body. FEBS Lett 249:195–198

15. Buckler KJ (1997) A novel oxygen-sensitive potassium current in rat carotid body type I cells. J Physiol 498:649–662

16. Kim D, Cavanaugh EJ, Kim I, Carroll JL (2009) Heteromeric TASK-1/TASK-3 is the major oxygen-sensitive background K$^+$ channel in rat carotid body glomus cells. J Physiol 587:2963–2975

17. Wyatt CN, Peers C (1995) Ca(2+)-activated K+ channels in isolated type I cells of the neonatal rat carotid body. J Physiol 483:559–565

18. Riesco-Fagundo AM, Pérez-García MT, González C, López-López JR (2001) O(2) modulates large-conductance Ca(2+)-dependent K(+) channels of rat chemoreceptor cells by a membrane-restricted and CO-sensitive mechanism. Circ Res 89:430–436

19. López-López J, González C, Ureña J, López-Barneo J (1989) Low pO$_2$ selectively inhibits K channel activity in chemoreceptor cells of the mammalian carotid body. J Gen Physiol 93:1001–1015

20. Grynkiewicz G, Poenie M, Tsien RY (1985) A new generation of Ca^{2+} indicators with greatly improved fluorescence properties. J Biol Chem 260:3440–3450

21. Becker PL, Fay FS (1987) Photobleaching of fura-2 and its effects on determination of calcium concentrations. Am J Phys 253:C613–C618

22. Duchen MR, Biscoe TJ (1992) Mitochondrial function in type I cells isolated from rabbit arterial chemoreceptors. J Physiol 450:13–31

23. Buckler KJ, Turner PJ (2013) Oxygen sensitivity of mitochondrial function in rat arterial chemoreceptor cells. J Physiol 591:3549–3563

Chapter 13

Testing Acute Oxygen Sensing in Genetically Modified Mice: Plethysmography and Amperometry

Patricia Ortega-Sáenz, Candela Caballero, Lin Gao, and José López-Barneo

Abstract

Monitoring responsiveness to acute hypoxia of whole animals and single cells is essential to investigate the nature of the mechanisms underlying oxygen (O_2) sensing. Here we describe the protocols followed in our laboratory to evaluate the ventilatory response to hypoxia in normal and genetically modified animals. We also describe the amperometric technique used to monitor single-cell catecholamine release from chemoreceptor cells in carotid body and adrenal medulla slices.

Key words Acute oxygen sensing, Hypoxia, Hyperventilation, Plethysmography, Single-cell secretion, Amperometry, Genetically modified mice

1 Introduction

Oxygen (O_2) is essential for life, and O_2 deficiency (hypoxia), even only transient, can have detrimental effects and critically contribute to the pathogenesis of severe and highly prevalent diseases in the human population. Adaptive responses, which can be acute or chronic, have evolved to minimize the effect of hypoxia on cells [1]. Chronic responses to hypoxia mainly depend on the prolyl hydroxylase/hypoxia-inducible factor pathway, which modulates the transcription of numerous "O_2-sensitive" genes involved in intermediary metabolism, angiogenesis, and red blood cell proliferation [2]. In mammals, hypoxia also triggers acute (in seconds) life-saving cardiorespiratory reflexes (hyperventilation and sympathetic activation) to increase gas exchange in the lungs and the delivery of O_2 to critical organs, such as the brain and heart. Changes in arterial O_2 tension (PO_2) are detected by peripheral chemoreceptors, which are highly irrigated sensory organs innervated by afferent nerve fibers terminating at brain centers involved in the control of respiration and autonomic function [3]. The most important chemoreceptor in mammals is the carotid body (CB), a small organ

L. Eric Huang (ed.), *Hypoxia: Methods and Protocols*, Methods in Molecular Biology, vol. 1742,
https://doi.org/10.1007/978-1-4939-7665-2_13, © Springer Science+Business Media, LLC 2018

strategically located in the carotid bifurcation. The CB is organized in clusters of cells called glomeruli, which contain O_2-sensing neuron-like glomus cells as well as other cell types, including blood vessels and nerve fibers. In response to hypoxia, glomus cells depolarize due to inhibition of "O_2-sensitive" K^+ channels. Cell depolarization leads to the opening of voltage-gated Ca^{2+} channels, Ca^{2+} influx, and release of transmitters that activate the sensory fibers [4, 5]. Although there are abundant evidence supporting the existence of autocrine and paracrine interactions among the different cell types in the CB glomerulus, several reports have shown that sensitivity to hypoxia is a cell autonomous process and that the hypoxic responses of single glomus cells are, in most aspects, similar to the input/output relationships in the entire CB [6].

The knowledge of CB function has advanced significantly in the past decades; however, understanding the mechanisms of acute O_2 sensing has been hampered by the tiny size of the CB (<30 mm³ volume in man) [7] and the small number of glomus cells (~1500/ CB/mouse). In addition, sensitivity to hypoxia is a labile phenomenon that, for unknown reasons, can sometimes disappear in apparently healthy cells. In the past few years, the study of responsiveness to hypoxia of genetically modified mice has been of great value to test several hypotheses of acute O_2 sensing and to identify cellular processes involved in sensory transduction by chemoreceptor cells [8–10]. The studies of genetically modified mice require the use of reliable methodologies to compare their responsiveness to hypoxia with respect to controls. We have found whole-body plethysmography in combination with amperometric recordings in CB slices the most appropriate techniques for in vivo and in vitro assessment of the alterations in acute O_2 sensing resulting from changes in the level of expression of specific genes. In this chapter, we describe these methodologies as adapted to our laboratory.

2   Materials

2.1 Plethysmography

1. The experimental setup for volume-constant whole-body plethysmography (Fig. 1a) consists of two methacrylate-walled chambers (one for the control mouse and the other for the test mouse) closed with an airtight seal, except for a small leak used

→

Fig. 1 (continued) Recorded signals are sent to a four-channel amplifier and acquired using iox2 software. Representative recordings of the respiratory flow signal obtained from a wild-type mouse in normoxia (blue) and hypoxia (red) are represented inside the box. (**b**) Top. Changes in respiratory frequency recorded from a control mouse in response to two sequential exposures to hypoxia (decrease in O_2 tension from 21 to 10% O_2) and a final exposure to hypercapnia (increasing CO_2 tension from 0.04 to 5% CO_2). Bottom. Changes in O_2 (blue points) and CO_2 (red points) tension during the experiment. (**c**) Average changes in respiratory frequency in response to hypoxia (see O_2 tension in the lower panel) of 17 wild-type mice. Each red point represents the mean ± standard error of the 17 measurements. See text for further explanation

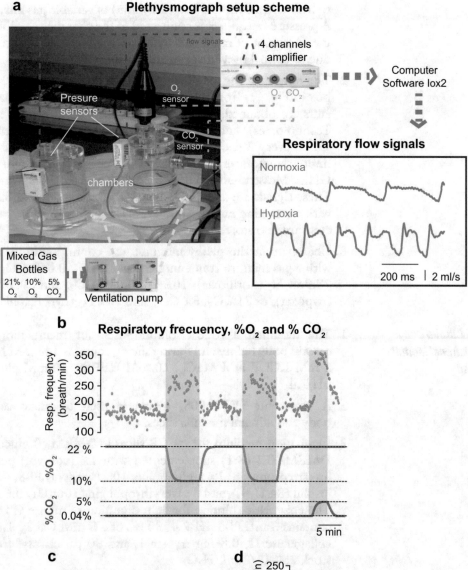

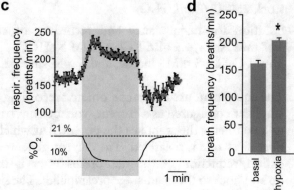

Fig. 1 Recording of respiratory parameters by plethysmography. (**a**) Image illustrating the two chambers used to perform whole-body plethysmography in mice. Chambers are connected to tubes with variable gas mixtures and contain individual pressure sensors. O_2 and CO_2 are monitored by sensors connected to one of the chambers.

to maintain a constant flow (1 L/min) of variable gas mixtures. A pressure transducer is connected to the chamber to allow the measurement of rapid and transient small changes of pressure due to animal movement or respiration.

2. The output of the amplifier provides a flux signal (mL/s) (*see* box in Fig. 1a) from which the appropriate respiratory parameters are obtained. To this end, we use iox2 software (Emka Technologies), and the calculated respiratory parameters are stored every 2 s. O_2 and CO_2 levels in the chambers are continuously monitored by individual sensors (oxy 3690 MP, GHM Messtechnik, and NS:MCO, Cibertec) (*see* Fig. 1a). These signals are amplified, digitized, and recorded unfiltered with a sampling rate of 1000 Hz. Data points of these parameters are also stored every 2 s.

3. Chambers in the plethysmograph are continuously perfused with a gas mixture containing either 21% O_2, 0.04% CO_2, and 78.96% N_2 (normoxia); 10% O_2, 0.04% CO_2, and 89.96% N_2 (hypoxia); or 21% O_2, 5% CO_2, and 74% N_2 (hypercapnia).

2.2 Carotid Body and Adrenal Medulla Slicing

1. The modified Tyrode's solution used for maintaining the carotid body before and during slicing contains 148 mM NaCl, 2 mM KCl, 3 mM $MgCl_2$, 10 mM HEPES, 10 mM glucose, pH 7.4.

2. A vibratome (VT1000S, Leica) is used to make carotid body and adrenal medulla slices.

3. Incubation medium for rat CB slices is DMEM (0 glucose)/ DMEM-F12 (3:1) supplemented with 1% (vol/vol) penicillin/streptomycin, 1% L-glutamine, 10% (vol/vol) FBS, 4.2 μL insulin (84 U/L), and 3 μL erythropoietin (1000UI/0.5 mL). The enzymatic solution for light digestion of mouse CB slices contains 3 mL PBS pH 7.4, 30 μL of a 5 mM $CaCl_2$, 1.8 mg collagenase II, 0.8 mg trypsin I, and 30 μL elastase from a stock 250 U/2 mL H_2O.

4. Solution for maintenance and recording of adrenal slices: 117 mM NaCl, 4.5 mM KCl, 23 mM $NaHCO_3$, 1 mM $MgCl_2$, 1 mM $CaCl_2$, 5 mM glucose, and 5 mM sucrose.

2.3 Amperometry

1. The equipment used for amperometric recording consists of an amplifier configured as a current/voltage converter connected to a voltmeter that supplies the potential at which the carbon-fiber electrode is polarized. The current generated by the oxidation/reduction of the neurotransmitter is driven by the carbon fiber to the head-stage preamplifier. The signal is amplified, filtered (low pass, 100 Hz), and digitized at a frequency of 250 Hz with an analog/digital converter (ITC-16, Instrutech Corporation). Data acquisition is done using Pulse/Pulsefit software (HEKA Electronics).

2. The control recording solution in amperometry experiments contains 117 mM NaCl, 4.5 mM KCl, 23 mM $NaHCO_3$, 1 mM $MgCl_2$, 2.5 mM $CaCl_2$, 5 mM glucose, and 5 mM sucrose, at ~35 °C. In high K^+ solutions, NaCl is replaced equimolarly with KCl. The "normoxic" solution is bubbled with a gas mixture of 5% CO_2, 20% O_2, and 75% N_2 (O_2 tension ~145 mmHg). The "hypoxic" solution is bubbled with 5% CO_2 and 95% N_2 to reach an O_2 tension in the chamber of ~15 mmHg. The "hypercapnic" solution is bubbled with 20% CO_2, 20% O_2, and 60% N_2. Osmolality of solutions is ~300 mosmol/kg and the pH 7.4.

3 Methods

3.1 Plethysmography: General Aspects of the Technique

1. Plethysmography is the standard method used for the study of lung ventilation in conscious unrestrained rodents [11–13]. In our laboratory, we use whole-body plethysmography to compare the hypoxic ventilatory response (HVR) recorded from genetically modified mice and their corresponding control littermates of the same gender (*see* **Note 1**). The barometric plethysmography technique measures changes in pressure that occur while the animal is breathing, in normal air and during exposure to hypoxia or hypercapnia. Each chamber pressure transducer is connected to the port of a four-channel amplifier where the signal is digitized and sent to a computer for recording and analysis (*see* Subheadings in Fig. 1a).

2. Plethysmographs are calibrated to allow the measurement of rapid and transient small changes of pressure due to animal movement or respiration. The calibration of the system to detect these small changes in volume, which are superimposed on the constant flow of gas mixtures, is done with syringes delivering an air volume (~1 mL) similar to tidal respiratory volume in mouse.

3. For acquisition of pressure changes in the chamber, we use a sample rate of 500 Hz (one data point every 2 s) and a band-pass filter (0.25–45 Hz). Signals from the O_2 and CO_2 sensors are also amplified, digitized, and recorded unfiltered with a sampling rate of 1000 Hz. Data points of these parameters are also stored every 2 s.

3.2 Plethysmography: Experimental Procedure and Data Analysis

1. To study the respiratory function, mice are placed inside chambers continuously perfused with normoxic, hypoxic, or hypercapnic gas mixtures (*see* Subheading in Fig. 1a). Before beginning the experiment, the mouse is left to calm and fall asleep in the chamber.

2. The experimental protocol normally consists of two cycles of normoxia/hypoxia (5 min at 10% O_2)/normoxia followed by a cycle of normoxia/hypercapnia (2 min at 5% CO_2)/normoxia (*see* Fig. 1b). The iox2 software includes an automatic system to differentiate between rhythmic respiration and pressure changes related to irregular animal movement, sniffing, sighs, or washing behavior allowing the elimination of wrong data from the analysis. The parameter "successful rate" (percentage of success among a number of attempts) indicates the number of breaths recorded without any abnormal movement. In our laboratory, we only use for analysis of respiratory function data obtained during 100% successful rate.

3. Although most available plethysmographs provide several respiratory variables (breathing frequency, minute volume, peak expiratory flow, peak inspiratory flow, tidal mid-expiratory flow, and tidal volume), we routinely use respiratory frequency as the most reliable and informative parameter, as the increase in minute ventilation during hypoxia is mainly due to the increase in respiratory frequency. In addition, changes in tidal volume are very small and not highly reproducible due, among other causes, to small leakage in the recording chambers.

4. Representative recordings of mouse ventilatory responses to hypoxia and hypercapnia are shown in Fig. 1b. In these and other recordings (*see* Fig. 2a, b), the mean of four consecutive data points (each one acquired every 2 s) was obtained and represented as a single data point (that represents 8 s of recording time). To calculate changes in respiratory frequency during an experiment (*see* Fig. 2c, d), basal respiratory frequency was estimated in each animal by averaging the values of 25 points (200 s) previous to hypoxia and 25 points after complete recovery. Respiratory frequency reached during exposure to hypoxia or hypercapnia was estimated by averaging 40 points (320 s at 10% O_2) or 20 points (160 s at ~4–5% CO_2) before returning to normoxia. We followed exactly the same time course in different experiments, thereby allowing us to calculate the mean and standard error of each point with values obtained from different groups of animals. A representative example illustrating the time course of the HVR measured in 17 control (wild-type) mice is shown in Fig. 1c. Alterations of the HPV in genetically modified mice are illustrated in Fig. 2 (*see* **Note 2**).

3.3 Amperometry: Carotid Body and Adrenal Medulla Slicing

1. Animals are sacrificed by intraperitoneal administration of a lethal dose of sodium thiopental (120–150 mg/kg). Complete carotid bifurcations and adrenal glands are quickly removed and placed on ice-cooled and O_2-saturated modified Tyrode's

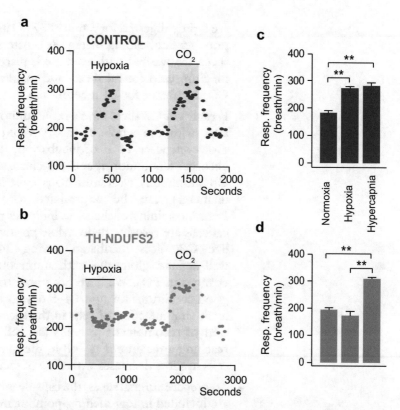

Fig. 2 Ventilatory response to hypoxia and hypercapnia of genetically modified animals. (**a**, **b**) Representative recordings of changes in respiratory frequency elicited by hypoxia (blue rectangle) and hypercapnia (red rectangle) in a control (**a**, blue dots) and a TH-NDUFS2 (Ndufs2-null) (**b**, red dots) mouse. (**c**, **d**). Quantification of the respiratory frequency during normoxia (21% O_2 and 0.04% CO_2), hypoxia (10% O_2), and hypercapnia (5% CO_2) in control and TH-NDUFS2 mice. *See* text for further explanation (reproduced from ref. [10] with permission from Elsevier)

solution. Carotid bodies are removed from the adjacent artery, cleaned, and included in 1% (wt/vol) low-melting-point agarose (FMC) in PBS. Small pieces (~1 cm length) of insulin syringe, perfectly polished and closed with Parafilm in one of the ends, are used to make agarose blocks. The inclusion is done at 42 °C, and quickly the agarose block is cooled on ice to slow down temperature.

2. The agarose block is glued with cyanoacrylate to the stage of a vibratome chamber and covered with the same cold, O_2-saturated Tyrode's solution. Slices of 150-μm thickness are cut with standard razor blades. The resulting slices are washed twice with cold sterile PBS and treated differently if they are from rat or mouse. Rat CB slices are normally placed on 35-mm Petri dishes with incubation medium. Mouse CB slices

are lightly digested for 5 min at 37 °C using an enzymatic solution (*see* Subheading in Fig. 3a). After this treatment, mouse slices are washed with cold PBS, placed in the same culture medium used for rat slices, and maintained at 37 °C in a 5% CO_2 incubator for 24 h before use.

3. Freshly cut CB slices have a rather uniform histological appearance with the whole surface covered by connective tissue. However, after 24 h of incubation, the texture of the slice becomes less uniform, and cell clusters (similar to the typical CB glomeruli) with numerous spherical cells (8–12 μm in diameter) can be appreciated (*see* Figs. 3a and 4a). Immunostaining of the slices indicates that most of the spherical cells are tyrosine hydroxylase positive [14]. We usually utilize CB slices incubated during 24–48 h, which show well-defined glomeruli with numerous cells responding to changes in PO_2. We have been able to maintain CB slices in good conditions for up to 4–5 days, which allowed us to perform viral infection of cells in the slices to induce the expression of recombinant proteins (*see* ref. [10]). However, after reaching a maximum at ~48 h, the number of O_2 responding cells in slices decreases with time of incubation.

4. To prepare adrenal slices, the capsule is removed and the glands are included in low-melting-point agarose at a temperature of 47 °C. Glands are cut into 200-μm slices following a protocol similar to that of CB slices. Adrenal slices are then washed with the recording solution and bubbled with carbogen (5% CO_2 and 95% O_2) at 37 °C in a water bath for 30 min. Fresh slices are used for the experiments during 4–5 h after preparation. Typical appearance of a fresh adrenal medulla slice is shown in Fig. 4d.

3.4 Measurement of Single-Cell Secretion by Amperometry (see Note 3)

1. The setup used for amperometric recording is illustrated in Fig. 3a.

2. Carbon-fiber electrodes (Fig. 3a) are prepared according to the protocol described in refs. [15, 16]. Briefly, synthetic graphite fibers, 10 μm in diameter (AMOCO), are inserted into a polyethylene tube (Portex) and the end of the tube heated to melt it with the carbon fiber to electrically insulate the fiber and allow the current to flow only through the fiber. The opposite end of the polyethylene fragment is inserted into a glass capillary (Kimax-51, Kimble products), to facilitate its connection to the head stage of the amplifier, and carefully sealed with Sylgard (SIGMA). Before used, the electrode is filled with a 3 M KCl solution to bring the carbon fiber into contact with the silver wire of the preamplifier.

3. To measure dopamine release from CB glomus cells, we normally use 10-μm carbon-fiber electrodes polarized to +750 mV

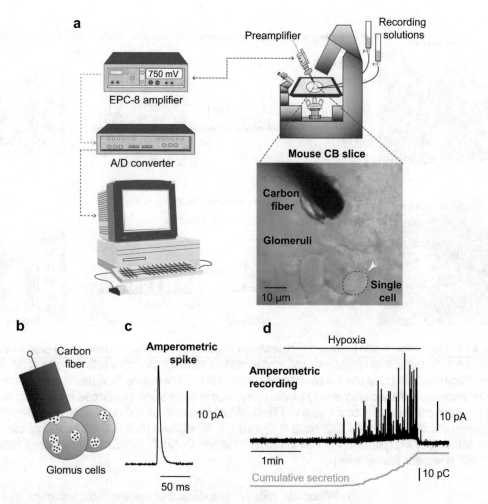

Fig. 3 Recording of cell secretory activity by amperometry. (**a**) Schematic representation of the equipment used in an amperometry setup. The microscope photograph shows the carbon-fiber electrode placed near a glomus cell in a carotid body slice with several cell clusters (glomeruli). (**b**) Schematic drawing of quantal dopamine release from a glomus cell recorded with the amperometric technique. (**c**) Amperometric spike due to the fusion of a single vesicle, showing the characteristic fast rising phase and slower exponential decay. (**d**) Typical amperometric recording of the secretory activity elicited by hypoxia in a glomus cell. The bottom panel shows the cumulative secretion signal of the recording. *See* text for details

and positioned near the cells under visual control (*see* Fig. 3a, b; *see* also Fig. 4a). Polarization voltage (+ 750 mV) is the redox potential, measured by cyclic voltammetry, needed for glomus cell dopamine oxidation [17]. A slightly higher polarization voltage (+800 mV) is used to record catecholamine secretion from adrenal chromaffin cells.

4. For experiments of testing responsiveness to hypoxia, a slice is transferred to a recording chamber and continuously perfused with normoxic and hypoxic solutions at ~35 °C.

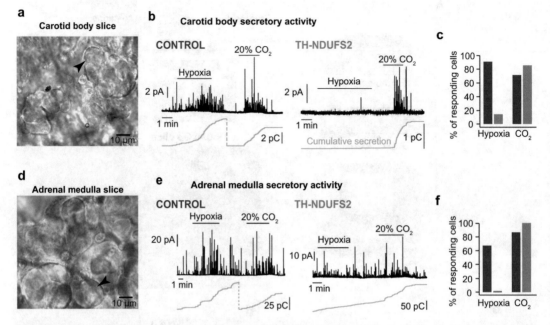

Fig. 4 Secretory responses to hypoxia and hypercapnia of genetically modified animals. (**a**, **d**) Representative microscope photographs of carotid body and adrenal medulla slices. Black arrows indicate some of the single glomus and chromaffin cells that can be clearly appreciated in the slice surface. (**b**, **e**) Representative ampero-metric recording of the secretory activity elicited in response to hypoxia and hypercapnia in cells from carotid body and adrenal slices from control (left) and TH-NDUFS2 (right) mice. (**c**, **f**) Quantification of the percentage of glomus and chromaffin cells responding to hypoxia and hypercapnia relative to the number of cells that were activated by high potassium (100%) in control (blue) and TH-NDUFS2 (red) mice (reproduced from ref. [10] with permission from Elsevier)

5. When the cells are stimulated to release catecholamine by exo-cytosis, oxidation of the neurotransmitter molecules results in the transfer of electrons to the carbon fiber, thereby generating a current that is recorded as a function of time. Single exocy-totic events appear as spike-like signals representing the release of either individual catecholaminergic vesicles or vesicles fused before exocytosis. An example of a large secretory spike is shown in Fig. 3c. The magnitude of the current generated is proportional to the number of molecules oxidized as defined by the equation:

$$\int I_t dt = Q = \eta FN$$

where I_t is the current generated at the surface of the sensing electrode, Q is the charge at the surface of the sensing electrode (calculated as the time integral of the amperometric current), n is the number of electrons transferred per molecule that is oxidized (e.g., 2 electrons in the case of dopamine), N is the number of moles of neurotransmitter oxidized, and F is the Faraday constant (96.500 coulomb/mol).

6. The typical secretory response to hypoxia of cells in CB slices is represented in Fig. 3d. Under normoxic conditions, most cells (as the one in this example) had no measurable secretory activity or some occasional exocytotic events at a very low frequency. However, some cells can be more active with a spontaneous secretory event frequency of 10 per minute or higher. After switching to the hypoxic solution, the cell responded with a progressive increase in the frequency and amplitude of the spikes that partially fused into a broad concentration envelope. After switching back to the normoxic solution, recovery was fast, and the cell returned to control conditions in less than 30 s. Reproducible responses to hypoxia can be normally observed in the cells if several minutes of rest are allowed for recovery between successive stimuli. Given the proximity of the amperometric electrode to the cell under study (*see* Fig. 3a, b), it seems that most of the spikes recorded are due to exocytotic events from the same cell. However, the possibility that some small spikes appearing in the recordings are due to secretory events occurring in neighboring cells cannot be excluded. The progressive increase in spike amplitude during exposure to hypoxia is a phenomenon observed repetitively in different experiments, which probably results from the intracytoplasmatic fusion of vesicles occurring before release (compound exocytosis) [14].

7. To quantify the secretory response induced by different stimuli, several parameters are calculated using the software Igor Pro (WaveMetrics). Secretion rate (femtocoulombs (fC)/min) is calculated as the amount of charge transferred to the recording electrode during a given period of time. Frequency of spiking is the number of spikes during the last minute of exposure to the stimulus. The cumulative secretion signal (in picocoulombs) is the sum of the time integral of successive spikes (green trace in Fig. 3d, bottom).

8. Representative examples of amperometric recordings performed on chemoreceptor cells in CB and AM slices of genetically modified mice are illustrated in Fig. 4 (*see* **Note 4**).

4 Notes

1. The hypoxic ventilatory response (HVR) is the first defense of mammals against environmental or systemic hypoxia. Hyperventilation is triggered by activation of the arterial chemoreceptors in response to a decrease in arterial PO_2. Depending on the duration of the hypoxic stimulus, several time domains in the HVR have been described [18–21]. Short-term exposures to hypoxia (lasting ~5 min) induce an increase in minute ventilation (breathing frequency × tidal volume)

called acute HVR. The design of experimental methods to measure HVR (in particular the acute component), with reliable results among different laboratories, has been a matter of debate during the last decades, and several attempts have been made to adopt a consensus methodology [22, 23]. In animal studies, quantitative differences in the magnitude of the HVR have been attributed to sex, age, body mass, and even to different inbred strains of the same species [23, 24]. It is well known that ventilation depends on both the activity of the central chemoreceptors and the peripheral chemoreflex drive to breathe [25]. When hypoxia stimulates the peripheral chemoreflex, its effect is assumed to be additive with the already present central chemoreflex drive [22, 26]. Determination of the peripheral contribution therefore requires subtraction of the central contribution from the measured ventilation. However, during hyperventilation at the same time that blood O_2 tension increases, a decrease in blood CO_2 tension occurs, which by itself alters the activity of central and peripheral chemoreceptors [27, 28]. This is one of the main reasons why the HVR can change depending on the conditions used in the different laboratories. Basically, the HVR can be measured with CO_2 tensions allowed to change (poikilocapnic) or fixed (isocapnic). The poikilocapnic measure is the one normally used to mimic ventilatory responses to hypoxic environments (with low CO_2 tension similar to air). Isocapnic HVR measures, which require the use of gas mixtures with a constant, relatively high, CO_2 tension (between 1.5 and 3% CO_2), are preferred to study the changes of chemoreceptor O_2 sensitivity, e.g., during hypoxic ventilatory decline when the low environmental O_2 tension is maintained for several hours [22].

2. Genetically modified mice are broadly used in biomedical research and are particularly useful to study acute O_2 sensing, as CBs are tiny organs in which large-scale biochemical or molecular biology techniques are difficult to apply. The functional role of specific genes can be studied in mice with selective disruption of genes targeted to a specific tissue or cell type that can be also controlled in time. One representative example of this type of studies is the analysis of the *Ndufs2* conditional knockout mice. *Ndufs2* encodes a 49 kDa subunit, which forms part of the ubiquinone binding site of mitochondrial complex I (MCI). Generation of this mice model was inspired by our previous studies indicating that rotenone selectively occludes responsiveness to hypoxia in CB glomus cells [29, 30]. Rotenone is a distal MCI blocker that binds to the ubiquinone binding site; therefore, our data suggested that MCI could be implicated in acute O_2 sensing (*see* ref. [29]). Using Cre-loxP technologies, we generated conditional knockout mice in which the *Ndufs2* gene was specifically deleted in

tyrosine hydroxylase (TH)-positive catecholaminergic cells (TH-NDUFS2 mice), which include the O_2-sensitive glomus cells and chromaffin cells in the CB and AM, respectively [10]. Animals were genotyped and the *Ndufs2* deletion confirmed by comparing *Ndufs2* mRNA levels in catecholaminergic tissues (CB and SCG) of Ndufs2-null and wild-type mice [10]. Fig. 2 shows plethysmographic recordings of a TH-NDUFS2 mouse and a control littermate performed in parallel. The wild-type (control) mouse responded to hypoxia and hypercapnia with the typical increase in respiratory frequency that rapidly returned to its basal values after disappearance of the stimuli (Fig. 2a). Responsiveness to hypoxia was practically abolished in the TH-NDUFS2 mouse, whereas the increase in respiratory frequency induced by hypercapnia was unaltered (Fig. 2b). Quantitative analyses of respiratory frequency in the differential experimental conditions are shown in Fig. 2c, d. These data demonstrate that a normal MCI function in CB glomus cells is necessary for a normal HVR (*see* further mechanistic details in refs. [10, 31]).

3. Amperometry is an electrophysiological technique used to measure changes in the concentration of electroactive molecules (catecholamines, O_2, etc.) that can be oxidized or reduced at an electrode surface. A small carbon-fiber electrode, carefully located near the cell and polarized at the redox potential of the molecule that is going to be detected, allows the experimenter the direct monitoring of neurotransmitter (e.g., catecholamine) released [16] or O_2 tension near a cell [32].

4. Cells in CB slices (Fig. 4a) respond to hypoxia and hypercapnia with a powerful secretory activity (Fig. 4b, left). The response to hypoxia is selectively abolished in TH-NDUFS2 mice (Fig. 4b, right), which as shown before (*see* Fig. 1) lack the HVR. Note that in TH-NDUFS2 mice responsiveness to hypercapnia remains unaltered. The percentage of cells responding to hypoxia or hypercapnia in each animal type is given in Fig. 4c. Chromaffin cells in adrenal slices (Fig. 4d) from wild-type mice also show characteristic secretory responses to hypoxia and hypercapnia (Fig. 4e, left). Responsiveness to hypoxia is also selectively abolished in TH-NDUFS2 mice (Fig. 4e, right) (*see* also Fig. 4f).

Acknowledgments

This work was supported by the Botín Foundation, the Spanish Ministry of Economy and Innovation (SAF2012-39343, SAF2016-74990-R), and the European Research Council (ERC Advanced Grant PRJ201502629).

References

1. López-Barneo J, Pardal R, Ortega-Sáenz P (2001) Cellular mechanism of oxygen sensing. Annu Rev Physiol 63:259–287

2. Semenza GL (2012) Hypoxia-inducible factors in physiology and medicine. Cell 148:399–408

3. Weir EK, López-Barneo J, Buckler KJ, Archer SL (2005) Acute oxygen-sensing mechanisms. N Engl J Med 353:2042–2055

4. López-Barneo J, Pardal R, Montoro RJ, Smani T, García-Hirschfeld J, Ureña J (1999) K+ and Ca2+ channel activity and cytosolic [Ca2+] in oxygen-sensing tissues. Respir Physiol 115:215–227

5. Kemp PJ, Peers C (2007) Oxygen sensing by ion channels. Essays Biochem 43:77–90

6. López-Barneo J, Ortega-Sáenz P, González-Rodríguez P, Fernández-Agüera MC, Macías D, Pardal R, Gao L (2016) Oxygen-sensing by arterial chemoreceptors: mechanisms and medical translation. Mol Asp Med 47-48:90–108

7. Ortega-Sáenz P, Pardal R, Levitsky K, Villadiego J, Munoz-Manchado AB, Duran R, Bonilla-Henao V, Arias-Mayenco I, Sobrino V, Ordonez A, Oliver M, Toledo-Aral JJ, López-Barneo J (2013) Cellular properties and chemosensory responses of the human carotid body. J Physiol 591:6157–6173

8. Ortega-Sáenz P, Pascual A, Piruat JI, López-Barneo J (2007) Mechanisms of acute oxygen sensing by the carotid body: lessons from genetically modified animals. Respir Physiol Neurobiol 157:140–147

9. Ortega-Sáenz P, Levitsky KL, Marcos-Almaraz MT, Bonilla-Henao V, Pascual A, López-Barneo J (2010) Carotid body chemosensory responses in mice deficient of TASK channels. J Gen Physiol 135:379–392

10. Fernández-Agüera MC, Gao L, González-Rodríguez P, Pintado CO, Arias-Mayenco I, García-Flores P, García-Perganeda A, Pascual A, Ortega-Sáenz P, López-Barneo J (2015) Oxygen sensing by arterial chemoreceptors depends on mitochondrial complex I signaling. Cell Metab 22:825–837

11. Aaron EA, Powell FL (1993) Effect of chronic hypoxia on hypoxic ventilatory response in awake rats. J Appl Physiol (1985) 74:1635–1640

12. Jacky JP (1978) A plethysmograph for long-term measurements of ventilation in unrestrained animals. J Appl Physiol Respir Environ Exerc Physiol 45:644–647

13. Mortola JP, Frappell PB (1998) On the barometric method for measurements of ventilation, and its use in small animals. Can J Physiol Pharmacol 76:937–944

14. Pardal R, López-Barneo J (2002) Carotid body thin slices: responses of glomus cells to hypoxia and K(+)-channel blockers. Respir Physiol Neurobiol 132:69–79

15. Chow RH, l von R (1995) Electrochemical detection of secretion from single cells. In: Sakmann B, Neher E (eds) Single-Channel recording, 2nd edn. Plenum Press, New York, pp 245–275

16. Gillis KD (1995) Techniques for membrane capacitance measurements. In: Sakmann B, Neher E (eds) Single-Channel recording, 2nd edn. Plenum Press, New York, pp 155–198

17. Ureña J, Fernández-Chacón R, Benot AR, Alvarez de Toledo GA, López-Barneo J (1994) Hypoxia induces voltage-dependent Ca2+ entry and quantal dopamine secretion in carotid body glomus cells. Proc Natl Acad Sci U S A 91:10208–10211

18. Easton PA, Slykerman LJ, Anthonisen NR (1986) Ventilatory response to sustained hypoxia in normal adults. J Appl Physiol (1985) 61:906–911

19. Liang PJ, Bascom DA, Robbins PA (1997) Extended models of the ventilatory response to sustained isocapnic hypoxia in humans. J Appl Physiol (1985) 82:667–677

20. Powell FL, Milsom WK, Mitchell GS (1998) Time domains of the hypoxic ventilatory response. Respir Physiol 112:123–134

21. Steinback CD, Poulin MJ (2007) Ventilatory responses to isocapnic and poikilocapnic hypoxia in humans. Respir Physiol Neurobiol 155:104–113

22. Duffin J (2007) Measuring the ventilatory response to hypoxia. J Physiol 584:285–293

23. Teppema LJ, Dahan A (2010) The ventilatory response to hypoxia in mammals: mechanisms, measurement, and analysis. Physiol Rev 90:675–754

24. Palmer LA, May WJ, deRonde K, Brown-Steinke K, Gaston B, Lewis SJ (2013) Hypoxia-induced ventilatory responses in conscious mice: gender differences in ventilatory roll-off and facilitation. Respir Physiol Neurobiol 185:497–505

25. Cunningham DJC, Robbins PA (1986) Wolff CB integration of respiratory responses to changes in alveolar partial pressures of CO_2 and O_2 and in arterial pH. In: Fishman AP, Cherniack NS, Widdicombe JG (eds) Handbook of physiology, section 3, The

Respiratory System. II. American Physiological Society, Bethesda, MD, pp 475–528

26. Clement ID, Bascom DA, Conway J, Dorrington KL, O'Connor DF, Painter R, Paterson DJ, Robbins PA (1992) An assessment of central-peripheral ventilatory chemoreflex interaction in humans. Respir Physiol 88:87–100

27. Lloyd BB, Cunningham DJC (1963) A quantitative approach to the regulation of human respiration. In: Cunningham DJC, Lloyd BB (eds) The regulation of human respiration. Blackwell, Oxford, pp 331–349

28. Mohan R, Duffin J (1997) The effect of hypoxia on the ventilatory response to carbon dioxide in man. Respir Physiol 108:101–115

29. Ortega-Sáenz P, Pardal R, García-Fernández M, López-Barneo J (2003) Rotenone selectively occludes sensitivity to hypoxia in rat carotid body glomus cells. J Physiol 548:789–800

30. García-Fernández M, Ortega-Sáenz P, Castellano A, López-Barneo J (2007) Mechanisms of low-glucose sensitivity in carotid body glomus cells. Diabetes 56:2893–2900

31. Gao L, González-Rodríguez P, Ortega-Sáenz P, López-Barneo J (2017) Redox signaling in acute oxygen sensing. Redox Biol 12:908–915

32. Ganfornina MD, López-Barneo J (1992) Potassium channel types in arterial chemoreceptor cells and their selective modulation by oxygen. J Gen Physiol 100:401–426

Chapter 14

Immunohistochemistry of the Carotid Body

Jayasri Nanduri and Nanduri R. Prabhakar

Abstract

Immunohistochemistry (IHC) enables the detection and distribution of proteins in cells of tissues. IHC has become an indispensable approach for studying oxygen sensing by the carotid body (CB). This chapter provides a detailed description of IHC of CB tissue and isolated CB cells.

Key words Carotid bifurcation, Primary and secondary antibodies, Fluorophores, Antigen retrieval, Blocking reagents

1 Introduction

Immunohistochemistry (IHC) is a commonly employed technique for studying localization of proteins in tissues/cells using antibodies capable of binding to proteins of interest. The detection method can be either direct or indirect. In the direct method, the antibody is linked to a reporter like fluorescent dye, or an enzyme, which will give rise to a signal, e.g., fluorescence or color from enzyme reaction, which is then detected by a microscope. Often, this approach is not sensitive enough to detect protein(s) of lower abundance. In the indirect method, a reporter-coupled secondary antibody, such as goat anti-mouse IgG, is tagged to the primary antibody which allows multiple binding of secondary antibodies to the same primary antibody, thereby increasing sensitivity to detect low-abundant proteins. The disadvantage of the indirect method is the risk of non-specific binding of the secondary antibody and longer experimental times. There are several reporter systems available. In the reporter system based on enzyme reaction, the secondary antibody is coupled to an enzyme such as horseradish peroxidase (HRP), which yields a brown reaction product after adding diaminobenzidine (DAB) as a final substrate. The brown color stain may be visualized using light microscopy. The sensitivity can be further amplified by using avidin-biotin complex (ABC) which involves using biotinylated conjugated secondary antibody

L. Eric Huang (ed.), *Hypoxia: Methods and Protocols*, Methods in Molecular Biology, vol. 1742,
https://doi.org/10.1007/978-1-4939-7665-2_14, © Springer Science+Business Media, LLC 2018

and HRP-conjugated avidin. The most popular reporter systems currently being used are the fluorophores which are fluorescent chemical compounds that reemit light upon light excitation and visualized using a fluorescence microscope. Depending on the spectral characteristics (excitation and emission), multiple fluorophores with different colors are available which can be combined in a single sample. This approach is most useful to study localization of a protein within a specific organelle using a marker protein or co-localization of two proteins within a given cell. IHC is commonly employed for visualizing the localization of the protein in a specific cell type within a given tissue section.

Immunocytochemistry (ICC) is used for determining the localization of proteins in cell cultures or in cells in suspension. IHC and ICC differ in sample processing with minimal differences in antibody staining procedures. ICC is useful for assessing the effect of knockout or overexpression of a gene on protein expression in cell cultures.

Carotid bodies (CBs) are sensory organs for monitoring the chemical composition of the arterial blood including O_2 levels [1]. Anatomically, CBs are located bilaterally at the bifurcation of the common carotid artery into external and internal carotid arteries. The CB is physically a small tissue weighing ~25 μg in mice and ~50–60 μg in rats [1]. The CB tissue is primarily composed of two cell types: type I (also called glomus cell) and type II or sustentacular cell. Type I cells are neuronal phenotype, and type II cells resemble glial cells of the nervous system [1]. A substantial body of evidence indicates that type I cells are the primary site of sensory transduction and they work in concert with the nearby afferent nerve ending as a sensory unit [1]. CB being physically small tissue, either IHC or ICC is widely employed to study cellular aspects of CB function [2–4].

The protocols for IHC for studying CB can be divided into four steps. The first two steps involve harvesting, mounting, and sectioning of CB. These steps are critical for generating good sections for reliable and reproducible staining of proteins. **Steps 3 and 4** include imaging and analysis of the images to identify cellular structures and quantitative assessment by morphometry.

2 Materials

2.1 Harvesting CBs

1. Adult rats or mice.
2. Urethane.
3. Forceps.
4. Catheters.
5. Syringe.
6. Sharp scissors.

7. 4% paraformaldehyde.

8. 30% sucrose.

9. Phosphate-buffered solution (PBS).

10. A dissecting microscope.

11. 15 mL polypropylene conical tubes.

12. Ice.

2.2 Mounting CB

1. Phosphate-buffered solution (PBS) (Gibco Invitrogen).

2. Tissue-Tek, OCT compound.

3. 60 mm Sylgard-coated dish.

4. Pins.

5. Plastic cryomolds.

6. Two pairs of straight tweezers.

7. Microdissecting spring scissors.

8. Dry ice.

9. Styrofoam container.

10. Cryostat.

11. Superfrost Plus slides.

2.3 Staining

1. Super PAP Pen HT (Research Products International).

2. Humidifier box (any plastic box covered with aluminum foil outside and a wet paper towel inside).

3. Phosphate-buffered solution.

4. Normal goat serum (NGS).

5. Triton X-100.

6. Microscope cover glass 24 × 50–1.

7. Mounting medium containing DAPI (4′,6-diamidino-2-phenylindole).

8. Clear nail polish.

9. Slide box.

10. Blocking reagent: 20% normal goat serum (NGS) + 0.2% Triton X-100 in PBS. It should be prepared fresh before use.

11. Permeabilization solution: 20% Triton X-100 in PBS.

12. Antibody solution: 1% NGS + 0.2% Triton X-100 in PBS. Spin down NGS solution prior to use to remove particle/sediment from suspension.

13. Primary and secondary antibodies: Dilute primary and secondary antibodies in 1% NGS as specified by manufacturer's instructions or published information.

14. Washing buffer: 0.05% Triton X-100 in PBS.

2.4 Imaging and Analysis

1. A fluorescent microscope.
2. ImageJ software (https://imagej.nih.gov/ij/).
3. Fiji image analysis software (http://fiji.sc/Fiji).

2.5 Immuno-cytochemistry of CB Glomus Cells

1. Dulbecco's Modified Eagle Medium (DMEM)/Ham F12 supplemented with 5% fetal bovine serum, 2% insulin-transferrin-selenium (ITS), and 1% penicillin/streptomycin.
2. Locke's solution: NaCl 154 mM, KCl 5.6 mM, Na_2HPO_4 2.15 mM, NaH_2PO_4 0.85 mM, glucose 10 mM, and HEPES 10 mM.
3. Dissociation solution I: Add 0.5 mL of collagenase (stock 8 mg/mL), 10 µL of DNase 1 (stock 10 mg/mL) solution, and 6 mg of BSA in 2 mL Locke's solution and sterilize by filtering.
4. Dissociation solution II: Add 6.67 µL of DNase I stock solution to 1 mL of Locke's solution.
5. A shaking water bath.
6. Fire-polished glass pipettes.
7. Collagen (type VII)-coated cover slips.

3 Methods

3.1 Harvesting CBs

1. Carotid bifurcations are harvested from adult rats or mice anesthetized with urethane (1.2 g/kg; IP).
2. Please change Animals' to Animal's chest is open, and a catheter is placed in the heart and perfused initially with PBS containing heparin (1 unit/mL) for 15 min or until no blood is seen in the perfusate coming out of the animal. Then the perfusion medium is switched to 4% paraformaldehyde for 30 min to fix the tissues.
3. Subsequently, animal's jaw and arms are taped and placed under a well-lit dissecting microscope.
4. Using scissors, a midline incision is made from the base of the neck up to the middle of the jaw.
5. Cut away the muscle until the trachea is visible.
6. Clip off small muscles posterior and lateral to the trachea.
7. The common carotid artery (CCA) is posterior and lateral to the trachea.
8. Remove the trachea allowing visualization of the CCA.
9. Lift the CCA with forceps and separate the CCA from the nearby connective tissue.

10. Identify where the CCA bifurcates into the interior carotid artery (ICA) and exterior carotid artery (ECA), and cut the ICA and ECA posterior and superior to the hypoglossal nerve (white nerve that goes across ECA and ICA).

11. Place the bifurcations in tubes containing PBS on ice.

12. Bifurcations are cryoprotected in 30% sucrose/PBS at 4 °C for 3–4 h or overnight.

13. *Carotid bifurcations from pups can be harvested directly after anesthesia without perfusion.* The harvested CB tissue from pups is then fixed in 4% paraformaldehyde for 4 h at room temperature or overnight at 4 °C and cryoprotected in 30% sucrose/PBS at 4 °C for 3–4 h. *It is not recommended to fix the tissue for more than 24 h because overfixation can cause poor antibody binding.*

3.2 Mounting Carotid Bifurcations

1. Add OCT compound to plastic cryomold by slowly filling up to the notch on the mold (about ¼ full). Make sure to avoid air bubbles. Let OCT settle in molds at room temperature while isolating the carotid bifurcation.

2. Label cryomold with sample ID.

3. Collect dry ice in Styrofoam container. Break ice into small chunks to create a flat level surface. This is an important step that allows placing the cryomold containing the tissue on flat surface of dry ice.

4. Wash dehydrated fixed tissue in PBS (3×) for 5 min per wash.

5. Place the bifurcation in 1 mL of PBS in a 60 mm Sylgard-coated dish (a pinch of carbon powder is added to Sylgard silicone, which provides the black background for easy visualization of the CB) and place beneath a well-lit microscope.

6. Orient the bifurcation so that the superior cervical ganglion (SCG) is visible from the top. The SCG is a large whitish, translucent tissue located between ECA and ICA.

7. Pin down the CCA, and pull laterally the ECA and ICA and pin them down on Sylgard. Using fine scissors, gently clean the connective tissue surrounding the CCA.

8. Remove the SCG by lifting its base and pulling it up slowly while carefully rubbing away the tissue beneath. Be careful not to accidentally pull off the CB.

9. Remove excess connective tissue and cut to expose the occipital artery, the carotid sinus nerve, and the CB. The CB can be visualized under the dissecting microscope as a small, brownish yellow structure. *Note: The occipital artery is the landmark and always protrudes from the opposite side of the CB.*

10. Trace the carotid sinus nerve where it joins the glossopharyngeal nerve.

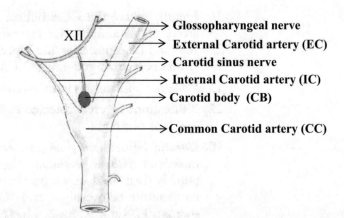

Fig. 1 Schematic representation of carotid body (CB) bifurcation. CBs are located bilaterally at the bifurcation of the common carotid artery (CC) into external carotid (EC) and internal carotid (IC) arteries

11. Gently clean the connective tissue off the CB. Again be very careful as to non-accidentally remove the CB (*see* **Note 1**).

12. Once sufficiently cleaned, the carotid bifurcation along with the CB and the carotid sinus is ready to be mounted (Fig. 1).

13. Check the cryomold containing OCT for any air bubbles. If there are any, use a pin to move them aside.

14. Unpin the bifurcation and place in the cryomold containing OCT. Place the CB bifurcation in the center of OCT avoiding any air bubbles (*see* **Note 2**).

15. After mounting the CB bifurcation, place the cryomold on a flat level bed of dry ice. Keep on dry ice for 10–15 min.

16. Frozen tissue can now be transferred to −80 °C until further sectioning.

17. Set up the cryostat as per manufacturer's instructions.

18. Cut 8–10 μm sections. The best CB sections are usually obtained when the block temperature is around −18 to −20 °C. *Do not allow mounted frozen tissue to thaw before cutting.*

19. Transfer sections to Superfrost Plus slides and collect two tissue sections per slide. Make sure the tissue sections are on the frosted side of the microscope slide.

20. Tissues can be stored at 4 °C for a few days; for longer-term storage, keep them at −80 °C.

3.3 Staining

1. Take out slides from −80 °C and place in the humidity box to bring them to room temperature.

2. Mark the tissue section boundaries with hydrophobic slide marker (PAP pen).

3. Wash the tissue section gently with 400 µL PBS. During this step, place the pipette tip to the side of the tissue section and not directly on top of the section to avoid dislodging the tissue section. Wash three times over a period of 2–3 min. *(Never leave the tissue dry.)*

4. Block the non-specific staining by incubating the tissue section in 200–400 µL of 20% NGS (blocking solution) for 30 min at room temperature (*see* **Note 3**).

5. Prepare primary antibody solution. Dilute the antibody in 1% NGS as specified by manufacturer's instructions or published information (*see* **Note 4**). Centrifuge the antibody before adding to the sample.

6. Remove blocking solution and add 200–400 µL of primary antibody solution to each tissue section sample. To study co-localization of two proteins in the same section, double immunofluorescence procedure can be carried out. Incubate the section with a mixture of two primary antibodies raised in different species (e.g., rabbit against rat protein 1 and mouse against rat protein 2 for a rat CB section).

7. Incubate the slides in primary antibody at 37 °C for 2 h or 4 °C overnight. For most antibodies, overnight incubation at 4 °C is recommended. However, the duration of incubation with primary antibody may need to be optimized for each antibody. *(In addition, it is important to include a negative control using the incubation buffer with no primary antibody to exclude non-specific staining.)*

8. Wash the tissue sections three times with washing buffer for 5 min.

9. Prepare secondary antibody by diluting in 1% NGS as specified by manufacturer's instructions or published information. If two primary antibodies are added in **step 6**, prepare a mixture of two secondary antibodies with two different fluorochromes, i.e., Alexa flour 488 (FITC) conjugated to rabbit to detect protein 1 and Alexa 555 (red) conjugated to mouse to detect protein 2. *Briefly spin the tube containing the secondary antibody before taking an aliquot* (*see* **Note 5**).

10. Add 200–400 µL of diluted secondary antibody to each sample and incubate at room temperature for 1 h (*see* **Note 6**). (All steps are done in the humidity box in dark.)

11. Wash with washing buffer five times, each wash for 5 min.

12. Immediately add 20 µL/slide of mounting medium containing DAPI, which binds to DNA and stains the nuclei. DAPI is excited by ultraviolet light and emits blue light. *DAPI is not recommended when visualizing protein localized in nuclei.*

13. Carefully lay the cover slip on the drop of mounting medium so that the medium spreads evenly with no air bubbles. Dab off the edges of the cover slip with tissue paper to remove excess medium.

14. Seal the edges of the cover slip with clear nail polish.

15. Visualize using a fluorescent microscope.

16. Store the stained slides in a slide box at −20 °C or −80 °C.

3.4 Imaging

Chromogranin A (CGA) and tyrosine hydroxylase (TH) are expressed within the glomus cells of the CB and often used as markers of these cells [5]. Shown in Fig. 2 are microscopic images of a mouse CB section stained with anti-CGA and anti-TH primary antibodies and biotinylated secondary antibody [6]. Staining was visualized in bright field microscopy by Vectastain Elite avidin-biotinylated enzyme and diaminobenzidine peroxidase as substrate.

Figure 3 shows images of a rat CB section stained for NADPH oxidase (Nox2) using anti-Nox2 antibody and fluorescent-conjugated secondary antibody [7]. Nox2-like immunoreactivity is localized to the cytoplasm of cells using fluorescein isothiocyanate (FITC)-conjugated secondary antibody. For identification of Nox2 to glomus cells, sections were doubled stained with polyclonal anti-CGA antibody followed by Alexa Fluor 555-conjugated secondary antibody.

3.5 Analysis

3.5.1 Morphometric Analysis of CB and Glomic Cell Volume

CB and glomic cell volumes can be analyzed using ImageJ software. IHC is performed on individual serial sections (minimum of 10) which are then individually imaged. For each image, CB area is measured manually by tracing the periphery of the CB. Glomic cell area can be calculated manually by tracing the periphery of glomus cells stained with TH or CGA (marker proteins). Glomic

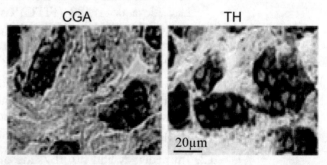

Fig. 2 Chromogranin A (CGA) and tyrosine hydroxylase (TH) expressed primarily within glomus cells of the mouse carotid body are stained using biotinylated secondary antibody. Staining was visualized in bright field microscopy by Vectastain Elite avidin-biotinylated enzyme and diaminobenzidine peroxidase as substrate

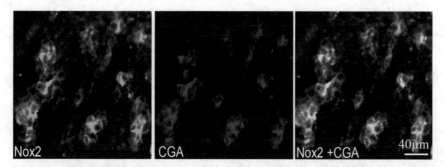

Fig. 3 Rat carotid body section stained using fluorescent-conjugated secondary antibodies. NADPH oxidase (Nox2)-like immunoreactivity is localized to the cytoplasm of cells using FITC-conjugated secondary antibody (Alexa Fluor 488; Molecular Probes). For identification of Nox2 to glomus cells, sections were doubled stained with polyclonal anti-Chromogranin A (CGA) antibody (established marker of glomus cells), followed by Alexa Fluor 555-conjugated secondary antibody (Molecular Probes)

cell volume is calculated by sum of each area multiplied by thickness and number of sections [8, 9].

3.5.2 Quantitative Analysis of co-Localization of Proteins

The degree of co-localization of two proteins can be determined using a "co-localization threshold" plug-in developed for Fiji image analysis software. This plug-in involves spatial correlation analysis along with an algorithm that permits automatic threshold identification [10, 11]. The areas for co-localization analysis are selected by drawing the region of interest stained with antibodies for two individual proteins. For each cell, co-localization coefficient for Protein 1 (P1) and for Protein 2 (P2) are calculated and analyzed using algorithms as described [12].

3.6 Immuno-cytochemistry (ICC) of CB Glomus Cells

3.6.1 CB Cell Isolation

1. Dissect carotid artery bifurcations from rats or mice anesthetized with urethane (1.2 g/kg, IP) and place them in ice-cold $Ca^{2+}-/Mg^{2+}$-free PBS solution.

2. Add 1 mL dissociation solution I to two rat CBs (1 mL to four mouse CBs) and incubate at 37 °C for 25 min, in a shaking water bath (100 rpm).

3. To release cells, the tissue fragments are gently triturated for 1 min at 5 min intervals during incubation using fire-polished glass pipette. The duration of trituration could be longer or shorter (hold the tube to the light to see that there are no pieces or clumps) (*see* **Note 7**).

4. After 25 min, add cold 5 mL of Locke's solution and centrifuge for 5 min at $150 \times g$ (~12,000 rpm).

5. Resuspend the cell pellet in dissociation solution II and incubate for an additional 10 min at 37 °C in the shaking water bath and repeat **step 4**. (Skip this step for dissociation of mouse CB cells and proceed to **step 6**.)

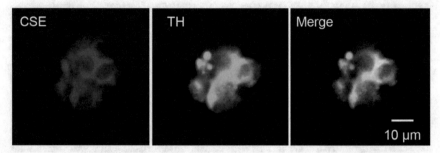

Fig. 4 Isolated glomus cells from a rat carotid body were stained using fluorescent-conjugated secondary antibodies. Cystathionine gamma-lyase (CSE)-like immunoreactivity is localized to the cytoplasm of cells using Alexa Fluor 555-conjugated secondary antibody (red; Molecular Probes). For identification of CSE to glomus cells, cells were doubled stained with polyclonal anti-tyrosine hydroxylase (TH) antibody (established marker of glomus cells), followed by FITC-conjugated (Alexa Fluor 488) secondary antibody (Molecular Probes)

6. Resuspend the isolated cells in culture medium and gently disperse them with a sterile pipette.

7. Plate cells on collagen (type VII)-coated cover slips and place them in a humidified incubator equilibrated with 21% O_2 + 5% CO_2 maintained at 37 °C.

3.6.2 Immuno-cytochemistry of Isolated CB Cells

1. Glomus cells plated on collagen-coated cover slips are fixed with 4% paraformaldehyde for 30 min at room temperature.

2. For staining procedure, follow the protocols described in Subheading 2.3. Shown in Fig. 4 is an example of isolated CB glomus cells stained with antibodies specific for cystathionine gamma-lyase (CSE) and TH (a marker of glomus cells).

4 Notes

1. Care should be taken to clean the CB bifurcation free of all connective tissue without accidentally removing the CB.

2. CB should be mounted horizontally in the cryomold to get good sections. In order to avoid any folding, a few drops of OCT compound can be added directly to the CB preparation at **step 14** in Subheading 3.2 before placing in the cryomold. When mounted properly, one gets approximately 16–18 good sections from a rat CB and 8–10 sections from a mouse CB. Before staining, it is recommended to check each section under a light microscope to ensure the intactness of the tissue.

3. NGS is used as the blocking agent, because the species of the secondary antibody used is usually from goat. Blocking reagent is 10–20% solution of serum from the species in which the secondary antibodies are raised. In our experience, the use of other blocking reagents such as BSA (3–5%) tends to give higher background.

4. Primary antibody concentrations should be optimized by testing dilutions from 1:50 to 1:500.

5. Vortex and centrifuge both primary and secondary antibody solutions before use as they can aggregate and add debris to fixed tissue preparation.

6. When using a mixture of antibodies/fluorophores to visualize different proteins, control experiment with each individual antibody/fluorophore needs to be routinely done to standardize the antibody concentration and check for tissue autofluorescence and for overlap of fluorophore spectra.

7. Dissociation of CB cells depends on the amount of trituration. Too little trituration results in clumps and too much triturating leads to dead cells.

Acknowledgments

This work was supported by the National Institutes of Health grant PO1-HL-90554.

References

1. Kumar P, Prabhakar NR (2012) Peripheral chemoreceptors: function and plasticity of the carotid body. Compr Physiol 2:141–219. https://doi.org/10.1002/cphy.c100069

2. Izal-Azcárate A, Belzunegui S, San Sebastián W, Garrido-Gil P, Vázquez-Claverie M, López B, Marcilla I, Luquin MA (2008) Immunohistochemical characterization of the rat carotid body. Respir Physiol Neurobiol 161:95–99. https://doi.org/10.1016/j.resp.2007.12.008

3. Gonzalez C, Almaraz L, Obeso A, Rigual R (1994) Carotid body chemoreceptors: from natural stimuli to sensory discharges. Physiol Rev 74:829–898

4. Heym C, Kummer W (1989) Morphology and immunocytochemistry of two endocrine cell types in the guinea-pig esophageal epithelium. Cell Tissue Res 256:635–643

5. Karasawa N, Kondo Y, Nagatsu I (1982) Immunohistocytochemical and immunofluorescent localization of catecholamine-synthesizing enzymes in the carotid body of the bat and dog. Arch Histol Jpn 45:429–435

6. Kline DD, Peng YJ, Manalo DJ, Semenza GL, Prabhakar NR (2002) Defective carotid body function and impaired ventilatory responses to chronic hypoxia in mice partially deficient for hypoxia-inducible factor 1 alpha. Proc Natl Acad Sci U S A 99:821–826. https://doi.org/10.1073/pnas.022634199

7. Peng YJ, Nanduri J, Yuan G, Wang N, Deneris E, Pendyala S, Natarajan V, Kumar GK, Prabhakar NR (2009) NADPH oxidase is required for the sensory plasticity of the carotid body by chronic intermittent hypoxia. J Neurosci 29:4903–4910. https://doi.org/10.1523/JNEUROSCI.4768-08.2009

8. Pawar A, Nanduri J, Yuan G, Khan SA, Wang N, Kumar GK, Prabhakar NR (2009) Reactive oxygen species-dependent endothelin signaling is required for augmented hypoxic sensory response of the neonatal carotid body by intermittent hypoxia. Am J Physiol Regul Integr Comp Physiol 296:R735–R742. https://doi.org/10.1152/ajpregu.90490.2008

9. Peng YJ, Makarenko VV, Nanduri J, Vasavda C, Raghuraman G, Yuan G, Gadalla MM, Kumar GK, Snyder SH, Prabhakar NR (2014) Inherent variations in CO-H2S-mediated carotid body O2 sensing mediate hypertension and pulmonary edema. Proc Natl Acad Sci U S A 111:1174–1179. https://doi.org/10.1073/pnas.1322172111

10. Manders EMM, Verbeek FJ, Aten JA (1993) Measurement of co-localization of objects in dual-colour confocal images. J Microsc 169:375–382

11. Costes SV, Daelemans D, Cho EH, Dobbin Z, Pavlakis G, Lockett S (2004) Automatic and quantitative measurement of protein-protein colocalization in live cells. Biophys J 86:3993–4003.   https://doi.org/10.1529/biophysj.103.038422

12. Makarenko VV, Usatyuk PV, Yuan G, Lee MM, Nanduri J, Natarajan V, Kumar GK, Prabhakar NR (2014) Intermittent hypoxia-induced endothelial barrier dysfunction requires ROS-dependent MAP kinase activation. Am J Physiol Cell Physiol 306:C745–C752. https://doi.org/10.1152/ajpcell.00313.2013

Chapter 15

Hypoxia Signaling and Placental Adaptations

Damayanti Chakraborty, Regan L. Scott, and Michael J. Soares

Abstract

Oxygen is an essential nutrient for cells. Oxygen is delivered to tissues via red blood cells through the vasculature. Molecular mechanisms mediating cellular responses to low oxygen tension have been identified. Hypoxia-inducible factors (HIFs) are activated by low oxygen and promote transcriptional regulation of downstream effector genes, which lead to cellular adaptations. Controlled hypoxia exposure is utilized as an experimental tool to investigate biological processes, regulating cellular adaptations. Here we describe detailed protocols for hypoxia exposure of pregnant rodent models and low oxygen exposure of trophoblast stem cells, utilizing gas-regulated chamber systems. The presentation also includes phenotypic analyses of the manipulated animal models and cells.

Key words Hypoxia signaling, Hypoxia chamber, Trophoblast cells, Pregnancy, Rat Models

1 Introduction

Hypoxia is a condition of decreased oxygen tension below a critical threshold, which brings about an alteration of normal functioning of organs, tissues, and cells [1]. The concept of hypoxia is context dependent. Normoxic conditions for embryonic or adult cells are variable. Certain regions in the thymus, medulla of the kidney, and bone marrow can exist in as low as 1% oxygen due to their specialized, atypical vascular networks [2]. Early embryonic development takes place in low oxygen (1–2%) [3]. Homeostatic responses to low oxygen tension are elicited through induction of specialized transcription factors known as hypoxia-inducible factors (HIFs) [4]. HIFα subunits form heterodimers with aryl hydrocarbon receptor nuclear translocator (ARNT, also known as HIF1β) and bind to specific DNA sequences known as hypoxia-responsive elements (HREs).

Placental and fetal development relies on oxygen availability. Oxygen concentrations at the maternal-fetal interface are low (1–2%) before chorioallantoic placenta formation [3]. Prior to gestation day (gd) 9.5, the mouse embryo is heavily dependent

L. Eric Huang (ed.), *Hypoxia: Methods and Protocols*, Methods in Molecular Biology, vol. 1742,
https://doi.org/10.1007/978-1-4939-7665-2_15, © Springer Science+Business Media, LLC 2018

on glycolysis for ATP generation [2]. However, by gd 10.5–11.5, establishment of placental circulation increases oxygen concentrations at the maternal-fetal interface. Once the hemochorial placenta is established, oxygen concentrations increase to approximately 6–8%. In vitro studies demonstrated that low oxygen tension can modulate trophoblast differentiation [5–11]. An oxygen gradient has also been proposed to be a primary effector regulating gestation-dependent uterine vascular remodeling [7, 8, 12–14]. Mutagenesis of components of HIF signaling in mice (*Arnt, Vhl, Egln1, HIF1a,* and *Epas1*) results in disruptions in trophoblast lineage commitment and aberrant placental phenotypes [5, 15–18]. An appropriate balance between trophoblast proliferation and differentiation is required to produce a functional placenta [14]. Low oxygen has an instructive role on trophoblast lineage commitment and placental development [19–22]. Low oxygen tensions promote HIF stabilization and alter the epigenetic landscape of rat trophoblast cells affecting structure and function of the hemochorial placenta [20]. Thus, manipulation of oxygen tension can be utilized as an experimental tool to interrogate mechanisms controlling hemochorial placentation. These trophoblast cell adaptive responses may be impaired in pregnancy associated disorders, such as preeclampsia syndrome [20].

In this chapter, we describe protocols for in vivo hypoxia exposure of pregnant rats and in vitro low oxygen exposure of trophoblast stem (TS) cell models and the respective phenotypic analyses of the manipulated placentation sites and TS cells.

2   Materials

2.1 Preparation of Rats for Timed Pregnancy

1. Holtzman Sprague-Dawley rats.
2. Plastic transfer pipette.
3. Multi-well glass.
4. Light microscope (×40–100 magnification).
5. Sterile saline solution (0.9% NaCl).

2.2 Calibration of the Gas-Regulated Chamber

1. In vivo hypoxia chamber with a ProOx 110 regulator (Biospherix).
2. Nitrogen gas cylinders with regulators.

2.3 Preparation and Monitoring the Gas-Regulated Chamber for Experimental Exposures

1. Ethanol solution (70%).
2. Green Line IVC Sealsafe PLUS Rat Cages, surface area: 890 cm^2 (Techniplast).

2.4   Collection of Tissue Specimens

1. Dissecting microscope (×10–20 magnification).
2. Sterile saline solution (0.9% saline).
3. Fine forceps and microdissecting spring scissors.
4. Heptane.
5. Liquid nitrogen and dry ice.

2.5   Immunohistochemical Analyses of Tissue Specimens

1. Microscope glass slides.
2. PAP Pen.
3. Cryostat.
4. StainTray humidified slide staining system.
5. 4% paraformaldehyde.
6. Phosphate-buffered saline (PBS).
7. 10% normal goat serum.
8. Primary antibodies: Cy3-conjugated anti-vimentin-Cy3 and FITC-conjugated anti-pan cytokeratin.
9. 4′,6-Diamidino-2-phenylindole, dihydrochloride (DAPI).
10. Fluoromount-G.
11. Glass coverslip.

2.6   RNA Extraction of Placental Tissue Samples

1. Polytron tissue homogenizer.
2. TRIzol reagent.
3. Chloroform.
4. Isopropanol.
5. Molecular biology grade water.
6. Ethanol (70%).

2.7   Protein Extraction from Frozen Tissue Samples

1. Razor blades.
2. RIPA buffer with protease inhibitor cocktail (10 mM Tris–HCl, pH 7.2, 1% Triton X-100 or 1% Nonidet P-40, 1% sodium deoxycholate, 0.1% SDS, 150 mM NaCl, 5 mM EDTA, 1 mM sodium orthovanadate, 1 mM phenylmethylsulfonyl fluoride, and 10 µg/mL aprotinin).
3. PBS.
4. Cell scrapers.
5. Bioruptor sonication system.
6. Tabletop temperature-regulated centrifuge.

2.8   Preparation of Rat TS Cells for Hypoxia Exposure

1. Rat TS cells [23].
2. Rat TS cell basal culture medium [RPMI 1640, 20% fetal bovine serum, 100 µM mercaptoethanol, 1 mM sodium pyru-

vate, 50 μM penicillin, and 50 U/mL streptomycin] supplemented with 70% rat embryonic fibroblast-conditioned medium, FGF4 (25 ng/mL), and heparin (1 μg/mL).

3. 0.25% trypsin solution.

2.9 Calibration of Gas-Regulated Cell Culture Incubator

1. Fyrite gas analyzer.
2. Cell culture incubator.
3. Nitrogen and carbon dioxide gas cylinders with regulators.

2.10 Immunocytochemistry of Rat TS Cells

1. Chamber Slide System.
2. Microscope glass slides.
3. StainTray humidified slide staining system.
4. 4% paraformaldehyde solution.
5. PBS.
6. 10% normal goat serum.
7. Primary antibodies: anti-lysine demethylase 3A (KDM3A; Novus Biologicals) and FITC-conjugated anti-pan cytokeratin (Sigma-Aldrich).
8. Secondary antibody: anti-rabbit Alexa 568 conjugate.
9. DAPI.
10. Fluoromount-G.
11. Glass coverslip.

2.11 RNA Extraction from Rat TS Cells

1. TRIzol reagent.
2. Chloroform.
3. Isopropanol.
4. Molecular biology grade water.
5. Ethanol (70%).

2.12 Protein Extraction from Rat TS Cells

1. RIPA buffer with protease inhibitor cocktail (*see* above).
2. PBS.
3. Cell scrapers.
4. Bioruptor sonication system.
5. Tabletop temperature-regulated centrifuge.

2.13 Analysis of Cell Movement through Extracellular Matrices

1. Extracellular matrix-coated BioCoat® Matrigel Invasion Chambers (Thermo-Fisher).
2. Diff-Quick stain (Allegiance Scientific Products).

3 Methods

3.1 Preparation of Rats for Timed Pregnancy

1. Animals are maintained in an environmentally controlled facility with lights on from 0600 to 2000 h (14 h light, 10 h dark cycle) and are allowed free access to food and water.

2. Female rats (8–10 weeks of age) are placed with male rats (>3 months of age) of the same strain.

3. Confirmation of mating is determined by inspection of vaginal lavages. Plastic pipettes are loaded with sterile saline solution. Saline is delivered to the vagina. Vaginal lavages are transferred to clean multi-well glass plates and observed under a microscope. The presence of sperm in the vaginal lavage is considered gd 0.5 of pregnancy.

3.2 Calibration of the Gas-Regulated Chamber (see Notes 1 and 2)

1. Calibrate the ProOx controller connected to the chamber 2–3 days before every experiment (Fig. 1a, b).

2. Connect the tubing of the calibration chamber to the sample port outlet on the ProOx controller (Fig. 1c, d).

3. Twist and open the bleeder knob completely, so that the nitrogen gas starts flowing into the calibration unit. Switch on the gas flow button.

4. Plug in the sensor tip into the calibration chamber and bathe the sensor in 100% nitrogen gas for 3–5 min.

5. Record the reading on the screen of the ProOx controller. It should read "0%." If the reading has drifted by ±0.3%, then proceed with "zero" calibration steps.

6. Hold "^" and "v" buttons for 3 s. The screen will read "tune off."

7. Press the "v" once—the screen will read "level 1."

8. Hold the "*" button and press the "^" button twice. The screen will read "level 3."

9. For calibrating the zero point, press the "^" button several times until the screen reads "zero." There will be an internal value associated with "zero."

10. Hold the "*" button and press the "^" button or the "v" to change the internal value.

11. Hold the "^" and "v" buttons for 3 s. The screen will show the new reading for the zero calibration. The reading should be as close as possible to 0%. If not, then repeat **steps 6–10**, changing to a different internal value as described in **step 10**.

12. After setting the zero calibration, proceed with "SPAN" (ambient oxygen) calibration.

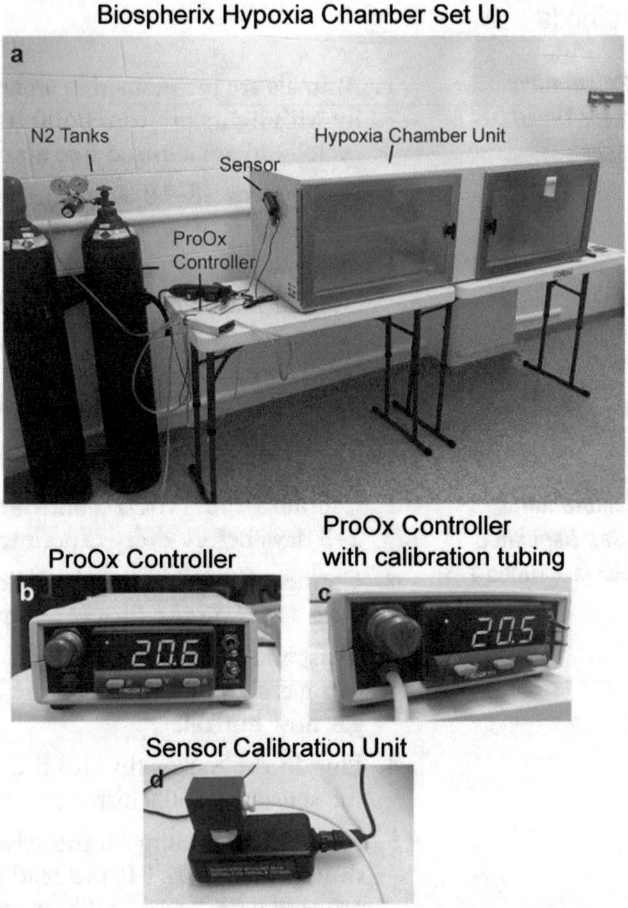

Biospherix Hypoxia Chamber Set Up

ProOx Controller

ProOx Controller with calibration tubing

Sensor Calibration Unit

Fig. 1 In vivo gas-regulated hypoxia chamber. (**a**) General setup of the gas-regulated hypoxia chamber for animal exposures. (**b**) The ProOx controller set at ambient oxygen tensions. (**c**) The yellow calibration tubing attached to the sample collection port of the ProOx controller connects to the sensor calibration unit (**d**)

13. Remove the sensor from the calibration chamber and expose the sensor tip to ambient oxygen.

14. At sea level, the ambient oxygen reading should be between 20.7% and 21%. If the reading varies by ±0.3%, please proceed with "SPAN" calibration.

15. Follow **steps 6–10**, with modification at **step 9**. Press the "^" button several times until the screen reads "SPAN."

16. At **step 11**, the reading should approximate 21% for the "SPAN" calibration. If not, then follow **steps 6–10**, changing to a different internal value as described in **step 10**.

17. After 48 h, the controller should be checked for calibration again before initiating the animal hypoxia exposure experiments.

3.3 Preparation and Monitoring the Gas-Regulated Chamber for Experimental Exposures

1. The gas-regulated chamber should be wiped clean with 70% ethanol and air-dried.

2. Cages containing gd 6.5 rats are placed into the chamber (*see* **Note 3**).

3. Open the nitrogen gas cylinders. Open valve and set the pressure reading to 20 psi.

4. Set the gas level to 10.5% by holding "*" button and pressing the "v" button until the desired nitrogen gas concentration is achieved. Gas will start flowing into the hypoxia chamber.

5. Verify that the ProOx controller reads the desired oxygen concentration.

6. The chamber should be opened every day during the experiment, for 1–2 min to relieve moisture accumulation and to permit any manipulation of the animals, food, and/or caging.

7. Cages, food, and water should be replaced every 2 days.

8. Buildup of pressure in the nitrogen tank tubing should be readjusted, by altering the regulator valve to 20 psi (*see* **Note 4**).

3.4 Collection of Tissue Specimens

1. At the end of the experiment, animals are removed from the chamber and euthanized.

2. Tissues are collected as quickly, as possible. Flash-freeze tissues for histological analysis in dry ice-cooled heptane and store at −80 °C or alternatively preserve the tissue in an appropriate fixative. Freeze tissue samples for protein or RNA extraction in liquid nitrogen and store at −80 °C until processed (*see* **Notes 5** and **6**).

3.5 Immunohisto-chemical Analyses of Tissue Specimens

1. The frozen conceptuses are equilibrated within a cryostat maintained at −20 °C.

2. Sections (10 μm) are cut and allowed to adhere on charged glass slides.

3. The sections are numbered serially and stored at −80 °C until processing.

4. For histological analyses, the sections should be taken out from −80 °C and placed on slide warmer for 10 min.

5. A PAP pen is used to draw a boundary line around the tissue section and air-dried at room temperature for another 5 min.

6. The slide is then dipped into freshly prepared paraformaldehyde solution for 12 min.

7. Slides are then washed three times with PBS (5-min/wash).

8. Normal goat serum (10%) blocking solution is added and the tissue section incubated in a humidified slide staining system for 30 min (*see* **Note 7**).

9. Primary antibodies (pan cytokeratin and vimentin) are diluted in blocking solution (1:200) and mixed thoroughly.

10. The blocking solution is decanted, and any excess solution is carefully removed from the slide.

11. The primary antibody solution is then added to the tissue section and incubated for 1 h at room temperature in a humidified slide chamber, protected from light (*see* **Note 7**).

12. After 1 h of incubation, the primary antibody solution is decanted. The slide is washed three times with PBS (5-min/wash) in the dark.

13. Reconstituted DAPI solution is added at a 1:50,000 dilution and incubated for 5 min at room temperature.

14. The slide is washed three times with PBS (5-min/wash) in the dark.

15. Fluoromount-G is added and a coverslip is carefully placed on the tissue section (*see* **Note 8**).

16. The slide is dried at room temperature overnight and subsequently inspected by fluorescence microscopy (Fig. 2).

3.6 RNA Extraction of Placental Tissue Samples

1. TRIzol reagent is added to the frozen tissue sample (10–20 mg of tissue/mL of TRIzol).

2. Disrupt the tissue at 4 °C with the polytron tissue homogenizer.

3. Add chloroform (200 μL) to the TRIzol-tissue mixture, mix thoroughly, and incubate at room temperature for 5 min.

4. Centrifuge at $15,700 \times g$ and 4 °C for 10 min.

5. Carefully remove the aqueous solution without disturbing the pink-colored organic phase.

6. Add equal amounts of isopropanol, mix thoroughly, and then transfer to −80 °C for at least 2 h.

7. Centrifuge as above at 4 °C for 20 min.

8. Decant the isopropanol without disturbing the pellet.

9. Add 70% ethanol and wash the pellet by centrifuging as above at 4 °C for 10 min.

10. Carefully remove the ethanol without disturbing the pellet.

11. Let the pellet air-dry for 10 min at room temperature.

12. Thoroughly suspend the dried pellet in the desired volume of molecular biology grade water and store at −80 °C.

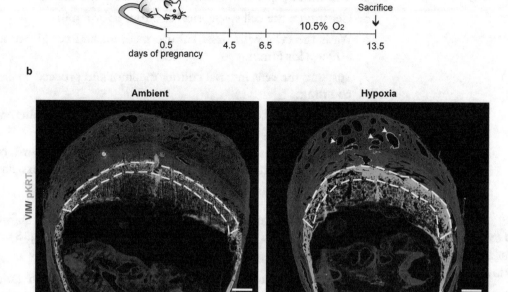

Fig. 2 Effects of hypoxia on hemochorial placentation in the rat. (**a**) Schematic representation of an experimental in vivo 10.5% oxygen exposure of pregnant rats. (**b**) Immunohistochemical analyses of vimentin (VIM) and pan cytokeratin (pKRT) within placentation sites from pregnant rats exposed to ambient or hypoxia (10.5%) oxygen. Note the deep endovascular trophoblast invasion in the hypoxia-exposed rats (white arrows). Scale bar = 1 mm

3.7 Protein Extraction from Frozen Tissue Samples

1. A frozen tissue specimen (20–30 mg) is minced into small pieces using a razor blade and transferred to a 1.5 mL Eppendorf tube.

2. Add RIPA buffer (400 µL) and mix thoroughly.

3. Place the tube in the bioruptor sonicator and sonicate for 5 min at a high-power setting with a 30 s on/30 s off cycle (*see* **Note 9**).

4. Centrifuge at $15,700 \times g$ and 4 °C for 20 min.

5. Remove the supernatant carefully without disturbing the pellet, aliquot as appropriate into new tubes, and store at −80 °C until further analysis.

3.8 Preparation of Rat TS Cells for Hypoxia Exposure

1. Plate rat TS cells in 25 cm² flasks and maintain in basal culture medium for 3 days, changing the medium every day.

2. On day 4, wash the cells twice with PBS and add 2 mL of trypsin solution and incubate at room temperature for approximately 2 min.

3. Tap the flask gently, to lift off the cells. Add basal culture medium to neutralize the trypsin.

4. Mix with a pipette to generate single-cell suspensions.

5. Centrifuge the cell suspension at $500 \times g$ for 5 min.

6. Wash the cells with basal culture medium and centrifuge at $500 \times g$ for 5 min.

7. Suspend the cells in basal culture medium and proceed to cell counting.

8. Plates 80,000 cells/well in a six-well culture plate or alternatively into a chamber slide (Day 0).

9. After 24 h, remove the medium completely and add 2 mL of fresh basal culture medium (Day 1). Repeat the next day (Day 2).

3.9 Calibration of the Gas-Regulated Cell Culture Incubator (see Notes 10 and 11)

1. Calibrate the gas-regulated cell culture incubator 2 days before initiation of the experimental low oxygen exposure (Fig. 3).

2. Set the incubator at the desired oxygen percentage.

3. Hold the Fyrite device containing oxygen evaluation fluid upright and depress the plunger valve to vent out Fyrite solution. Release the plunger valve.

4. Invert the device and then hold the device angularly to drain in fluid into the top reservoir.

5. Turn the device upright and hold it at a 45° angle to drain the fluid into bottom reservoir. Repeat **steps 1–3** twice.

6. Depress the plunger valve. Note the fluid level in the column.

7. Adjust the 0-level mark, so that the fluid meniscus matches with the zero mark.

8. To check calibration of the oxygen sensor, insert the other end of the sampling tube into the sample port nozzle of the incubator (Fig. 3c).

9. Hold the Fyrite device in upright position and place the rubber connecter tip over the Fyrite plunger valve (Fig. 3d).

10. Depress the plunger valve with the rubber connecter tip.

11. Pump sample gas by squeezing and releasing aspirator bulb 18 times. During the 18th bulb squeeze, release the connecter tip and plunger valve.

12. Invert the Fyrite device until the fluid drains into the top reservoir. Then set the device upright to drain the fluid into the bottom reservoir. Repeat this mixing step three times.

13. Hold the Fyrite device at 45° angle to allow the fluid to drain completely into bottom reservoir.

14. Stabilize the flow and hold the Fyrite device upright allowing the column to stabilize.

Hypoxia Incubator

Fyrite O₂ Analyzer

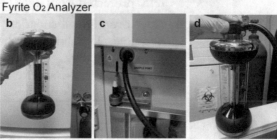

Fig. 3 In vitro gas-regulated cell culture incubator. (**a**) General setup of gas-regulated hypoxia incubator for cell culture exposures. (**b**) Fyrite oxygen analyzer held at upright position showing ambient oxygen (21% at sea level) reading. (**c**) The rubber tubing connecting the oxygen analyzer to the incubator gas sampling port. (**d**) The rubber nozzle set up on the Fyrite gas analyzer. Note the other end of the nozzle tube connects to the incubator sample port

15. Immediately read the oxygen (%) on the scale at the reading mark corresponding to the top of the fluid column.

16. If the oxygen concentration reading does not match the concentration at which the incubator is set, then press the "mode" button a few times until the reading shows "Calibration." Press "Enter" and press the "mode" button again until the oxygen offset setup is reached. Click the arrows to set oxygen offset value to the Fyrite reading.

17. This calibration process should be repeated on the day the experiment is initiated.

18. Initiating the hypoxia exposure experiment:

    (a) Press the "mode" button on the incubator panel until the panel reads "set point."

    (b) Press the ">" button until the oxygen set point is achieved.

(c) Set the desired oxygen concentration and press the "enter" button.

(d) Open the nitrogen gas tank.

(e) Nitrogen gas will flow in the incubator until the desired oxygen level is achieved.

3.10 Immunocyto-chemistry of Rat TS Cells (Fig. 4)

1. Cells plated in chamber slides are fixed with freshly prepared paraformaldehyde solution for 12 min and washed.

2. Proceed through **steps 7–10** of Subheading 3.5 (*see* above). Anti-KDM3A can be used to monitor hypoxia TS cell responses [21].

3. The primary antibody is then added to the tissue section and incubated overnight at 4 °C in a humidified slide staining system, protected from light (*see* **Note 7**).

4. After overnight incubation, the primary antibody solution is decanted. The slide is washed three times with PBS (5-min/wash) in the dark.

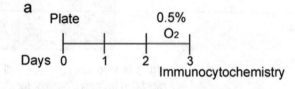

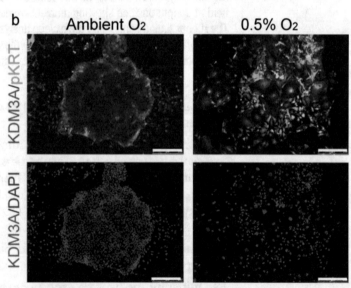

Fig. 4 Effects of low oxygen on TS cell expression of KDM3A. (**a**) Schematic representation of in vitro 0.5% oxygen exposure of rat TS cells. (**b**) Immunocytochemical localization of KDM3A (red) and pan cytokeratin (pKRT, green) on rat TS cells exposed to ambient or 0.5% oxygen for 24 h. Note the increased expression of KDM3A (a hypoxia-responsive protein) in the hypoxia-exposed rat TS cells. Scale bar = 250 μm

5. PBS is decanted and excess PBS is removed.

6. The secondary antibody is added at a 1:500 dilution and incubated at room temperature for 45 min.

7. The slide is washed three times with PBS (5-min/wash) in the dark.

8. Reconstituted DAPI solution is added at a 1:50,000 dilution and incubated for 5 min at room temperature.

9. The slide is washed three times with PBS (5-min/wash) in the dark.

10. Fluoromount-G is added and a coverslip is carefully placed on the tissue section (*see* **Note 8**).

11. The slide is dried at room temperature overnight and subsequently inspected by fluorescence microscopy.

3.11   RNA Extraction from Rat TS Cells (See Note 12)

1. Remove the cell culture plate from the incubator and place on ice.

2. Wash the adherent cells once with PBS and then decant the PBS.

3. Add 1 mL TRIzol/well of the six-well culture plate. Gently swirl the plate for few min to evenly distribute the TRIzol solution.

4. Collect the TRIzol solution in 1.5 mL Eppendorf tube.

5. Proceed with **steps 4–17** of Subheading 3.6 (*see* above).

3.12   Protein Extraction from Rat TS Cells

1. Remove the cell plate from the incubator and place on ice.

2. Wash the adherent cells once with PBS and then decant the PBS.

3. Add 200 μL of RIPA buffer to each well. Gently swirl the plate for a few min to evenly distribute the RIPA buffer.

4. With a cell scraper, lift the cells into the RIPA buffer. Add an extra 100 μL of RIPA buffer and collect the solution in an Eppendorf tube.

5. Proceed with **steps 5–8** of Subheading 3.7 (*see* above).

3.13   Analysis of Cell Movement through Extracellular Matrices (See Note 13)

The invasive phenotype of trophoblast cells is enhanced by hypoxia and can be assessed by determining the movement of cells through an extracellular matrix [20].

1. Rat TS cells are seeded at 5×10^4 per 3 mL in basal culture medium on the upper chamber of an extracellular matrix-coated (BioCoat Matrigel) transwell chamber.

2. Cells are incubated in oxygen-regulated cell culture chambers as described above for 24 h.

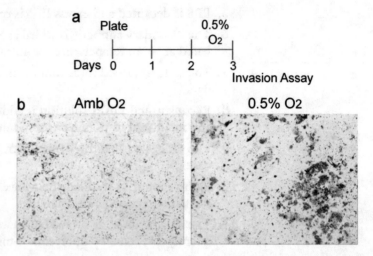

Fig. 5 Effects of low oxygen on TS cell invasion through extracellular matrix-coated transwells. (**a**) Schematic representation of the assessment of TS cell invasion through Matrigel-coated transwells. (**b**) Membranes stained with Diff-Quick to identify invasive TS cells exposed to ambient (Amb O_2) or 0.5% O_2. Note increased TS cell invasion in low oxygen-exposed rat TS cells

3. Cells from the upper surface are removed by scraping, and cells on the under surface are fixed and stained with Diff-Quick.

4. Membranes are then removed, placed on slides, and cells attached to the under surface of the transwell chamber counted with the aid of a microscope ocular grid (Fig. 5).

4   Notes

1. Careful monitoring and calibration of the in vivo hypoxia chamber are required to ensure maintenance of accurate oxygen concentrations.

2. The oxygen sensor for the ProOx controller is functional between 1 and 1.5 years. $A > 0.3\%$ deviation within a week of calibration is a good indicator of a failing oxygen sensor and the need for its replacement before performing any experiments. An efficient way to checking the accuracy of the ProOx oxygen sensor is using a portable oxygen monitor (BW CLIP 2 Single Gas Detector; PK Safety, Alameda, CA, cat. no BWC2-X) and placing it inside the chamber. The display on the portable device and the ProOx monitor should read the same oxygen concentration. When installing the new sensor, please follow the standard calibration procedure before setting up an experiment.

3. Avoid housing more than two animals per cage and limit the number of cages per chamber to not more than three. Exceeding the recommended animal numbers in the chamber is problematic. They increase humidity in the chamber and adversely affect animal housing conditions.

4. There is a tendency for gas to accumulate in tubing connected to the nitrogen gas cylinder, which affects the pressure. Maintain a pressure reading of <20 psi. At higher pressures, the tubing may detach and compromise the experiment.

5. For in vivo hypoxia exposure experiments, it is most optimal to keep the time for tissue collection short and the dissections efficient. Extended exposure to ambient oxygen prior to euthanasia can compromise hypoxia manipulation.

6. Cut tissues into small pieces and individually freeze prior to transferring to tubes and storage at −80 °C. This procedure facilitates recovery of tissues for subsequent biochemical analyses.

7. Duration of primary antibody incubation for immunohisto-chemistry and immunocytochemistry should be standardized for each antibody. Incubation for 1 h at room temperature or overnight incubation at 4 °C is appropriate for many antibodies. When different secondary antibodies are used, diluting the secondary antibody in blocking solution helps reduce non-specific background. Alternatively, a prolonged blocking step for 1 h also helps to reduce non-specific background signal.

8. Air bubbles beneath the coverslip can be disruptive for observing the immunostained images. To remove air bubbles, the slide should be completely immersed in PBS for 1.5 h at room temperature. Then gently tap the slide side to side to remove the coverslip. The stained section should be washed twice with fresh PBS. Excess PBS should be carefully removed. Place a new coverslip on the slide with a drop of Fluoromount-G.

9. If a Diagenode bioruptor sonicator is not available, then a conventional probe-based sonicator (e.g., Branson Sonifier, Thermo-Fisher) can be used during protein extraction. Use a setting of 10% duty cycle, output control of 2 with 30 s pulse, and 30 s ice cooling. Perform 3–4 cycles of sonication. Alternatively, if a sonicator is not available, then following cell lysis in RIPA buffer, the solution can be extensively passed through a syringe (at least 10 times) to breakdown nuclear material.

10. When oxygen calibration fluid in the Fyrite unit turns dark blue-green, it is insensitive to changing oxygen concentrations. This will lead to erroneous calibration of the oxygen sensor. To assess efficacy of the fluid, check oxygen percentage reading of ambient air. In a clean environment, pump in ambi-

ent air into the Fyrite device following the procedure as mentioned in Subheading 3.6, then check the reading. Repeat the process three times to obtain three independent readings. Each reading should be 21%. If the reading is not close to 21%, then the calibration fluid should be replaced.

11. Some gas-regulated incubators cannot be set for oxygen tensions less than 1%. To achieve less than 1% oxygen concentration (e.g., 0.5%), calibrate the sensor for 1% oxygen using the Fyrite device, but set the incubator oxygen at 1.5%. In that case, when the set point is entered as 1%, the effective oxygen tension inside the incubator will be 0.5%. After, the reading stabilizes at 1%, follow the procedure described in Subheading 3.6 to check the oxygen concentration of the incubator.

12. Depending on the cell type, low oxygen exposure can alter cell proliferation and viability. This may result in lower RNA yields. To visualize RNA pellets effectively, glycogen solution may be added before the isopropanol centrifugation step. The resulting pellet will be a visible white pellet. This will ensure proper RNA extraction and final suspension from small numbers of cells.

13. Standardizing appropriate cell numbers is critical for Matrigel transwell invasion assays. High cell numbers can lead to erroneous data. Commercially available Matrigel can be utilized in 1:10 dilution, to coat cell culture inserts. The inserts can then be placed in cell culture incubators for 2.5 h to uniformly coat the well. While, preparing the insert for Diff-Quick staining, extra care should be taken to completely remove the cells from the upper Matrigel-coated surface. If this step is forgotten or inefficiently performed, the cell counts will be spurious.

Acknowledgments

We acknowledge present and past members of the Soares Laboratory for their contributions to the development of these research model systems. This work was supported by NIH HD020676.

References

1. Hockel M, Vaupel P (2001) Tumor hypoxia: definitions and current clinical, biologic, and molecular aspects. J Natl Cancer Inst 93(4):266–276

2. Simon MC, Keith B (2008) The role of oxygen availability in embryonic development and stem cell function. Nat Rev Mol Cell Biol 9(4):285–296

3. Rodesch F, Simon P, Donner C et al (1992) Oxygen measurements in endometrial and trophoblastic tissues during early pregnancy. Obstet Gynecol 80(2):283–285

4. Semenza GL (2010) Oxygen homeostasis. Wiley Interdiscip Rev Syst Biol Med 2(3):336–361

5. Adelman DM, Gertsenstein M, Nagy A et al (2000) Placental cell fates are regulated in vivo by HIF-mediated hypoxia responses. Genes Dev 14(24):3191–3203

6. Alsat E, Wyplosz P, Malassine A et al (1996) Hypoxia impairs cell fusion and differentiation process in human cytotrophoblast, in vitro. J Cell Physiol 168(2):346–353

7. Caniggia I, Mostachfi H, Winter J et al (2000) Hypoxia-inducible factor-1 mediates the biological effects of oxygen on human trophoblast differentiation through TGFbeta(3). J Clin Invest 105(5):577–587

8. Genbacev O, Joslin R, Damsky CH et al (1996) Hypoxia alters early gestation human cytotrophoblast differentiation/invasion in vitro and models the placental defects that occur in preeclampsia. J Clin Invest 97(2):540–550

9. James JL, Stone PR, Chamley LW (2006) The effects of oxygen concentration and gestational age on extravillous trophoblast outgrowth in a human first trimester villous explant model. Hum Reprod 21(10):2699–2705

10. Jiang B, Kamat A, Mendelson CR (2000) Hypoxia prevents induction of aromatase expression in human trophoblast cells in culture: potential inhibitory role of the hypoxia-inducible transcription factor Mash-2 (mammalian achaete-scute homologous protein-2). Mol Endocrinol 14(10):1661–1673

11. Nelson DM, Johnson RD, Smith SD et al (1999) Hypoxia limits differentiation and upregulates expression and activity of prostaglandin H synthase 2 in cultured trophoblast from term human placenta. Am J Obstet Gynecol 180(4):896–902

12. Burton GJ (2009) Oxygen, the Janus gas; its effects on human placental development and function. J Anat 215(1):27–35

13. Burton GJ, Jaunaiux E (2001) Maternal vascularisation of the human placenta: does the embryo develop in a hypoxic environment? Gynecol Obstet Fertil 29(7–8):503–508

14. Fryer BH, Simon MC (2006) Hypoxia, HIF and the placenta. Cell Cycle 5(5):495–498

15. Cowden Dahl KD, Fryer BH et al (2005) Hypoxia-inducible factors 1alpha and 2alpha regulate trophoblast differentiation. Mol Cell Biol 25(23):10479–10491

16. Gnarra JR, Ward JM, Porter FD et al (1997) Defective placental vasculogenesis causes embryonic lethality in VHL-deficient mice. Proc Natl Acad Sci U S A 94(17):9102–9107

17. Maltepe E, Krampitz GW, Okazaki KM et al (2005) Hypoxia-inducible factor-dependent histone deacetylase activity determines stem cell fate in the placenta. Development 132(15):3393–3403

18. Takeda K, Ho VC, Takeda H et al (2006) Placental but not heart defects are associated with elevated hypoxia-inducible factor alpha levels in mice lacking prolyl hydroxylase domain protein 2. Mol Cell Biol 26(22):8336–8346

19. Rosario GX, Konno T, Soares MJ (2008) Maternal hypoxia activates endovascular trophoblast cell invasion. Dev Biol 314(2):362–375

20. Chakraborty D, Cui W, Rosario GX et al (2016) HIF-KDM3A-MMP12 regulatory circuit ensures trophoblast plasticity and placental adaptations to hypoxia. Proc Natl Acad Sci U S A 113(46):E7212–E7221

21. Chakraborty D, Rumi MA, Konno T et al (2011) Natural killer cells direct hemochorial placentation by regulating hypoxia-inducible factor dependent trophoblast lineage decisions. Proc Natl Acad Sci U S A 108(39):16295–16300

22. Chakraborty D, Rumi MA, Soares MJ (2012) NK cells, hypoxia and trophoblast cell differentiation. Cell Cycle 11(13):2427–2430

23. Asanoma K, Rumi MA, Kent LN et al (2011) FGF4-dependent stem cells derived from rat blastocysts differentiate along the trophoblast lineage. Dev Biol 351(1):110–119

Chapter 16

Evaluation of Erythrocyte Changes After Normoxic Return from Hypoxia

Jihyun Song and Josef T. Prchal

Abstract

Hypoxia increases erythropoiesis by hypoxia-inducible factors (HIF), principally by HIF-2, which upregulates erythropoietin transcription. This results in an increase of red blood cell (RBC) production and delivery of more oxygen to tissues. Upon rapid return to normoxia, hypoxia-induced polycythemia is overcorrected by neocytolysis, a transient destruction of preferentially young RBCs bearing low catalase (downregulated by hypoxia-stimulated microRNA(miR)-21) caused by reactive oxygen species (ROS) from expanded mitochondria. In order to study molecular mechanism of neocytolysis, it is critical to differentiate life span of young and old RBCs and to measure the hematological changes before and after hypoxia treatment. Here we describe the methodological aspects of these measurements.

Key words Hypoxia, Erythrocytes, Biotin labeling, Reactive oxygen species, Mitochondrial mass, Evans blue staining, HIF, Catalase, Bnip3 L, MicroRNA-21

1 Introduction

Measurement of hematocrit levels is a rapid and convenient assessment of red blood cell (RBC) mass and changes of erythropoiesis. However, hematocrit levels are also altered by plasma volume [1]. Thus, for an estimate of RBC mass, hematocrit should be adjusted by the plasma volume. A convenient nonradioactive way to measure plasma volume is Evans blue staining method. Intravenously administrated Evans blue dye binds to plasma albumin. The injected Evans blue is diluted by the volume of plasma. After its plasma concentration is determined by spectrophotometer, the plasma volume is calculated [2, 3]; for an estimate of RBC survival, RBC biotin is measured in time. Here we describe the modified Evans blue and RBC and reticulocyte survival methodologies [4].

Neocytolysis is a preferential destruction of young red blood cells to correct hypoxia-induced polycythemia upon return to normoxia [5]. To evaluate neocytolysis, we describe the biotin-labeling

L. Eric Huang (ed.), *Hypoxia: Methods and Protocols*, Methods in Molecular Biology, vol. 1742,
https://doi.org/10.1007/978-1-4939-7665-2_16, © Springer Science+Business Media, LLC 2018

method of RBCs that differentiates young RBCs from old RBCs and measures their life span [6]. Rapid change of oxygen tension from hypoxia to normoxia temporarily but dramatically decreases HIFs, increases ROS from excessive reticulocytes' mitochondrial mass by Bnip3L downregulation, and decreases catalase activity via miR-21 increase [7]; the methods to measure reticulocytes' mitochondrial mass, ROS, and transcript levels of *Bnip3L*, *Cat* (encodes catalase), and *miR-21* in reticulocytes are described here.

2   Materials

1. **Preparation of Evans Blue Solution:** Add 1 mg of Evans blue to 100 μL of PBS and then perform 1:5 dilution (such as adding 20 μL of 1% of Evans blue to 80 μL of PBS). 100 μL of 0.2% Evans blue solution is used for one mouse.

2. **Preparation of MitoTracker** (MitoTracker® Deep Red FM): Add 92 μL of DMSO into one vial of tube to make 1 mM of stock solution and store at −20 °C with protection from light. Dilute stock solution at 1:100 to make 10 μM of working solution when you use. Avoid freeze and thaw cycles.

3. **Preparation of ROS Detector** (CM-H2DCFDA, Molecular probes): Add 86 μL of DMSO into one vial to make 1 mM of working solution and store at −20 °C with protection from light. Avoid freeze and thaw cycles.

4. **Preparation of MitoSOX** (MitoSOX™ Red mitochondrial superoxide indicator, molecular probes): Add 65 μL of DMSO into one vial to make 1 mM of MitoSOX and store at −20 °C with protection from light. Avoid freeze and thaw cycles.

5. Thiazole Orange Dye (BD Bioscience).

6. Polyacryl Carrier (Molecular Research Center).

7. SuperScript™ VILO™ MasterMix (Invitrogen).

8. Gene-Specific TaqMan Probes: *Bnip3L* (Assay ID: Mm00786306_s1), *Cat* (Assay ID: Mm00437992_m1), and normalize transcript levels against 18S (Assay ID: Hs03003631_g1) (Applied Biosystems).

9. TaqMan® MicroRNA Reverse Transcription Kit.

10. MicroRNA-specific TaqMan probes: *miR-21* (Assay ID: 000397) and *U6snRNA* (Assay ID: 001973).

3 Methods

3.1 Measurement of Hematocrit Levels

1. Collect whole blood into heparinized microcapillary tube (commercially available from multiple sources) and seal one end of tube with Cha-seal tube sealing compound.

2. Centrifuge at $11,200 \times g$ for 3 min.

3. Measure total blood volume and packed blood volume using microcapillary reader.

4. Calculate hematocrit levels using equation in Fig. 1.

3.2 Measurement of Plasma Volume

1. Inject 100 μL of 0.2% Evans blue solution per one mouse by either tail vein or retro-orbital injection.

2. After 10 min, collect 30 μL of whole blood every 5 min in EDTA-coated tube.

3. Separate plasma by centrifugation for 10 min at $1000 \times g$ at room temperature.

4. Dilute plasma at 1:100 with 1× PBS (add 5 μL of plasma to 495 μL of PBS).

5. Measure absorbance of diluted plasma and 1:250 diluted injection solution at 620 nm using spectrophotometer.

6. Calculate plasma volume using equation in Fig. 2.

7. Measure the weight of mouse and express plasma volume as μL/g.

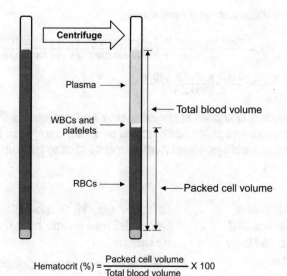

$$\text{Hematocrit (\%)} = \frac{\text{Packed cell volume}}{\text{Total blood volume}} \times 100$$

Fig. 1 Measurement of hematocrit levels. After centrifuge microcapillary tube with whole blood, hematocrit level is calculated by measuring total blood volume and packed cell volume via microcapillary reader

a

Absorbance of Evans blue in plasma

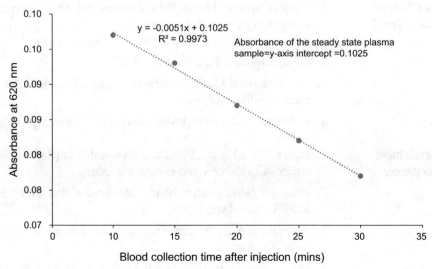

Blood collection time after injection (mins)

b Absorbance of injection solution

Absorbance of injection solution	
Dilution factor	Absorbance at 620 nm
250	0.607

c Calculation of plasma volume

$$PV = \frac{\text{absorbance of the injection solution * its dilution factor * injection volume}}{\text{absorbance of the steady state plasma sample * dilution factor}}$$

$$= \frac{0.607 \times 250 \times 100}{0.1025 \times 100} = 1480.5 \ \mu l$$

Fig. 2 Example of calculation of plasma volume. (**a**) Absorbance of Evans blue in plasma was measured at 620 nm at various time points. Absorbance of the steady state (y-axis intercept) was calculated by regression line. (**b**) Absorbance of injection solution was measured using 1:250 diluted solution. (**c**) Equation of calculating plasma volume

3.3 Measurement of Hypoxia-Born Red Blood Cells Life Span by Biotin Labeling

1. Inject 40 mg/kg of EZ-Link Sulfo-NHS-LC-Biotin (Thermo Scientific) into mouse by intravenous injection before hypoxia treatment.

2. Place the mice in a hypobaric chamber for 10 days at 12% oxygen.

3. After 10 days' hypoxia treatment, collect 10 μL of whole blood from mice every 4 days after hypoxia treatment in EDTA-coated tube.

4. Prepare and label following tubes: "unstained," "TER119 only," "streptavidin only," and "samples").

5. Incubate 2 μL of whole blood in 500 μL of 1% BSA with 1× PBS for 30 min at 4 °C.

6. Add antibodies to each tube: Add following antibodies into each tube. For *unstain* tube, add nothing. For TER119 only tube, add 0.8 μL of 0.5 mg/mL of FITC rat anti-mouse TER119. For streptavidin only tube, add 2 μL of 0.2 mg/mL of PE-Cy5-conjugated streptavidin. For sample tube, add 0.8 μL of 0.5 mg/mL of FITC rat anti-mouse TER119 and 2 μL of 0.2 mg/mL of PE-Cy5-conjugated streptavidin.

7. Incubate for at least 30 min at 4 °C on rotator in the dark and centrifuge at 2300 × g for 5 min at room temperature and resuspend pellet with 500 μL of 1× PBS.

8. Analyze the percentage of PE-Cy5-positive cells among FITC-positive mature red blood cells by FACS.

3.4 Measurement of Life Span of Hypoxic Reticulocytes by Biotin Labeling

1. Inject 40 mg/kg of EZ-Link Sulfo-NHS-LC-Biotin (Thermo Scientific) into mouse by intravenous injection at a day before return to normoxia, as done for "red blood cells life span, above in Subheading 3.3."

2. Maintain the mice in hypobaric chamber for one more day.

3. Collect 10 μL of whole blood every 12 h after hypoxia treatment in EDTA-coated tube.

4. Prepare and label following tubes: "unstain," "thiazole orange only," "PE-Cy5 conjugated streptavidin only," and "samples" and prepare appropriate gates for FAC analysis.

5. Add 250 μL thiazole orange dye (Cat. No. 349204, BD Bioscience) into thiazole orange only and sample tubes.

6. Add 250 μL of 1× PBS into unstain, PE-Cy5 conjugated streptavidin only.

7. Add 1.25 μL of whole blood into each tube and incubate for at least 30 min at room temperature at dark.

8. Add 5 μL of PE-Cy5-conjugated streptavidin into PE-Cy5 conjugated streptavidin only and sample tubes and incubate for at least 30 min at 4 °C on rotator at dark.

9. Centrifuge at 2300×g for 5 min at room temperature and remove supernatant and resuspend pellet with 300 μL of 1× PBS.

10. Analyze the percentage of PE-Cy5-positive biotin-labeled cells among FITC-positive reticulocytes by FACS.

3.5 Measurement of Mitochondrial Mass in Reticulocytes and Mature Red Cells by FACS

1. Collect 10 μL of whole blood in EDTA-coated tube.

2. In order to make proper controls for FACs analysis, please prepare six tubes "unstain," CD71 only, TER119 only, MitoTracker only, CD71/TER119, and sample.

3. Add 2 μL of whole blood in each tube to 500 μL of 4% BSA in PBS and vortex.

4. Incubate for 30 min at 4 °C on rotator.

5. Add antibodies in each tube. For *unstain* tube, add nothing. For CD71 only tube, add 2 μL of 0.2 mg/mL of PE rat anti-mouse CD71. For TER119 only tube, add 0.8 μL of 0.5 mg/mL of FITC rat anti-mouse TER119. For MitoTracker only, add no antibody. For CD71/TER119 and sample tube, add 2 μL of 0.2 mg/mL of PE rat anti-mouse CD71 and 0.8 μL of 0.5 mg/mL of FITC rat anti-mouse TER119.

6. Incubate for 30 min at 4 °C on rotator in the dark.

7. Add 1.25 μL of 10 μM of MitoTracker® Deep Red FM (final concentration is 25 nM) in MitoTracker only and sample tube.

8. Incubate for 30 min at 4 °C on rotator in the dark.

9. Centrifuge for 5 min at 2300 × g at room temperature.

10. Discard supernatant and resuspend cells with 500 μL of 1× PBS.

11. Determine mean fluorescence of APC (MitoTracker) in reticulocytes (CD71⁺/TER119⁺) and percentage of MitoTracker-positive reticulocytes (CD71⁺/TER119⁺) by FACS (Fig. 3a).

3.6 Measurement of ROS in Reticulocytes, Mature Red Cells by FACS

1. Collect 10 μL of whole blood in EDTA-coated tube.

2. Add 2 μL of whole blood in each tube into 300 μL of 1× PBS and vortex. In order to make proper controls for FACs analysis, please make six tubes, as described in above Subheading 3.5.

3. Incubate for 30 min at 4 °C on rotator in the dark.

4. Add antibodies in each tube like below table.

   For unstain tube, add nothing. For CD71 only tube, add 2 μL of 0.2 mg/mL of PE rat anti-mouse CD71. For TER119 only tube, add 2 μL of 0.2 mg/mL of PE-Cy7 rat anti-mouse TER-119. For ROS only, add no antibody. For CD71/TER119 and sample tube, add 2 μL of 0.2 mg/mL of PE rat anti-mouse CD71 and 2 μL of 0.2 mg/mL of PE-Cy7 rat anti-mouse TER-119.

5. Incubate for 30 min at 4 °C on rotator in the dark.

6. Add 1.5 μL of 1 mM of *ROS detector* to ROS only and sample tube.

7. Incubate for 30 min at 4 °C on rotator in the dark.

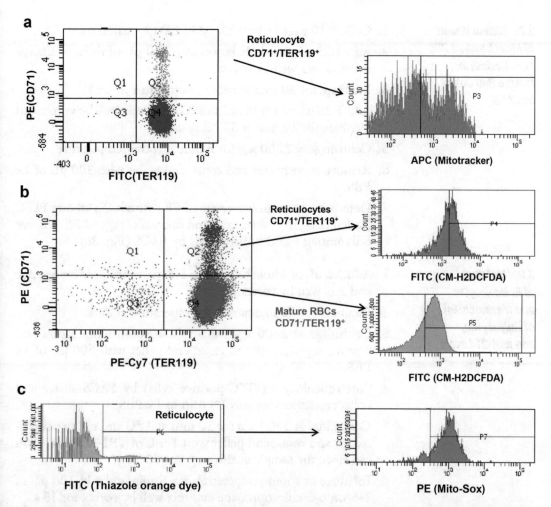

Fig. 3 Examples of FACS analysis. (**a**) Mitochondrial mass measurement in reticulocytes (CD71$^+$/TER119$^+$). Mean fluorescence of APC (MitoTracker) and percentage of APC positive cells from reticulocytes are measured by FACS analysis. (**b**) ROS measurement in reticulocytes (CD71$^+$/TER119$^+$) and mature RBCs (CD71$^-$/TER119$^+$). Mean fluorescence of FITC (CM-H2DCFDA, ROS indicator) and percentage of FITC positive cells in reticulocytes and mature RBCs are accessed by FACS analysis. (**c**) Mitochondrial ROS measurement in reticulocytes. Mean fluorescence of PE (MitoSOX) and PE positive cells among FITC positive cells (reticulocytes) are determined by FACS analysis

8. Centrifuge for 5 min at 2300 × *g* at RT.

9. Discard supernatant and resuspend cells with 500 μL of 1× PBS.

10. Determine mean fluorescence of FITC (CM-H2DCFDA) in reticulocytes (CD71$^+$/TER119$^+$) and mature RBCs (CD71$^-$/TER119$^+$) and percentage of ROS-positive reticulocytes and mature RBCs by FACS (Fig. 3b).

3.7 Measurement of Mitochondrial ROS in Reticulocytes, Mature Red Cells by FACS

1. Collect 10 μL of whole blood in EDTA-coated tube.

2. Add 1.25 μL of whole blood into 250 μL of thiazole orange dye and mix well by vortex.

3. Incubate for 30 min at room temperature at dark.

4. Add 1.25 μL of 1 mM of MitoSOX and mix well by vortex and incubate for 10 min at 37 °C at dark.

5. Centrifuge at $2300 \times g$ for 5 min at room temperature.

6. Remove supernatant and resuspend pellet with 300 μL of 1× PBS.

7. Determine mean fluorescence of PE (MitoSOX) among FITC (reticulocytes)-positive cells and the percentage of PE-positive cells among FITC-positive cells by FACS (Fig. 3c).

3.8 Isolation of Reticulocytes and Measurement of Bnip3L, Cat, and miR-21 Levels

1. Add 2.5 μL of whole blood into 500 μL of thiazole orange dye and mix well by vortex.

2. Incubate for 30 min at room temperature at dark.

3. Centrifuge at $2300 \times g$ for 5 min at room temperature and remove supernatant and resuspend pellet with 500 μL of 1× PBS.

4. Sort reticulocyte (FITC-positive cells) by FACS sorter and collect reticulocytes into 1% BSA in 1× PBS.

5. Centrifuge at $2300 \times g$ for 10 min at 4 °C and remove supernatant and resuspend pellet with 1 mL of TRI-Reagent. (You can store the sample at this step at −80 °C.).

6. Incubate at room temperature for 5 min and add 100 μL of 1-bromo-3-chloropropane and mix well by vortex for 15 s.

7. Incubate at room temperature for 15 min and centrifuge at $12,000 \times g$ for 15 min at 4 °C.

8. Add 0.5 mL of isopropanol to new tube and transfer aqueous.

9. Add 3 μL of Polyacryl carrier (Molecular research center) to visualize pellet and mix well by vortex.

10. Incubate at room temperature for 10 min and centrifuge at $12,000 \times g$ for 10 min at 4 °C.

11. Discard supernatant and add 1 mL 75% ethanol and centrifuge at $7500 \times g$ for 5 min at 4 °C.

12. Remove the ethanol and air-dry RNA pellet until no more ethanol smell.

13. Dissolve RNA in 20 μL of DEPC-treated water and incubate for 5 min at 60 °C.

14. Store RNA at −80 °C or use them for measuring transcript levels.

15. In order to measure transcript levels of *Bnip3L* and *Cat*, use 5 µL of total RNA to make cDNA using SuperScript™ VILO™ MasterMix.

16. Measure transcript levels of *Bnip3L* and *Cat* by real-time PCR using gene-specific TaqMan probes: *Bnip3L* and *Cat* and normalize transcript levels against 18S.

17. In order to measure *miR-21* levels, use 10 ng of total RNA to make cDNA using TaqMan® MicroRNA Reverse Transcription Kit.

18. Measure transcript level of *miR-21* by real-time PCR using microRNA-specific TaqMan probes (*miR-21*, and *U6snRNA*, and normalize against U6snRNA).

4 Notes

1. 10 µl of whole blood, anticoagulated in EDTA-coated tube, is sufficient to measure mitochondrial mass, ROS, and mitochondrial ROS experiment [7].

2. The biotin is administered just prior to hypoxic exposure; thus biotin-labeled RBCs are normoxia born, and hypoxia- and post-hypoxia-born RBCs are unlabeled [7].

3. Biotin-labeled reticulocytes are hypoxia-born reticulocytes, since reticulocytes are in circulation for 1–2 days; injected biotin at one day before return to normoxia binds to reticulocytes produced during hypoxia [7].

4. Reticulocytes should be incubated with thiazole orange dye for at least 30 min but no more than 4 h. During the incubation, every 10 min, samples should be mixed by vortexing [7].

5. ROS signal is not time sensitive but Mitotracker signal increases in a time-dependent manner; thus Mitotracker signal should be measured within 30 min after incubation [7].

6. Mito-Sox signal slowly decreases in time. It should be analyzed within 15 min after incubation [7].

7. The transcript levels in reticulocytes, 10,000 reticulocytes, are sufficient to measure transcript levels of *Bnip3L*, *Cat*, and *miR-21* [7].

References

1. Dill DB, Costill DL (1974) Calculation of percentage changes in volumes of blood, plasma, and red cells in dehydration. J Appl Physiol 37(2):247–248
2. Crooke AC, Morris CJ (1942) The determination of plasma volume by the Evans blue method. J Physiol 101(2):217–223
3. Kalra PR, Anagnostopoulos C, Bolger AP, Coats AJ, Anker SD (2002) The regulation and measurement of plasma volume in heart failure. J Am Coll Cardiol 39(12):1901–1908
4. Vogel J, Kiessling I, Heinicke K, Stallmach T, Ossent P, Vogel O, Aulmann M, Frietsch T, Schmid-Schonbein H, Kuschinsky W, Gassmann M (2003) Transgenic mice overexpressing erythropoietin adapt to excessive erythrocytosis by regulating blood viscosity. Blood 102(6):2278–2284. https://doi.org/10.1182/blood-2003-01-0283

5. Alfrey CP, Rice L, Udden MM, Driscoll TB (1997) Neocytolysis: physiological downregulator of red-cell mass. Lancet 349(9062): 1389–1390. https://doi.org/10.1016/S0140-6736(96)09208-2
6. Sandoval H, Thiagarajan P, Dasgupta SK, Schumacher A, Prchal JT, Chen M, Wang J (2008) Essential role for Nix in autophagic maturation of erythroid cells. Nature 454(7201):232–235. https://doi.org/10.1038/nature07006
7. Song J, Yoon D, Christensen RD, Horvathova M, Thiagarajan P, Prchal JT (2015) HIF-mediated increased ROS from reduced mitophagy and decreased catalase causes neocytolysis. J Mol Med (Berl) 93(8):857–866. https://doi.org/10.1007/s00109-015-1294-y

Chapter 17

Hypoxic Treatment of Zebrafish Embryos and Larvae

Hiroyasu Kamei and Cunming Duan

Abstract

Zebrafish has emerged as an informative animal model to study the biological impact and molecular mechanisms of hypoxia. Here we describe a simple method to induce hypoxia in zebrafish embryos and larvae. This protocol is easy and reproducible and does not require expensive equipment or specialized devices. It can be adapted in large, medium, and small scales. This protocol is also well-suited for experiments requiring chemical drug treatment and can be applied to other fish and amphibian species.

Key words Dissolved oxygen, Hypoxic water, Zebrafish, Embryo, Developmental timing

1   Introduction

Oxygen is a vital factor for fetal tissue development, growth, and adult physiology. Hypoxia, or low oxygen tension, imposes strains on fetal tissue growth and organogenesis and is a cause of intra-uterine growth restriction in human fetus [1]. This effect is observed in a wide variety of vertebrate species, including human, amphibians, and fish [2–7]. Hypoxia is also involved in the pathogenesis of other human diseases such as tumorigenesis.

Zebrafish is an excellent vertebrate model organism in developmental biology. Zebrafish embryos fertilize and develop externally. The rapidly developing and transparent embryos make it possible to manipulate environmental oxygen tension and observe the phenotypic changes in real time. Furthermore, major components of the zebrafish hypoxia-inducible factor system have been characterized and their genes identified, and they are highly similar to those found in humans and mammals [2, 4–6, 8–10]. Therefore, knowledge gained from studying zebrafish embryos and larvae will not only provide new insight into fish physiology but will also deepen our understanding of hypoxia physiology in general.

In this chapter, we describe a simple method to induce hypoxia in zebrafish embryos and larvae. This protocol is easy, reproducible, and does not require expensive equipment or specialized devices.

L. Eric Huang (ed.), *Hypoxia: Methods and Protocols*, Methods in Molecular Biology, vol. 1742,
https://doi.org/10.1007/978-1-4939-7665-2_17, © Springer Science+Business Media, LLC 2018

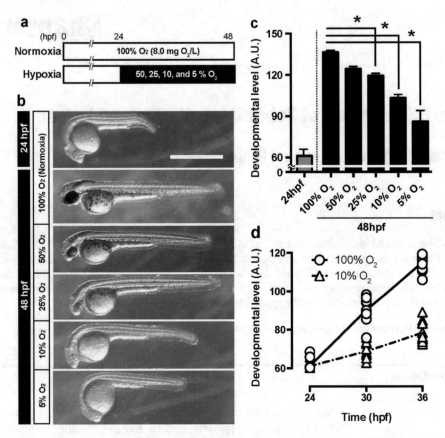

Fig. 1 Hypoxia decreases zebrafish embryo growth and development in a concentration-dependent manner. The result of an experiment carried out using the medium-scale system is shown. (**a**) A diagram illustrating the experimental design. Wild-type 24 hpf zebrafish embryos were transferred to $1 \times$ E3 solution with the indicated O_2 levels. (**b**) Representative embryo images after 24-h treatment. Scale bar = 1.0 mm. (**c**) Changes in developmental rate. Developmental rate is defined by head-trunk angle changes [4]. Data is shown as means $\pm$ SEM ($n = 10$). *, $P < 0.05$, One-way ANOVA followed by Dunnett's multiple comparison test. (**d**) Time-dependent effect of hypoxia. Each embryo is indicated as one dot

The effectiveness of hypoxic treatment can be simply assessed by embryo morphology and quantitative parameters [11]. A representative experiment result is shown in Fig. 1. This method can be easily applied to other fish and amphibian species.

2 Materials

2.1 Reagents

The following is a list of chemicals/reagents required. All can be obtained from local dealers.

1. Routine chemicals (NaCl, KCl, $CaCl_2 \cdot 2H_2O$, $MgSO_4 \cdot 7H_2O$, $NaHCO_3$, and dimethyl sulfoxide (DMSO)) can be purchased from a local chemical dealer.

2. Penicillin-Streptomycin solution (×100), cell culture-grade.

3. 0.5% methylene blue solution, microbiology-grade.

4. Pronase (EC Number 3.4.24.4), non-sterile.

5. Nitrogen gas: Purity is more than 99.99 vol.%.

2.2 Preparation of Solutions

1. Stock E3 solution (60×): Add 17.2 g NaCl, 0.76 g KCl, 2.9 g $CaCl_2 \bullet 2H_2O$, and 4.9 g $MgSO_4 \bullet 7H_2O$ to about 800 mL double-distilled water. Fill it up to 1 L. The solution should be filter sterilized and keep in an autoclaved bottle.

2. Working E3 solution (1×): Add 16.6 mL 60× E3 stock solution and 100–200 μL 0.5% methylene blue in ~ 900 mL double-distilled water. Adjust the pH to 7.4 by adding appropriate amount of $NaHCO_3$ solution. Add the Penicillin-Streptomycin solution if necessary. Fill up to 1 L with distilled water (*see* **Note 1**).

3. Pronase solution: Add 100 mg pronase to 10 mL E3 solution. This is the 10× stock solution and can be kept at −20 °C until use. Prepare pronase working solution on the day of experiment by diluting the above 10× stock solution with 1× E3 working solution.

2.3 Equipment/ Apparatus

1. 0.45 μm vacuum filter unit.

2. Dissolved oxygen (D.O.) meter: YSI/Nanotech Inc./Xylem Inc., YSI Model 5000, YSI ProODO, or equivalent meter with detection range from 0.1 mg O_2/L to 0.01 mg O_2/L). Make sure that your D.O. meter is functioning normally before each use.

3. Airstone: AQUAmunch, Fish Tank Air Stone Cylinder Shape Aerator Pump Bubble Airstone (model BS022).

4. Air tube.

5. Tanks: 500 mL tanks from Duran® Jar and 3 L tanks from Kotobuki Crystal cube 150 H/B) or any equivalent tanks.

6. Glass bottles: Media, Storage, Screw Cap, 500, 1000, and 2000 mL.

7. Tubes: Centrifuge tubes, 50, 15, and 1.7 mL.

2.4 Zebrafish Embryos

1. Adult zebrafish are maintained and fed following standard protocols [12]. Embryos are obtained by natural cross, and the embryos are kept in 1× E3 solution at 28.5 °C and staged according to the standard protocols [12].

2. Before the treatment, embryos are subjected to pronase treatment to remove chorion. For this, embryos of the desired developmental stage are placed in the working pronase

solution. After 5–10 min incubation at 28.5 °C, chorion becomes brittle. Embryos are easily released from the chorion by gentle pipetting. If chorion is not easily broken within a certain period (e.g., 5 min), wait a few minutes longer. After the dechorionation, embryos are washed with fresh 1× E3 solution and kept at 28.5 °C until use.

3. Depending on the stage of the embryos used, the pronase treatment length varies. In general, 5–10 min is recommended for 24-h post-fertilization (hpf) embryos. Please note that too long of a treatment or pipetting too strongly may damage or even kill embryos.

3   Hypoxic Treatment Protocol

We have designed and tested protocols of three different scales. The large-scale system protocol is shown in Fig. 2, and this system is suitable for handling large numbers of embryos (~500 embryos/group) for a relatively long period (~several days). The medium-scale system (Fig. 3) can handle ~50 embryos/group. This system is well-suited for the experiment combining hypoxia and drug treatment. The small-scale system is suitable for experiments handling 10–15 embryos/group (Fig. 4). It is particularly useful when you need to test expensive or valuable chemicals/drugs.

Due to space limitation, we will describe the large-scale system protocol step by step (Fig. 2). Technical details pertaining to the medium- and small-scale systems are also mentioned.

Step 1: Prepare several 500 mL tanks. Only two tanks are shown in Fig. 2 (1) for illustration simplicity. Transfer embryos to Tank A with minimal normoxic water. This is the experiment tank. Transfer the same amount of normoxic water to Tank B, which is used to measure the D.O. levels (*see* **Note 2**). It is also acceptable to use 500 mL bottles.

Step 2: Set up a 3-L tank as shown in Fig. 2 (2). Fill the tank with 80–90% E3 solution and babbling with N_2 gas until it reaches the desired D.O. levels. Place a cover on the top of the tank to reduce the air contact while bubbling N_2 (*see* **Note 3**). We usually bubble N_2 for 10–20 min at the flow rate of 5–10 L gas/min.

Step 3: Gently transfer the hypoxic water from the large tank to Tanks A and B. Please minimize aeration. Make sure that there are no air bubbles because they will increase the D.O. levels.

Step 4: Use Tank B to make sure that the D.O. levels are correct. If the final D.O. levels are too high, please repeat **step 1** by performing additional N_2 bubbling. If the final D.O. levels are too low, please adjust upward by adding normoxic water.

Step 5: Cover the tank with a lid and seal the tank. The samples are incubated at 28.5 °C until the end of experiment.

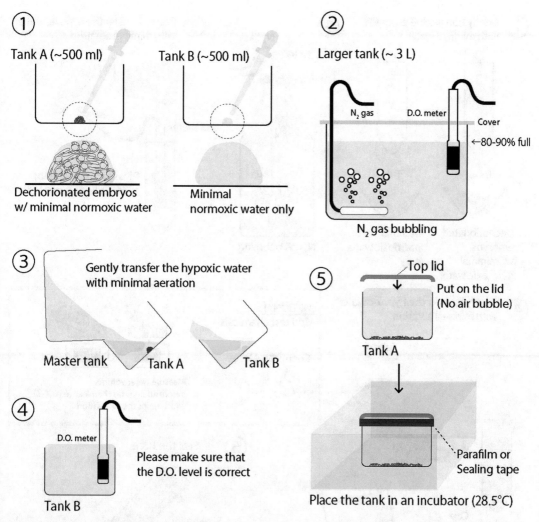

Fig. 2 A schematic diagram of the large-scale protocol

3.1 Several Issues Pertaining to the Medium-Scale System (Fig. 3)

Depending on the D.O. probe size, it may be difficult to monitor the D.O. levels while bubbling N_2 in a small bottle. In such case, perform N_2 bubbling first, and measure the D.O. level immediately after the bubbling (Fig. 3 (2)).

In Fig. 3 (4), if you need to combine the hypoxic treatment with pharmacological inhibitors, the water volume can be measured by weighing the tube before and after the addition of 1× E3 solution.

Please close the cap tightly to eliminate the air bubble/excess water (Fig. 3 (4)).

In Fig. 3 (5), make sure to place the tube laterally to spread embryos. This will help to avoid local hypoxia by overcrowding.

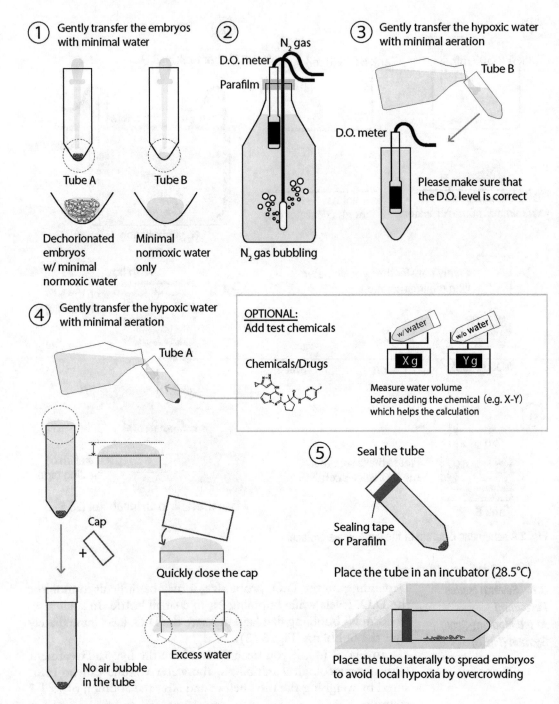

Fig. 3 A schematic diagram of the medium-scale protocol

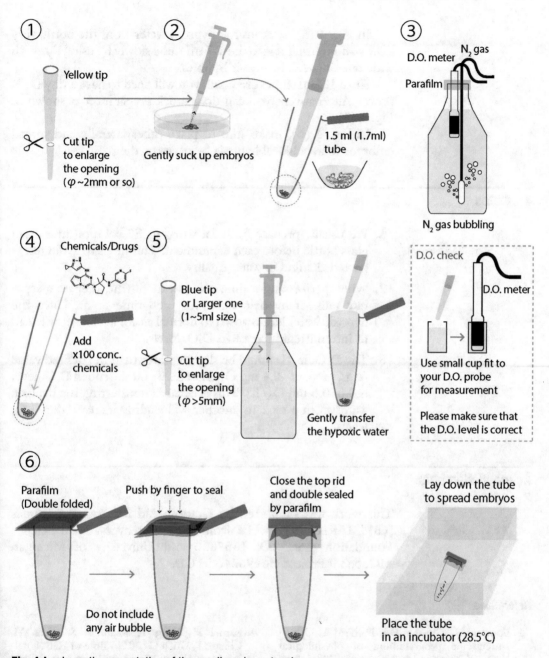

Fig. 4 A schematic presentation of the small-scale protocol

3.2 Several Issues Pertaining to the Small-Scale System (Fig. 4)

Cut tips to widen the opening. Without cutting, tips can damage embryos.

Depending on the D.O. probe size, it may be difficult to monitor the D.O. levels while bubbling N_2 in a small bottle. In such case, perform N_2 bubbling first, and measure the D.O. level immediately after the bubbling (Fig. 4 (3)).

In Fig. 4 (5), suck up the hypoxic water from the bottle very slowly and transfer it into the 1.7 mL tube slowly by using tips with wide opening. Avoid any air bubbles.

Since 1.7 mL tubes are small, you will need to have a tiny D.O. probe. Alternatively, you can do a mock preparation as shown in Fig. 4 (5).

In Fig. 4 (6), make sure to place tubes laterally and spread embryos to avoid local hypoxia by overcrowding.

4　Notes

1. We usually prepare fresh 1× working E3 solution in a clean glass bottle before each experiment. This helps to avoid unexpected changes in water quality.

2. When performing a time-course experiment, prepare a separate tank of hypoxic water for each time point. Once the hypoxic water is reexposed to normal air for a while, it will lead to intermittent increase in D.O. levels.

3. The D.O. level should be slightly lower than the desired value (e.g., prepare ~0.4 mg O_2/L water if your intended D.O. level is 0.5–0.6 mg O_2/L). This is because transferring the hypoxic water from a tank to another will slightly increase the D.O. levels.

Acknowledgments

This work was supported by Grant-in-Aid for Young Scientists [(B)# 15K18799] to HK from JSPS and by National Science Foundation grant IOS-1557850 and University of Michigan MCube2.0 Project U0496246 to CD.

References

1. Fowden AL, Giussani DA, Forhead AJ (2006) Intrauterine programming of physiological systems: causes and consequences. Physiology (Bethesda) 21:29–37

2. Kajimura S, Aida K, Duan C (2005) Insulin-like growth factor-binding protein-1 (IGFBP-1) mediates hypoxia-induced embryonic growth and developmental retardation. Proc Natl Acad Sci U S A 102:1240–1245

3. Moore LG, Shriver M, Bemis L, Hickler B, Wilson M, Brutsaert T, Parra E, Vargas E (2004) Maternal adaptation to high-altitude pregnancy: an experiment of nature—a review. Placenta 25(Suppl A):S60–S71

4. Kamei H, Ding Y, Kajimura S, Wells M, Chiang P, Duan C (2011) Role of IGF signaling in catch-up growth and accelerated temporal development in zebrafish embryos in response to oxygen availability. Development 138:777–786

5. Kamei H, Lu L, Jiao S, Li Y, Gyrup C, Laursen LS, Oxvig C, Zhou J, Duan C (2008) Duplication and diversification of the hypoxia-inducible IGFBP-1 gene in zebrafish. PLoS One 3:e3091

6. Zhang P, Yao Q, Lu L, Li Y, Chen PJ, Duan C (2014) Hypoxia-inducible factor 3 is an oxygen-dependent transcription activator and

regulates a distinct transcriptional response to hypoxia. Cell Rep 6:1110–1121

7. Hidalgo M, Le Bouffant R, Bello V, Buisson N, Cormier P, Beaudry M, Darribere T (2012) The translational repressor 4E-BP mediates hypoxia-induced defects in myotome cells. J Cell Sci 125:3989–4000

8. Duan C (2016) Hypoxia-inducible factor 3 biology: complexities and emerging themes. Am J Physiol Cell Physiol 310:C260–C269

9. Kajimura S, Aida K, Duan C (2006) Understanding hypoxia-induced gene expression in early development: in vitro and in vivo analysis of hypoxia-inducible factor 1-regulated zebra fish insulin-like growth factor binding protein 1 gene expression. Mol Cell Biol 26:1142–1155

10. Zhang P, Bai Y, Lu L, Li Y, Duan C (2016) An oxygen-insensitive Hif-3alpha isoform inhibits Wnt signaling by destabilizing the nuclear beta-catenin complex. elife 5:e08996

11. Kimmel CB, Ballard WW, Kimmel SR, Ullmann B, Schilling TF (1995) Stages of embryonic development of the zebrafish. Dev Dyn 203:253–310

12. Westerfield M (2007) The zebrafish book. A guide for the laboratory use of zebrafish (Danio rerio), 5th edn. University of Oregon Press, Eugene

Chapter 18

Microinjection of Antisense Morpholinos, CRISPR/Cas9 RNP, and RNA/DNA into Zebrafish Embryos

Yi Xin and Cunming Duan

Abstract

In this chapter, we describe a stepwise protocol of microinjection. Using this method, antisense morpholinos, CRISPR-Cas9 ribonucleoprotein complexes, capped mRNA, and DNA can be delivered into fertilized zebrafish eggs to manipulate gene expression during development. This protocol can also be adapted for microinjection in other fish and amphibian species.

Key words Zebrafish embryos, Microinjection, Morpholinos, CRISPR/Cas9, mRNA, Gene knockdown, Transgenesis

1 Introduction

Zebrafish has become a popular model organism in modern biology. The fast developing embryos and larvae are transparent, thus making it easy to follow the impact of genetic manipulation during development in real time. One pair of adult fish can produce hundreds of offspring, therefore providing ample supply of embryos for experiments. One of the commonly used approaches in zebrafish research is microinjection. This method employs a glass needle filled with the desired transformation reagent(s). The needle is attached and controlled by an apparatus that forces the solution out of the pipet by pressure. By adjusting the pulse duration, the injection volume can be controlled, thus ensures the reproducibility.

Microinjection is routinely used to deliver in vitro transcribed capped mRNA to increase the global expression of a gene product in a developing embryo [1]. It is also used to inject artificial chromosome-like bacterial (BAC) DNA or P1-derived artificial chromosome (PAC) DNA to express target gene(s) in a spatiotemporal-controlled manner [2]. Microinjection is also frequently used to deliver morpholinos (MOs) into zebrafish embryos [3]. MOs are synthetic oligonucleotides that can bind to

L. Eric Huang (ed.), *Hypoxia: Methods and Protocols*, Methods in Molecular Biology, vol. 1742,
https://doi.org/10.1007/978-1-4939-7665-2_18, © Springer Science+Business Media, LLC 2018

specific regions in RNAs by complementary pairing, thus blocking the access of other molecules. When injected to embryos, it can reduce the expression of a specific gene product either by blocking the translation or modify the pre-mRNA splicing [3]. Recently, the clustered regularly interspaced short palindromic repeats (CRISPR) and CRISPR-associated protein-9 nuclease (Cas9) system have emerged as powerful tool for genome editing in many organisms [4]. It has been shown that microinjection of preassembled Cas9 and single-molecule guide RNA (sgRNA) ribonucleoprotein complexes (RNPs) into the one-cell stage zebrafish embryos can cause mutagenesis within the first few cell divisions and permit phenotype assessment in the F0 generation [4].

In this chapter, we describe a stepwise microinjection protocol. Although this protocol is written for zebrafish, it can also be used for injecting transforming reagents into other fish and amphibian species.

2   Materials and Equipment

2.1   Preparation of Solutions

1. E3 embryo rearing solution: To make stock E3 solution (60×), add 17.2 g NaCl, 0.76 g KCl, 2.9 g $CaCl_2 \cdot 2H_2O$, and 4.9 g $MgSO_4 \cdot 7H_2O$ to about 800 mL double-distilled water. Adjust the pH to 7.4 by adding ~500 µL 2 N NaOH solution. Fill it up to 1 L. The solution should be filter sterilized and kept in an autoclaved bottle. Working E3 solution (1×) should be made on the day of the experiment. Add 16.6 mL 60× E3 stock solution and 100–200 µL 0.5% methylene blue in ~900 mL double-distilled water. Fill up to 1 L with distilled water.

2. Danieau's solution: To make 100 mL 30× stock solution, add 10.17 g NaCl, 0.156 g KCl, 0.296 g $MgSO_4 \cdot 7H_2O$, 0.425 g $Ca(NO_3)_2$, and 2.575 g HEPES in 80 mL double-distilled water, adjust pH to 7.6, and fill up to 100 mL. Store it in 4 °C. Working Danieau's solution (1×) is made by diluting the stock solution using double-distilled water.

3. Phenol red solution: Dissolve 0.5 g phenol red in 100 mL double-distilled water to obtain 0.5% phenol red stock solution. A final concentration of 0.025–0.050% phenol red is often used as an injection tracer.

2.2   Equipment and Tools Needed

1. Microinjection system: A typical microinjection system includes a stereomicroscope, a micromanipulator (World Precision Instruments or WPI) connecting to an injection needle, and a microinjector (WPI) that connects to a nitrogen gas tank (Fig. 1).

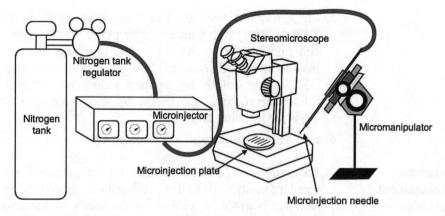

Fig. 1 A schematic diagram showing a typical microinjection setup. A microinjection plate with rows of zebrafish embryos is placed on the stage of a stereomicroscope. A micromanipulator is positioned on one side of the microscope, allowing the adjustment of angle and position of a microinjection needle. The microinjection needle is connected to a pneumatic microinjector, which is attached to a nitrogen tank through a nitrogen tank regulator and tubes

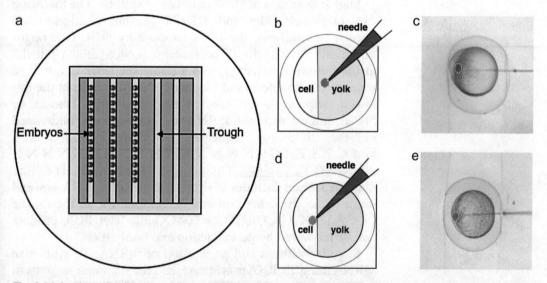

Fig. 2 (**a**) A schematic diagram of the microinjection plate. (**b**) A schematic diagram showing the site of morpholino or mRNA injection. (**c**) Top view of an egg injected with morpholino. The injection bolus is marked by a dashed line. (**d**) A schematic diagram showing the site of RNPs or BAC DNA injection. (**e**) Top view of an egg injected with BAC DNA. The injection bolus is marked by a dashed line

2. Microinjection plates: Add 1.5 g agarose to 100 mL 1× E3 solution in a microwavable flask. Microwave around 2 min until agarose completely dissolved. Pour the molten agarose into a 100× 15 mm petri dish (approximately 20 mL per dish). Place a plastic mold (Eppendorf) on the top of the agarose (*see* Fig. 2a). Take the plastic mold out after the agarose is solidified. Cover the plate surface with 1× E3 medium to preventing drying and store in 4 °C.

3. Microinjection needles: Pull Thin-Wall Single-Barrel Glass Tubing (WPI) with a micropipette puller (WPI) to obtain long and fine-tipped needles. The glass capillaries chosen should have an internal filament to allow the liquid to be carried to the injection tip quickly by capillary action. Immediately before the injection, break the needle tip with tweezers to create a ~2–3 μm opening end.

4. P10 Micropipettor and Microloader tips.

2.3 Preparation of MO, CRISPR/Cas9 RNP, and mRNA for Injection

1. Preparation of MO: MOs targeting the gene of interest and standard control MO can be ordered commercially. Stock MO solution is made by adding 1× Danieau's solution to a final concentration of 2 mM (~16 μg/μL) and stored in −80 °C in small aliquots. Before each injection, dilute the stock MO solution to a desired concentration using 1× Danieau's solution and phenol red (*see* **Note 1**).

2. Preparation of CRISPR/Cas9 RNPs:
   Step 1: Synthesis of DNA template of sgRNA: The following two oligonucleotides and DR274 plasmid (Addgene) are needed to synthesize the DNA template by PCR. One oligonucleotide includes the T7 promoter sequence (shown in italics in the following sequence) and a clamp sequence at the 5′ to stabilize the double-strand fragment (shown in bold in the following sequence), the selected target sequence (shown by "N"s), and the common sgRNA scaffold sequence (underlined in the following sequence) (5′-**GAAA***TAATACGACT CACTATAG*GNNNNNNNNNNNNNNNNNN NNNNGTTTTAGAGCTAGAAAT-3′) (*see* **Note 2**). The other oligonucleotide includes a short portion of sgRNA scaffold sequence in the other end and is common for all the targets (5′-AAAAGCACCGACTCGGTGCC-3′). The PCR product will be recovered by gel extraction and purification.

   Step 2: Synthesis and purification of sgRNA: To synthesize sgRNA using T7 RNA polymerase, add the following reagents to a 0.5 mL tube: ~30 ng DNA template, NTPs (0.5 mM each final), RNase inhibitor (1 U/μL final), DTT (5 mM final), 10× Reaction Buffer (2 μL), and H_2O (nuclease-free, fill up to 20 μL). Place the tube in a 37 °C incubator for 5 h to overnight. Stop the reaction by adding 1 μL Turbo DNase and 15 min incubating at 37 °C. Add 2 μL 8 M LiCl and 75 μL prechilled ethanol and place the tube in −80 °C for 1 h to overnight to precipitate the sgRNA. Next, spin at $13,000 \times g$ for 15 min at 4 °C, and remove the supernatant. Wash the pellet with prechilled 75% ethanol, spin the same speed for 5 min at 4 °C, and remove the supernatant. Dry the pellet to remove residual ethanol. Resuspend the pellet in 30 μL H_2O (nuclease-free) and store the sgRNA in −80 °C in small aliquots. The quality and quantity of synthesized sgRNA can be examined by electrophoresis.

Step 3: Mixing Cas9 protein with sgRNA: Cas9 protein containing a nuclear localization signal can be purchased from PNA Bio, Inc. (Thousand Oaks). Mix pure Cas9 protein with sgRNA solutions and incubate on ice for 10 min. Add KCl and phenol red to make the final concentrations as follows: Cas9 protein (~300 ng/μL), sgRNA (~40 ng/μL), KCl (300 mM) [3], and phenol red (~0.05%).

3. Preparation of capped mRNA: Capped mRNA is synthesized in vitro using mMessage mMachine SP6 kit (Ambion) and DNA template following the manufacturer's instruction. Purify the synthesized RNA sample using phenol:chloroform extraction followed by ethanol precipitation. Resuspend in 10 μL H_2O (nuclease-free) and store in −80 °C in small aliquots. Before injection, dilute stocks to a working concentration in 0.1 M KCl solution and phenol red.

3 Zebrafish Embryos

Adult zebrafish are maintained and fed following standard practice [5]. Embryos are obtained by natural cross, and the embryos are kept in 1× E3 solution at 28.5 °C (*see* **Note 3**). Line up eggs in a microinjection plate submersed in 1× E3 medium. To speed up the microinjection process, it is recommended to gently adjust the egg position in the trough by forceps to ensure all the eggs facing the same direction before the injection.

4 Microinjection Method

Load 2–4 μL solution of interest into the injection needle using a microloader. Attach the needle to the micromanipulator.

Adjust the angle and position of the needle through micromanipulator (Fig. 2). Typically, the angle between the needle and plate is approximately 45°, and the tip of needle is in the center.

Open the valve of nitrogen tank slowly and adjust the nitrogen tank regulator knob to set the outlet pressure approximately 20 psi. The injection pressure can be further adjusted using the microinjector knob.

Break the tip of the needle using a pair of forceps under the stereomicroscope (*see* **Note 4**).

In one smooth stroke, penetrate the chorion and insert the needle in the egg. For mRNA and MOs, the injection site is in the yolk (Fig. 2b, c). When injecting sgRNA, the injection site is in the cell (Fig. 2d, e).

The injection volume can be changed by adjusting the pressure pulse duration knob on the microinjector.

After injecting one embryo, pull the needle out and move the microinjection plate forward to inject the next embryo in line.

After finishing injection of one group, transfer the injected embryos into a petri dish containing 1× E3 medium and place the dish in the incubator (28.5 °C).

Two to three hours after the injection, check the injected embryos and remove the unfertilized eggs and dead embryos [6] (*see* **Note 5**).

5 Technical Notes

1. The optimal concentration for a particular MO or mRNA should be titrated based on the purpose of your study. High concentrations of transforming reagents have non-specific toxicity.

2. If your target sequence starts with G or GG, replace the GG at the end of T7 promoter with the target sequence. The optimal sgRNA length is 20 nt [7].

3. To ensure the quantity and quality of embryos for injection, dedicated breeding fish should be used. Regular mating is necessary to maintain the productivity of the breeding stocks.

4. The size of the needle tip is critical for a successful injection. If a needle opening is too large, it may rupture the embryo. If it is too small, it is easier to get clogged.

5. Figure 3 shows a representative mRNA and MO co-injection experiment to validate the effectiveness and specificity of MO knockdown. In this experiment, a GFP reporter is placed under the control of the 5′UTR and coding sequence of a gene of interest. When the mRNA is co-injected with a targeting MO that blocks the translation of this (TB), no GFP signal was observed. In contrast, the control MO (CT) did not affect the GFP expression (Fig. 3).

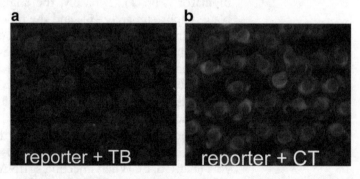

Fig. 3 Morphorlino-mediated knockdown of GFP expression. (**a**) Images of a group of 24 hpf embryos co-injected with a GFP reporter plasmid and a targeting morpholino (TB). (**b**) Images of a group of 24 hpf embryos co-injected with the same GFP reporter plasmid and a control morpholino

Acknowledgments

We thank Ms. Allison Malick, University of Michigan, for proofreading this manuscript. This work was supported by National Science Foundation grant IOS-1557850 and University of Michigan MCube2.0 Project U0496246 to CD.

References

1. Hogan BM, Verkade H, Lieschke GJ, Heath JK (2008) Manipulation of gene expression during zebrafish embryonic development using transient approaches. Methods Mol Biol 469:273–300

2. Suster ML, Abe G, Schouw A, Kawakami K (2011) Transposon-mediated BAC transgenesis in zebrafish. Nat Protoc 6:1998–2021

3. Burger A, Lindsay H, Felker A (2016) Maximizing mutagenesis with solubilized CRISPR-Cas9 ribonucleoprotein complexes. Development 143:2025–2037

4. Robu ME, Larson JD, Nasevicius A, Beiraghi S, Brenner C, Farber SA, Ekker SC (2007) p53 activation by knockdown technologies. PLoS Genet 3(5):e78

5. Westerfield M (2007) The zebrafish book. A guide for the laboratory use of zebrafish (Danio rerio), 5th edn. University of Oregon Press, Eugene

6. Kimmel CB, Ballard WW, Kimmel SR, Ullmann B, Schilling TF (1995) Stages of embryonic development of the zebrafish. Dev Dyn 203(3):253–310

7. Brocal I, White RJ, Dooley CM, Carruthers SN, Clark R, Hall A, Busch-Nentwich EM, Stemple DL, Kettleborough RN (2016) Efficient identification of CRISPR/Cas9-induced insertions/deletions by direct germline screening in zebrafish. BMC Genomics 17:259

Chapter 19

Western Blot Analysis of *C. elegans* Proteins

Dae-Eun Jeong, Yujin Lee, and Seung-Jae V. Lee

Abstract

C. elegans has been widely used as a model organism for basic biological research and is particularly amenable for molecular genetic studies using a broad repertoire of techniques. Biochemical approaches, including Western blot analysis, have emerged as a powerful tool in *C. elegans* biology for understanding molecular mechanisms that link genotypes to phenotypes. Here, we provide a protocol for Western blot analysis using protein extracts obtained from *C. elegans* samples.

Key words *C. elegans*, Protein extraction, Western blot, SDS-PAGE

1 Introduction

The roundworm *Caenorhabditis elegans* is one of the most important model animals in biological research. *C. elegans* is easy to culture in the laboratory, is genetically tractable, and has a short life cycle, high reproduction rates, a fully sequenced genome, and a completed cell lineage [1]. Many powerful genetic and biochemical approaches in *C. elegans* studies have yielded significant advances in developmental biology, neuroscience, and aging biology.

Despite the need and importance of biochemistry-based approaches, many *C. elegans* studies have relied primarily on molecular genetic and cell biological approaches for several reasons. First, expression patterns and levels of proteins tagged with a fluorescent protein can be easily observed in living, transparent *C. elegans* [2], which has been used as a popular substitute for Western blot analysis. Researchers can express tagged proteins by employing straightforward transgenesis techniques, including simple microinjection [3, 4]. In addition, single-copy insertions of tagged exogenous genes can be performed using MosSCI [5, 6]. Genome editing of *C. elegans* via the CRISPR/Cas9 system has also rapidly progressed, and provides the great advantage of labeling endogenous proteins with a fluorescent protein or a small peptide tag [7, 8]. Overall, these genetic techniques for protein tagging

L. Eric Huang (ed.), *Hypoxia: Methods and Protocols*, Methods in Molecular Biology, vol. 1742,
https://doi.org/10.1007/978-1-4939-7665-2_19, © Springer Science+Business Media, LLC 2018

paradoxically account for the scarcity of protein biochemical data in conventional *C. elegans* research papers. Moreover, the thick cuticle structure at the outer surface of *C. elegans* makes protein extraction challenging. The number of available *C. elegans*-specific antibodies is limited, although antibodies that target *C. elegans* homologues in other species can be tested for cross-reactivity. If antibodies of interest are not available, *C. elegans* protein-specific antibodies can be generated from rabbits or mice and subsequently purified using conventional affinity purification methods [9–11]. However, researchers should be aware of the fact that antibodies generated from animals infected with various parasitic nematodes may be less specific than other antibodies and may generate many background signals during Western blotting. Finally, *C. elegans* usually utilizes bacteria (mostly *E. coli* in laboratory conditions) as a food source, and several proteins expressed in bacteria share a high sequence similarity to *C. elegans* proteins. Thus, antibodies targeting *C. elegans* proteins of interest can sometimes bind orthologous bacterial proteins. In this case, bacteria should be thoroughly removed before worm samples are harvested for subsequent biochemical experiments. Despite these limitations, protein-based biochemical experiments have emerged as valuable tools for *C. elegans* research [9, 12, 13].

Western blot analysis detects specific proteins of interest from mixed proteins and typically includes three steps [14]: molecular-weight-dependent protein separation by sodium dodecyl sulfate-polyacrylamide gel electrophoresis (SDS-PAGE), electrophoretic transfer of proteins from the SDS-PAGE gels to nitrocellulose or polyvinylidene difluoride (PVDF) membranes, and detection of proteins of interest by using specific antibodies. Western blot analysis, in combination with other biological assays, helps researchers uncover molecular mechanisms by which specific proteins physiologically function. For example, Western blotting is useful for examining if protein levels are influenced by changes in genetic and/or environmental conditions. Researchers can also explore the physical interactions between two proteins by performing Western blot analysis following immunoprecipitation. In addition, subcellular localization of proteins can be determined by fractionation together with Western blotting. Here, we describe a protocol for protein extraction and Western blot analysis using *C. elegans* samples.

2   Materials

The standard culture temperature for *C. elegans* is 20 °C (*see* detailed methods for *C. elegans* culture in [15]). Prepare and store reagents at room temperature unless otherwise specified. We recommend all buffers listed below be made as stock solutions and then diluted to working concentrations presented below.

2.1 Protein Extraction

1. Worm samples: proteins can be extracted from various strains of *C. elegans*, including mutants or transgenic animals under diverse experimental conditions. For a simple Western blot analysis, 30–40 μL wet worm sample (i.e., approximately 4000 young adult wild-type worms) is sufficient for one assay. Developmental stages of animals for control and experimental groups should be tightly synchronized for minimizing experimental bias.

2. 1× M9 buffer: 42 mM Na_2HPO_4, 22 mM KH_2PO_4, 85 mM NaCl, and 1 mM $MgSO_4$. We recommend 1 mM $MgSO_4$ be added after autoclave sterilization of the M9 buffer.

3. 1× phosphate-buffered saline (PBS): 137 mM NaCl, 2.7 mM KCl, 4.3 mM Na_2HPO_4, and 1.47 mM KH_2PO_4.

4. 1× protease inhibitors (or commercially available protease inhibitor cocktails): 1 mM phenylmethane sulfonyl fluoride (PMSF), 10 μM pepstatin, 2.34 μM leupeptin, 1 mM dithiothreitol (DTT), 1 mM benzamidine, and 1.5 μM aprotinin. Protease inhibitors should be stored at −20 °C and added to PBS buffer right before experiments are performed.

5. Available protein quantitation kits, such as Bradford or bicinchoninic acid (BCA) protein assay kits.

6. 2× sample buffer (for boiling samples): 0.125 M Tris–HCl (pH 6.8), 4% SDS, 20% glycerol, and 0.05% bromophenol blue. Right before use, 10% β-mercaptoethanol (BME) or 10–20 mM DTT should be added to the 2× sample buffer (*see* **Note 1**).

7. 5× sample buffer (for sonication, homogenization, and grinding samples using a mortar): 0.325 M Tris–HCl (pH 6.8), 15% SDS, 50% glycerol (or 0.825 M sucrose), and 0.05% bromophenol blue. Right before use, 10% BME or 10–20 mM DTT should be added to the 5× sample buffer.

8. Heat block for boiling worm samples, sonicator for breaking worm samples by generating ultrasonic frequencies, glass homogenizers (pestles and tubes) for mechanical homogenization of worm samples, liquid nitrogen, and a mortar and pestle for grinding worm samples to make frozen worm powders.

9. Syringe filters (0.45 μm pore size) for discarding debris after grinding worm samples with mortar and pestle.

2.2 SDS-PAGE

1. SDS-polyacrylamide gel: the SDS-polyacrylamide gel consists of two different gel layers, the lower resolving gel (running gel or separating gel) and the upper stacking gel. SDS is required for denaturing and coating proteins with a negative charge, so proteins can migrate in a voltage- and molecular-weight-dependent manner. The stacking gel, which has a low concen-

tration of polyacrylamide and a low pH (6.8), provides a platform at which proteins with different sizes are stacked as a line at the boundary between stacking and resolving gels (*see* **Note 2**). Proteins are then separated in the resolving gel, which has a high concentration of polyacrylamide and a high pH (8.8), by electrophoretic force. The resolving gel should be prepared first. After polymerization of acrylamide in the resolving gel is complete, the stacking gel is subsequently cast at the upper region of the resolving gel.

(a) Resolving gel: 6–15% polyacrylamide, 0.375 M Tris–HCl (pH 8.8), 0.1% SDS, 0.1% ammonium persulfate, and tetramethylethylenediamine (TEMED). The percentage of polyacrylamide should be determined based on sizes of the proteins of interest (*see* **Note 3**). The proper volume of TEMED is also variable depending on the percentage of polyacrylamide in the resolving gel (*see* **Note 4**). Ammonium persulfate and TEMED should be added last to avoid premature acrylamide polymerization.

(b) Stacking gel: 5% polyacrylamide, 0.125 M Tris–HCl (pH 6.8), 0.1% SDS, 0.1% ammonium persulfate, and 0.1% TEMED.

2. 1× running buffer (pH 8.3): 25 mM Tris base, 250 mM glycine, and 0.1% SDS.

2.3 Protein Transfer from SDS-PAGE Gels to Membranes

1. Commercially available PVDF membrane or nitrocellulose membrane (*see* **Note 5**).

2. 1× transfer buffer (pH 8.3): 25 mM Tris base, 192 mM glycine, and 0.025–0.1% SDS (optional, *see* **Note 6**). Methanol (20%) should be added just before use.

3. Ponceau S solution for rapid detection of proteins on a membrane (optional).

2.4 Membrane Blocking

1. 1× PBS or Tris-buffered saline (TBS): 50 mM Tris–Cl (pH 7.5) and 150 mM NaCl. The pH of the buffer should be adjusted to 7.5 with HCl.

2. TBS-T or PBS-T: 1× TBS or PBS with 0.1% Tween 20.

3. Blocking solution: TBS-T or PBS-T containing 5% skim milk or bovine serum albumin (BSA).

4. A shaker (or rocker) for gentle agitation of membranes.

2.5 Antibodies

1. Primary antibodies that target specific proteins of interest. Primary antibodies may be preferentially *C. elegans* protein-specific. Alternatively, antibodies targeting orthologous proteins in other species whose epitopes are well conserved in the *C. elegans* proteins can be tested.

2. Secondary antibodies: target specificity of secondary antibodies depends on host species from which the primary antibodies were generated and purified. Commonly used secondary antibodies for Western blot analysis are conjugated to certain enzymes, such as horseradish peroxidase (HRP) and alkaline phosphatase (AP). Secondary antibodies specifically bind to primary antibodies and are used to visualize interactions between proteins of interest and primary antibodies through enzymatic reactions that convert invisible substrates to luminescent or colorimetric compounds. Secondary antibodies conjugated to fluorophores that do not require an enzymatic reaction or substrate treatment can also be used for protein detection. Moreover, multiple secondary antibodies conjugated to fluorophores with various colors can be used for detecting different proteins on a single membrane. Radioisotope (^{125}I)-labeled secondary antibodies can be used for generating very sensitive and quantitative autoradiographic signals. However, the autoradiography method for Western blot analysis is rarely used because of precautions associated with ^{125}I radioisotopes.

2.6 Detection of Proteins

1. Chemiluminescent HRP substrates or chromogenic AP substrates (pNPP or BCIP/NBT). Substrate types should be determined based on enzymes that are conjugated to the secondary antibodies. Below, we mention specific options for signal detection.

2a. Detection of proteins using X-ray films: X-ray films, film cassette, and developer and fixer solutions are required for detecting chemiluminescent or autoradiographic signals.

2b. Detection of proteins using a bioluminescence detector: proper imaging device that can detect chemiluminescent signals (e.g., ChemiDoc™ imager or ImageQuant™ LAS 2000).

2c. Detection of proteins with fluorescent signals: device for fluorescence imaging (e.g., Odyssey CLx or VersaDoc™ MP 5000).

---

3 Methods/Protocols

Perform experiments as follows (Fig. 1) at room temperature unless otherwise specified. Protein samples should be stored on ice to prevent degradation.

3.1 Preparation of Worm Samples

1. Wash worms synchronized at specific growth stages and in genetic backgrounds at least three times with M9 buffer to remove bacteria.

2. After the worms form a sediment, remove as much supernatant as possible.

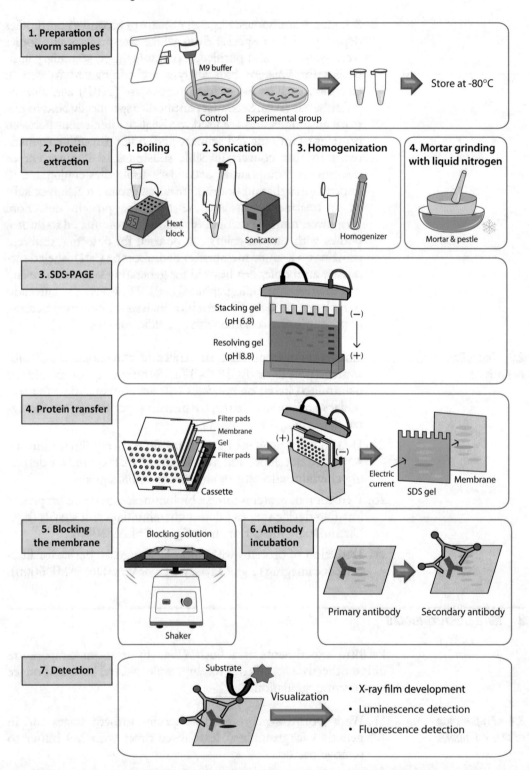

Fig. 1 Summary of protocols for Western blot analysis of *C. elegans* proteins. Shown are seven steps of the Western blot procedure using *C. elegans* protein samples. Numbers 1 through 7 indicate steps described in the protocol 3.1–3.7. Subheadings 3.2.1–3.2.4 are alternative protein extraction procedures described in Subheadings 3.2.1–3.2.4

3. Freeze the worm samples immediately by using liquid nitrogen or by placing in at −80 °C. The samples can be stored at −80 °C until subsequent experiments are performed. For simple boiling-based protein extraction (*see* Subheading 3.2.1), add an equivalent volume of 2× sample buffer containing freshly added 10% BME or 10–20 mM DTT to the worm samples prior to freezing.

3.2 Protein Extraction

Four alternative methods for protein extraction are presented below.

3.2.1 Protein Extraction with Simple Boiling

1. Thaw the frozen worm samples at room temperature, and then place them immediately on ice.

2. Boil the worm samples at 95 °C for 10 min.

3. Gently vortex the worm samples for 10 min.

4. Centrifuge (16,000 × *g*) the worm samples for 30 min at 4 °C.

5. Transfer clean supernatants (crude protein extracts) to new 1.5 mL microcentrifuge tubes, and keep the samples on ice. Use the supernatants as protein samples for SDS-PAGE (*see* **Note 7**).

3.2.2 Protein Extraction with Sonication

1. Thaw the frozen worm samples at room temperature.

2. Resuspend the worm pellets with 500 µL 1× PBS, and keep them on ice. Protease inhibitors in the PBS will help prevent rapid degradation of proteins.

3. Prepare and wash the microtip of the sonicator with distilled water (dH₂O).

4. Sonicate the samples. Insert the microtip into a 1.5 mL microcentrifuge tube that contains the worm sample on ice, and pulse the worm sample for 1 s. The number of pulse repeats depends on the amplitude and intensity of the sonication. Continuous cooling of the microcentrifuge tube by placing the sample on ice is required during the repeated pulses because sonication generates heat that may affect protein stability. Confirming sufficient sonication by observing the samples microscopically will help optimize the number and amplitude of the pulses.

5. Quantify proteins in the worm samples by using Bradford or BCA protein assays [16].

6. Prepare loading samples for SDS-PAGE by normalizing the concentrations of proteins in the samples using 1× PBS and then adding 5× sample buffer.

3.2.3 Protein Extraction with Homogenization

1. Thaw the worm samples at room temperature, and store the thawed samples on ice.

2. Resuspend the worm pellets with 500 μL 1× PBS, and keep them on ice. Protease inhibitors in the PBS will help prevent rapid protein degradation.

3. Prepare and wash the glass homogenizer with sufficient dH$_2$O.

4. Transfer the worm samples to the glass homogenizer tube.

5. Homogenize the samples. After 20 strokes for each worm sample, transfer the samples to new 1.5 mL microcentrifuge tubes.

6. Centrifuge (800 × g) the homogenized samples at 4 °C for 10 min.

7. Transfer the supernatants (protein extracts) to a new 1.5 mL microcentrifuge tubes on ice.

8. Resuspend the remaining pellets with 500 μL 1× PBS.

9. Repeat **steps 4–7**.

10. Combine the two sets of supernatants (homogenized proteins) and quantify total proteins by using Bradford or BCA protein assays.

11. Prepare loading samples for SDS-PAGE by normalizing the concentrations of proteins in the samples using 1× PBS and by adding 5× sample buffer.

3.2.4 Protein Extraction by Grinding with Mortar and Pestle

1. Thaw the worm samples at room temperature, and keep the thawed samples on ice until use.

2. Prepare a cold mortar and pestle using liquid nitrogen.

3. Remove buffer in the worm samples, and transfer the samples into the cold mortar. The worm samples will be frozen immediately after the transfer.

4. Grind the frozen worm samples with the cold pestle thoroughly to form a powder. This step may need additional liquid nitrogen to prevent the frozen samples from melting.

5. Dissolve the powder in 1× PBS that contains protease inhibitors.

6. Filter the samples with syringe filters (0.45 μm pore size) to discard debris.

7. Quantify proteins in the samples using Bradford or BCA protein assays.

8. Prepare loading samples for SDS-PAGE by normalizing the concentrations of proteins in the samples using 1× PBS and by adding 5× sample buffer.

3.3 SDS-PAGE

1. Prepare 1× running buffer and the SDS-polyacrylamide gel.

2. Before loading, boil the samples at 95 °C for 2 min, and briefly spin down the samples by centrifugation.

3. Load each protein extract and a protein molecular-weight ladder into the wells of SDS-polyacrylamide gel.

4. Electrophorese the samples at 80 V for approximately 20 min until the samples are stacked and reach the running gel.

5. Electrophorese the samples at 120–130 V for approximately 40–60 min until the loading dye reaches the bottom of the running gel. Running time differs depending on the percentage of polyacrylamide gel and the molecular weights of the target proteins.

3.4 Protein Transfer to the Membranes

1. Prepare cold 1× transfer buffer and PVDF membrane.

2. Treat the PVDF membrane with 100% methanol for 1–2 min to activate the membrane before use. The PVDF membrane should then be immersed in transfer buffer to remove air bubbles.

3. After electrophoresis, assemble the gel and membrane with a proper cassette in cold transfer buffer. Prevent the membrane from drying.

4. Transfer proteins from the gel to the membrane at 4 °C for 1 h at 300 mA or overnight at 30 mA.

5. Verify a successful transfer process by detecting pre-stained protein ladder on the membrane or by using Ponceau S staining. For Ponceau S staining, immerse the membrane in Ponceau S solution for 1–2 min and wash the membrane with dH_2O until background signals disappear. Ponceau S staining is reversible and can be removed following membrane blocking and washing.

3.5 Membrane Blocking to Reduce Background Signals

1. Prepare TBS-T- or PBS-T-containing 5% BSA or skim milk. PBS-T should not be used for detecting phosphorylated proteins with phospho-specific primary antibody.

2. Transfer the membrane into TBS-T- or PBS-T-containing 5% BSA or skim milk and agitate on a shaker for 1 h.

3.6 Treatment of Membranes with Antibodies

1. Prepare solution containing specific primary antibodies targeting proteins of interest and appropriate secondary antibodies, which are selected based on the animal from which the primary antibodies are generated. For example, anti-mouse IgG secondary antibodies should be used with anti-α-tubulin primary antibodies generated from mice.

2. Dilute a chosen specific primary antibody with TBS-T or PBS-T. To reduce background signals from nonspecific binding of the primary antibody, 5% BSA or skim milk is usually added to TBS-T or PBS-T for diluting antibody (*see* **Note 8**).

3. Transfer the membrane into TBS-T or PBS-T with diluted primary antibody, and incubate the membrane on a shaker for 1 h at room temperature. Incubation of the membrane with the primary antibody can also be done at 4 °C overnight (12–16 h) to maintain stability of the primary antibody for multiple uses.

4. Wash the membrane five times with TBS-T or PBS-T for 5 min on a shaker at room temperature.

5. Incubate the membrane in TBS-T or PBS-T with diluted secondary antibody with gentle shaking for 1 h at room temperature.

6. Wash the membrane five times with TBS-T or PBS-T for 5 min on a shaker at room temperature.

3.7 Detection of Proteins

Methods for detecting proteins should be determined based on secondary antibodies used.

3.7.1 Protein Detection with Chemiluminescent Substrates

1. Treat the membrane with chemiluminescent HRP substrate for 1–2 min with gentle agitation. The signal is the strongest immediately after treatment and declines over time due to reaction kinetics of the secondary antibodies conjugated with HRP and substrates. For visualization of protein signals, follow **step 1a** with X-ray film or **step 1b** with a bioluminescence detector:

   (a) Transfer the membrane into a film cassette, and expose the membrane to X-ray film in a dark room. Optimal exposure time depends on the concentration and affinity of antibodies to target proteins.

   (b) Transfer the chemiluminescent HRP substrate-treated membrane to a CCD camera device that detects luminescent signals. Determine the optimal exposure time and intensity by following the manufacturer's protocol to obtain images of protein bands.

3.7.2 Protein Detection with Chromogenic Substrates

1. Stain the membrane with commercially available chromogenic substrates of AP or HRP, depending on the types of secondary antibodies in use, for 5–10 min by following the manufacturer's protocol. Do not shake the membrane during the enzyme-substrate reaction.

2. When the color signals are optimally visible, rinse the membrane with dH_2O to stop the reaction.

3. Air-dry the membrane for 10–20 min.

4. Obtain a photograph of the membrane immediately after air-drying of the membrane, as the color signals become weak over time.

Fluorophore-based protein detection does not require an enzyme-substrate reaction but does require a device for fluorescent signal detection. Photographs of the membrane should be taken using a fluorescent detector by following the device manufacturer's protocol.

4 Notes

1. SDS-PAGE samples typically include a reducing agent (e.g., BME or DTT) for denaturing disulfide bonds of proteins. This denaturation is critical for proper separation of proteins based on molecular weight. DTT is more effective but less stable than BME. Either reducing agent should be added to sample buffer right before use.

2. The stacking gel and resolving gel should be pH 6.8 and pH 8.8, respectively. Working concentration 1× running buffer (pH 8.3) includes glycine, whose charge varies by pH. For exemple, weakly negative charge of glycine in the pH 8.3 running buffer allows it to enter the stacking gel upon applying electric currents. Glycine becomes neutral in the stacking gel (pH 6.8), which impedes its movement. Compared to glycine, Cl^- ions in the stacking gel move rapidly. These different mobilities of glycine and Cl^- in the stacking gel form a steep voltage gradient because of the temporary absence of electrolytes. The voltage gradient then allows glycine to move rapidly, and subsequently prompts proteins to migrate into the gel. Thus, in the stacking gel, proteins are mobilized between Cl^- and glycine, regardless of their molecular weights, and will migrate evenly. In the resolving gel (pH 8.8), negatively charged glycine moves more quickly, and the proteins can be separated by their molecular weights.

3. Determining the optimal concentration of polyacrylamide for SDS-PAGE is critical for clear separation of the proteins of interest. A protein migration chart is helpful for this determination, although in general, optimal concentrations of acrylamide for detecting proteins with different ranges of molecular weights are as follows: 7% for 50–500 kDa, 10% for 20–300 kDa, 12% for 10–200 kDa, and 15% for 3–100 kDa proteins.

4. TEMED generates free radicals by reacting with ammonium persulfate and promotes polymerization of acrylamide and bisacrylamide. Optimal TEMED concentrations differ depending on concentrations of the acrylamides as follows: 0.08% for 6%, 0.06% for 7–9%, and 0.04% for 10–15% acrylamides [14].

5. Two representative membrane types, nitrocellulose and PVDF, are often used for Western blot analysis. Nitrocellulose mem-

branes are relatively inexpensive and do not require an activation step using methanol prior to usage, which is required for PVDF membranes. However, the binding affinity of PVDF membranes to proteins is higher than that of nitrocellulose membranes because they are more hydrophobic. Thus, PVDF membranes are preferred when multiple stripping procedures are expected.

6. SDS in the transfer buffer helps proteins move out of the gel and increases transfer efficiency of proteins with high molecular weights but inhibits binding of proteins to membranes. Addition of SDS to the transfer buffer is not recommended for protein transfer using nitrocellulose membranes.

7. The boiling method with 2× sample buffer is straightforward for protein extraction from *C. elegans*, but cannot be coupled with protein quantitation methods based on chromogenic reactions (BCA and Bradford protein assays). Therefore, control Western blot analysis that determines the level of loading control proteins (e.g., α-tubulinand, β-actin) is recommended for normalization prior to actual experiments.

8. Optimizing the ratio of antibody dilution for detecting specific protein signals with minimal background signals should be established before performing actual experiments.

5 Closing Remarks

Biochemical experiments using *C. elegans* have been previously regarded as technically difficult. However, rapid advances in biological techniques have helped Western blot analysis gain popularity in *C. elegans* research. For example, the recent revolution of CRISPR/Cas9 genome editing technology has made epitope tagging of endogenous *C. elegans* proteins easier and subsequently has enabled Western blot analysis of such proteins. High-quality *C. elegans* research will ideally combine protein biochemistry, including Western blot analysis, with powerful genetic techniques to elucidate molecular mechanisms by which proteins function in *C. elegans* and higher animals.

Acknowledgments

This work was supported by the Science Research Center grant (NRF-2017R1A5A1015366) funded by the Ministry of Science, ICT, and Future Planning of South Korea to S.-J.V.L.

References

1. Corsi AK, Wightman B, Chalfie M (2015) A transparent window into biology: a primer on *Caenorhabditis elegans*. Genetics 200:387–407

2. Hutter H (2012) Fluorescent protein methods: strategies and applications. Methods Cell Biol 107:67–92

3. Evans TC (2005) Transformation and microinjection. WormBook:1–15

4. Praitis V, Maduro MF (2011) Transgenesis in *C. elegans*. Methods Cell Biol 106:161–185

5. Frokjaer-Jensen C, Davis MW, Hopkins CE, Newman BJ, Thummel JM, Olesen SP, Grunnet M, Jorgensen EM (2008) Single-copy insertion of transgenes in *Caenorhabditis elegans*. Nat Genet 40:1375–1383

6. Frokjaer-Jensen C, Davis MW, Sarov M, Taylor J, Flibotte S, LaBella M, Pozniakovsky A, Moerman DG, Jorgensen EM (2014) Random and targeted transgene insertion in *Caenorhabditis elegans* using a modified Mos1 transposon. Nat Methods 11:529–534

7. Dickinson DJ, Goldstein B (2016) CRISPR-based methods for *Caenorhabditis elegans* genome engineering. Genetics 202:885–901

8. Kim HM, Colaiacovo MP (2016) CRISPR-Cas9-guided genome engineering in *C. elegans*. Curr Protoc Mol Biol 115:31.37.1–31.37.18

9. Duerr JS (2006) Immunohistochemistry. WormBook:1–61

10. Hadwiger G, Dour S, Arur S, Fox P, Nonet ML (2010) A monoclonal antibody toolkit for *C. elegans*. PLoS One 5:e10161

11. Arur S, Schedl T (2014) Generation and purification of highly specific antibodies for detecting post-translationally modified proteins in vivo. Nat Protoc 9:375–395

12. Fonslow BR, Moresco JJ, Tu PG, Aalto AP, Pasquinelli AE, Dillin AG, Yates JR 3rd (2014) Mass spectrometry-based shotgun proteomic analysis of *C. elegans* protein complexes. WormBook:1–18

13. Walhout AJ (2006) Biochemistry and molecular biology. WormBook:1–10

14. Sambrook J, Russell DW (2001) Molecular cloning: a laboratory manual, 3rd edn. Cold Spring Harbor Laboratory Press, Cold Spring Harbor, NY

15. Stiernagle T (2006) Maintenance of *C. elegans*. WormBook:1–11

16. Krohn RI (2011) The colorimetric detection and quantitation of total protein. Curr Protoc Cell Biol Appendix 3:3H

Chapter 20

In Vivo Manipulation of HIF-1α Expression During Glioma Genesis

Patricia D.B. Tiburcio, Séan B. Lyne, and L. Eric Huang

Abstract

Hypoxia has long been recognized as a driving force of tumor progression and therapeutic resistance, and the transcription factor HIF-1α is believed to play a crucial role in these processes. Here we describe an efficient RCAS/Nes-TVA system that allows for in vivo manipulation of HIF-1α expression in the mouse neural progenitor cells. Simple production of the recombinant avian virus RCAS enables quick delivery of gene of interest through injection into the neural progenitors of transgenic mice expressing the viral cognate receptor TVA under the nestin promoter. By crossing with various commercially available genetically engineered mouse strains, a repertoire of mouse models can be created to study gene-specific effects on glioma genesis. This chapter provides details of plasmid construction, viral production, and intracranial delivery of transgenes, a methodology that can be easily adapted to a specific purpose.

Key words Gene transfer, Glioma, HIF-1α, Hypoxia, Nestin, RCAS/TVA

1 Introduction

There is a great utility to stably expressing or disrupting genes in specific cell types and cellular groups in biological research. Recombinant retroviruses have become one such means of genetic manipulation. The avian leukosis and sarcoma viral family contains the Rous sarcoma virus (RSV-A), which has been engineered to express transgenes up to 2.8 kb in size [1]. The integration of the modified RSV-A, known as RCAS (replication-competent ALSV long terminal repeat with a splice acceptor), with its cognate retroviral receptor TVA (tumor virus A) to be expressed in mammalian cells has created the RCAS/TVA system for gene function analysis in vivo [2]. With various available transgenic mice that express TVA under tissue-specific promoters, the RCAS/TVA system has been successfully used for the study of a variety of genes. Here we will discuss the use of the *nestin* promoter-driven TVA transgenic line (*Nes-TVA*) for investigating hypoxic effects on glioma genesis.

L. Eric Huang (ed.), *Hypoxia: Methods and Protocols*, Methods in Molecular Biology, vol. 1742,
https://doi.org/10.1007/978-1-4939-7665-2_20, © Springer Science+Business Media, LLC 2018

The RCAS/Nes-TVA system has the advantage of generating spontaneous gliomas in mice [3], and platelet-derived growth factor B (PDGFB) has been demonstrated to induce gliomas from low grade to high grade [4]. Furthermore, genetic crossing with commercially available mouse lines, e.g., *Hif1α* floxed or *HIF1A* conditionally expressed, can be used to expand the repertoire of the system for the study of tumor hypoxia in vivo. The major advantages of the RCAS/Nes-TVA system are that (1) it allows for efficient transfer of multiple genes into the TVA-expressing neural progenitors in the mouse brain through site-directed intracranial injection, a process taking a total of 4 weeks from virus production to in vivo transduction [2]; (2) only a small number of cells are targeted for gene expression or deletion in contrast to the conventional genetic approach that potentially affects the entire organ with tissue-specific promoters; (3) glioma development is within the normal tissues and in an immunocompetent environment; (4) RCAS/Nes-TVA has been used successfully especially in the study of glioblastoma [3]; and (5) mammalian cells are incompetent for RCAS replication and therefore incapable of viral spreading.

We describe here how to use the RCAS/Nes-TVA system to investigate the effect of HIF-1α on spontaneous glioma genesis driven by PDGFB. We have used a HIF-1α variant (HIF1αΔODD) that is deficient of the oxygen-dependent degradation domain [5, 6], which allows for stable expression of HIF-1α. We have also generated the *Nes-TVA;HIF1A^{lsl/lsl}* strain through genetic crossing with our *HIF1A^{lsl/lsl}* strain where a stable HIF-1α(P402A, P564A) [7] is controlled by a floxed stop cassette in the ROSA26 locus. In combination with the glioma driver PDGFB, these two systems allow for the study of glioma initiation and progression in the context of HIF-1α induction.

We begin by detailing the generation of RCAS vectors to express the gene of interest, production of high-titer retrovirus in the chicken fibroblast DF1 cells, and intracranial delivery of virus-producing cells into the TVA-positive glial progenitor cells. This methodology allows for effective manipulation of HIF-1α expression in spontaneously developed glioma, with an easy adaption to explore a variety of genes of interest in different scopes.

2   Materials

2.1 DF-1 Cell Culture and Retrovirus Production

1. Chicken fibroblast DF-1 cells.

2. CO_2 cell-culture incubator.

3. Biosafety class II laminar flow cabinet.

4. Inverted fluorescence microscope, with appropriate filters for fluorophores used.

5. 37 °C water bath.

6. DF-1 growth medium: Dulbecco's Modified Eagle's Medium (DMEM) composed of 4.9 g/L glucose, 4 mM L-glutamine, 1 mM sodium pyruvate, and 15 mg/L phenol red and supplemented with 10% v/v fetal bovine serum (FBS) and 1× minimum essential medium nonessential amino acid solution (MEM NEAA).

7. DF-1 freezing medium: 70% v/v DMEM, 20% FBS, 10% v/v dimethyl sulfoxide (DMSO).

8. Dulbecco's phosphate buffered saline (D-PBS) without calcium and without magnesium.

9. 0.25% Trypsin-EDTA.

10. Transfection reagent: Superfect (Qiagen) or others compatible with RCAS retroviral vector and DF-1 cells.

11. Gateway cloning vector system: Donor vector (pDNOR221 or others), BP clonase kit, and LR clonase kit (Invitrogen), and the destination vector RCAS-Y DV-A [8, 9].

12. Restriction enzymes.

13. DNA sequencing facility.

14. Plasmid extraction kit.

15. Tissue culture plates and flasks (100-mm dishes, T-25 flasks, T-75 flasks, 12-well plates).

16. Sterile 1.5-mL microcentrifuge tubes.

17. Sterile 1.8-mL cryotubes.

18. −1 °C/min cryotube freezing containers.

19. Sterile 15-mL conical tubes.

20. Sterile, filtered pipette tips (10, 20, 200, 1000 μL).

21. Sterile serological pipettes (2, 5, 10, 25 mL).

22. Pipettors (0.5–10, 2–20, 20–200, 100–1000 μL).

23. Disposable syringes for filtration.

24. 0.42-μm syringe filters.

2.2 Intracranial Gene Delivery

1. Nes-TVA mice: Tg(NES-TVA)J12Ech/J mice (Jackson Laboratory) or Nes-TVA genetically crossed with conditional inducible oxygen-stable HIF-1α mice (*LSL-HIF1dPA*) or conditional *Hif1a* knock-out mice (*Hif1a^flox^*).

2. Hank's balanced salt solution (HBSS).

3. Hemocytometer or automated cell counter.

4. Glass barrel syringe with a 0.2-mm gauge needle with limited dead volume from which a minimum of 0.5 μL and a maximum of at least 5 μL of sample can be accurately dispensed.

5. 70% ethanol.

6. Sterile 1.7-mL microcentrifuge tubes.

7. Sterile alcohol prep pads.

8. 10 mg/mL tamoxifen in sterile corn oil.

3 Methods

The following steps have been adapted from a previously published general protocol [2] and tailored to gene delivery into the murine neural progenitor cells.

3.1 RCAS Plasmid Construction

1. For convenience, we recommend the use of Gateway cloning by adding the *att*B1 and *att*B2 homologous recombination sequences to the 3′ and 5′ ends of the gene of interest via PCR amplification [8, 10]. (*see* **Notes 1** and **2**).

2. Clone the DNA fragment into a donor vector (e.g., pDNOR221) and subsequently shuttle into the RCAS-Y DV-A destination vector. For general information regarding retroviral vectors, refer to the general protocol [2] (Fig. 1).

3. Perform restriction enzyme digestions and DNA sequencing to confirm plasmid identity.

4. Prepare endotoxin-free, high-concentration ($\geq$500 ng/µL) plasmids using appropriate plasmid extraction kit of choice.

3.2 RCAS Virus Production

1. Thaw out a cryotube of parental DF-1 cells in a 37 °C water bath for 1 min. Transfer DF-1 cells to a 10-cm dish or T-75 flask with growth medium. Incubate the culture at 39 °C with 5% v/v CO_2 (*see* **Note 3**).

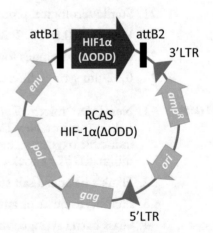

Fig. 1 Schematic diagram of RCAS-HIF1α(ΔODD) plasmid constructed through Gateway cloning. *att*B1 and *att*B2, homologous recombination sites; *amp*R, ampicillin resistance gene; *ori*, origin of replication; and LTR, long terminal repeats

2. DF-1 cells are expected to have a doubling time of about 48 h. Split cells as needed in ratios from 1:2 up to 1:10.

3. For each transfection, seed DF-1 2.5×10^5 cells in a T-25 flask. Include an additional for RCAS-GFP transfection control. Incubate the cells at 39 °C, 5% CO_2 v/v overnight (*see* **Note 4**).

4. Next day, perform individual transfection with RCAS vectors expressing of genes of interest and RCAS-GFP as transfection control according to the protocol of transfection reagent selected. The amount of plasmids also depends on the corresponding protocol (*see* **Note 3**).

5. On day 4–6 post transfection, expand cells to a T-75 flask if they become confluent. By day 7–10, 100% fluorescence of DF-1 GFP cells would indicate successful transfection and high-titer viral production.

6. By day 10–14, expand cells to a total of three T-75 flasks. If desired, determine viral titer beginning post-transfection day 9–11 (*see* **Note 4**). The cells are ready for injection beginning at day 10 and should not be used beyond day 30.

7. Freeze cell stocks beginning at day 10 up to day 21 by pelleting cells with centrifugation ($250 \times g$, 5 min at room temperature). Remove medium and resuspend the pellet in DF-1 freezing medium at a density of about 3×10^6 cells per mL. Aliquot 1 mL of the suspension to each 1.8-mL cryotube and transfer to a freezing container that has been warmed up to room temperature. Store the freezing container at −80 °C for overnight before transferring cryotubes to liquid nitrogen for long-term storage.

3.3 HIF-1a Overexpression in Spontaneous Glioma Genesis

Perform the following in compliance and with approval from relevant animal welfare regulatory agencies.

1. The transfected DF-1 cells between day 10 and day 30 are ready for injections. Alternatively, thaw out a frozen vial of DF-1 PDGFB, DF-1 HIF1α(ΔODD), and parental DF-1 or DF-1 GFP to 10-cm plates a week prior to the expected birth date of Nes-TVA (*see* **Notes 5** and **6**). Periodically monitor the DF-1 cells and split as needed prior to injection. Each 10-cm plate should yield approximately 10^7 DF-1 cells.

2. Once Nes-TVA mouse pups (day 0–2) become available, detach cells with trypsin dissociation.

3. Dissociate cells well before adding 10 mL of growth medium to stop the enzymatic reaction.

4. Transfer cell suspension into 15-mL conical tubes and centrifuge at $250 \times g$ for 5 min at room temperature.

5. Remove the medium and wash the cell pellet with 10 mL of D-PBS.

6. Count the DF-1 viral producer cells with a hemocytometer or with an automated cell counter.

7. Calculate the volume of cell suspension on the basis of $1-5 \times 10^4$ DF-1 PDGFB cells per injection. For DF-1 HIF1α(ΔODD), we recommend 2–5 times more cells to increase the probability of coinfection. The same number of parental DF-1 or DF-1 GFP cells is used as negative control of DF-1 HIF1α(ΔODD) (*see* **Notes 7–9**).

8. Centrifuge the combined cell suspension at $250 \times g$ for 5 min at room temperature.

9. Remove the D-PBS and resuspend the cell pellet in HBSS for an injection of both cell types between 2 and 3 μL (*see* **Note 8**).

10. Sterilize the glass barrel syringe to be used for neonatal injection by drawing 70% ethanol in and out of the entire length of the syringe at least 3 times. Draw sterile D-PBS in and out of the entire length of the syringe at least 3 times to rinse.

11. Draw in 2–3 μL of the cell suspension.

12. Hold the head of the pup between your thumb and index finger, with the mouse facing away, to stabilize the dorsal head region and wipe the injection site with an alcohol prep pad.

13. The tip of the syringe needle should be aimed at the lambda, at an approximately 90° angle from the surface. Target the cerebral cortex of either side by aiming toward the anterior and slightly to one side to inject the virus (Fig. 2).

14. Insert the needle to a depth of ~2 mm and quickly dispense the virus by pressing down the plunger.

15. Apply gentle pressure to the injection site for 5 s.

16. Depending on the purpose of the experiment and the nature of the injection, monitor mice for any signs of sickness (i.e., seizures, hydrocephalus, emaciation, lethargy) (*see* **Note 7**).

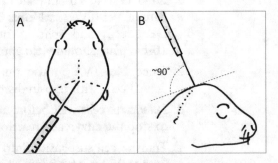

Fig. 2 Dorsal (**a**) and side (**b**) view of a Nes-TVA mouse neonate traced with relevant cranial sutures. The needle to be inserted at the lambda is directed to the right to inject the viral-producing DF-1 to the right cerebral cortex

3.4 Temporal Regulation of HIF-1α Expression During Glioma Genesis

Nes-TVA can be easily genetically crossed with LSL-HIF1dPA or Hif1a^{flox} mice to generate *Nes-TVA;LSL-HIF1dPA* or *Nes-TVA;Hif1a^{flox}* mice. The use of Cre recombinase variant CreER^{T2} and tamoxifen injection allows for temporal control of HIF-1α expression in the spontaneous PDGFB glioma model. Perform the following in compliance and with approval from relevant animal welfare regulatory agencies.

1. Proceed as indicated in Subheading 3.3, **steps 1–16**, and use 2×10^4 DF-1 PDGFB and 1×10^5 of tamoxifen-inducible CreER^{T2} for each mouse (*see* **Note 8**).

2. To activate HIF-1α expression or *Hif1a* disruption, 10 mg/mL tamoxifen in sterile corn oil is administered by intraperitoneal injection for 3–5 consecutive days. Sterile corn oil without tamoxifen serves as a control. While we have administered tamoxifen at 3 weeks post-viral infection, timing of tamoxifen administration and termination of the experiment will depend on the research question and experimental design.

4 Notes

1. Depending on the investigator's preference, fluorescent proteins and/or tags may be fused to or coexpressed with gene of interest as long as the entire insert is within 2.8 kb.

2. For the RCAS plasmid, use a high concentration (at least 500 ng/μL) of endotoxin-free plasmid. Supercoiled plasmids are required for the generation of high-titer viruses.

3. Use healthy, low-passage (<20) DF-1 cells. Check periodically for signs of stress (vacuoles, stasis) or contamination. Depending on the investigator's preference, antibiotics may be used, and this should not affect viral production. However, we recommend avoiding this as antibiotics can mask early signs of bacterial contamination.

4. Viral titration may be done to verify viral infectivity. At day 9–11, check if the T-75 flasks are about 80% confluent and replace the medium at a reduced volume (6–8 mL) to generate the viral supernatant. Seed 1×10^4 non-transfected DF-1 cells in a 12-well plate (eight wells for each virus generated, two additional for negative control). After 24 h, collect the viral supernatant and filter through a 0.42-μm syringe filter. Make a series of tenfold dilutions in DF-1 growth medium up to 10^{-8}. Remove the medium from the wells and replace with dilutions 10^{-5}, 10^{-6}, 10^{-7}, and 10^{-8} in duplicate. Replace the negative control wells with DF-1 growth medium. Incubate at 39 °C, 5% v/v CO_2 for 1 week, and assay with the appropriate method such as semiquantitative RT-PCR of gene of interest or fluo-

rescence microscopy. A producer cell line with viral titer $\geq 10^{-7}$ is considered ideal.

5. The ability to plan ahead of the birth of mice is crucial to timing the preparation of RCAS-producing DF-1. When setting up mating cages, we recommend weighing the female at day 0, around day 9–12, and finally day 17–20. We have found that a weight gain of at least 3 g at around day 10 is an indication of successful mating, thus allowing ample time to plan for an injection. Furthermore, obtaining the weight closer to the date of expectancy can help spot potential miscarriages and predict the size of the upcoming litter. While this may be dependent on the condition and nuances of the mouse strain used, we have found in Nes-TVA mice that an additional weight gain of 7 g or higher generally predicts a litter of eight or more pups.

6. Thawing transfected DF-1 cells at least a week prior to injection provides sufficient time for cells to recover, grow, and produce virus particles. Use of DF-1 cells that are ≥ 30 days from the date of transfection is not recommended because of reduced titer and possible accumulation of mutations on the gene of interest.

7. Depending on the specific purpose of the investigation, the number of DF-1 transfected cells to be injected per mouse may vary. For the purpose of generating PDGFB-induced glioma in Nes-TVA mice, we recommend using $1-10 \times 10^4$ of DF-1 with RCAS-PDGFB to generate grade II–III gliomas between 3 and 6 weeks (Fig. 3). A higher cell number may be used to minimize latency period or increase tumor grade. We have

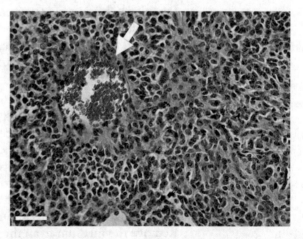

Fig. 3 Representative hematoxylin–eosin stained section of the cerebral cortex of a Nes-TVA mouse injected with 3×10^4 DF-1 PDGFB and 1×10^5 DF-1 HIF1$\alpha\Delta$ODD at day 0 and terminated at 6 weeks. Note hypercellularity, nuclear pleomorphism, and hyperchromasia of the tumor tissue. Intratumoral hemorrhage is indicated. Scale bar: 50 μm

observed that Nes-TVA mice are able to tolerate injections of *HIF1A(ΔODD)* at 10^5 DF-1 cells per injection. Tamoxifen injection at 3 weeks post injection or later ensures glioma genesis before induction or deletion of HIF-1α. For injections that may lead to tumor acceleration, we lean toward injecting a lower number of DF-1 PDGFB to produce grade II gliomas to give room for the tumor to progress further following tamoxifen injection.

8. When deciding the amount of two types of viral-producing DF-1 cells for injection, note that the ratio of each type injected can affect the degree of neural progenitor cells that are coinfected. For example, a ratio of 1:5 of DF-1 PDGFB to DF-1 CreER[T2] would increase the likelihood that PDGFB-infected neural progenitor cells are also infected with CreER[T2].

9. Nes-TVA is a randomly integrated transgene and may be lost at some point in a subset of pups in each litter. We recommend genotyping for the presence of the *TVA* gene using extracted genomic DNA from ear or tail clippings.

Acknowledgments

This work was supported in part by an NIH Grant CA084563 from the National Cancer Institute and by the University of Utah Funding Incentive Seed Grant. The authors wish to thank Kristin Kraus for editorial assistance.

References

1. Federspiel MJ, Hughes SH (1997) Retroviral gene delivery. Methods Cell Biol 52:179–214

2. von Werder A, Seidler B, Schmid RM et al (2012) Production of avian retroviruses and tissue-specific somatic retroviral gene transfer in vivo using the RCAS/TVA system. Nat Protoc 7:1167–1183

3. Holland EC, Varmus HE (1998) Basic fibroblast growth factor induces cell migration and proliferation after glia-specific gene transfer in mice. Proc Natl Acad Sci U S A 95:1218–1223

4. Dai C, Celestino JC, Okada Y et al (2001) PDGF autocrine stimulation dedifferentiates cultured astrocytes and induces oligodendrogliomas and oligoastrocytomas from neural progenitors and astrocytes in vivo. Genes Dev 15:1913–1925

5. Huang LE, Gu J, Schau M et al (1998) Regulation of hypoxia-inducible factor 1alpha is mediated by an O2-dependent degradation domain via the ubiquitin-proteasome pathway. Proc Natl Acad Sci U S A 95:7987–7992

6. Koshiji M, Kageyama Y, Pete EA et al (2004) HIF-1α induces cell cycle arrest by functionally counteracting Myc. EMBO J 23:1949–1956

7. Kageyama Y, Koshiji M, To KKW et al (2004) Leu-574 of human HIF-1alpha is a molecular determinant of prolyl hydroxylation. FASEB J 18:1028–1030

8. Loftus SK, Larson DM, Watkins-Chow D et al (2001) Generation of RCAS vectors useful for functional genomic analyses. DNA Res 8:221–226

9. Holmen SL, Williams BO (2005) Essential role for Ras signaling in glioblastoma maintenance. Cancer Res 65:8250–8255

10. Katzen F (2007) Gateway®recombinational cloning: a biological operating system. Expert Opin Drug Discovery 2:571–589

Chapter 21

In Vitro Assays of Breast Cancer Stem Cells

Debangshu Samanta and Gregg L. Semenza

Abstract

Aldehyde dehydrogenase and mammosphere assays enable the cost-effective quantification and characterization of cancer stem cells (CSCs) from cancer cell lines as well as cancer tissue. Here we describe the quantification of CSCs in breast cancer cell lines using aldehyde dehydrogenase and mammosphere assays under hypoxic (1% O_2) and non-hypoxic (20% O_2) culture conditions. Using this method, a significant enrichment of CSCs compared to bulk populations is observed when breast cancer cells are exposed to 1% O_2 for 72 h.

Key words Cancer stem cells (CSC), Tumor-initiating cells, Breast cancer, Hypoxia, Self-renewal, FACS, Hypoxia-inducible factors (HIFs)

1 Introduction

Breast cancer mortality usually occurs when cancer cells metastasize and become resistant to chemotherapy. Breast cancer stem cells (BCSCs), which are also called tumor-initiating cells, represent a subpopulation that is central to both of these processes. Although many breast cancer cells enter the circulation, only BCSCs are capable of forming a secondary tumor [1, 2]. Intratumoral hypoxia is common in most solid tumors, including breast, cervical, and head/neck cancers [3]. Hypoxia induces metastatic properties and the BCSC phenotypes through transcriptional activation of target genes by hypoxia-inducible factor 1 (HIF-1) and HIF-2 [4]. The gold standard assay of CSC self-renewal is the demonstration that in vivo transplantation with limiting cell numbers results in tumor formation in recipient mice, an assay that is time-consuming and expensive. Reliable in vitro assays are necessary to efficiently and cost-effectively quantify stem cells. In this chapter, we describe two methods that are used routinely to quantify BCSCs, namely, the aldehyde dehydrogenase [5] and mammosphere [6] assays. These assays identify populations of cells that are enriched by several orders of magnitude for tumor-initiating cells as compared to the bulk cancer cell population.

L. Eric Huang (ed.), *Hypoxia: Methods and Protocols*, Methods in Molecular Biology, vol. 1742,
https://doi.org/10.1007/978-1-4939-7665-2_21, © Springer Science+Business Media, LLC 2018

One widely accepted method for identifying cancer stem cells in vitro is based on the enzymatic activity of aldehyde dehydrogenase 1 (ALDH1) [7], which was first used to identify hematopoietic stem cells and which may regulate stem cell differentiation through the metabolism of retinol to retinoic acid [8, 9]. The commercially available ALDEFLUOR™ assay employs BODIPY-aminoacetaldehyde (BAAA), which freely diffuses into intact and viable cells and, in the presence of ALDH1, is converted to the fluorescent product BODIPY-aminoacetate, which is retained inside the cells. The amount of fluorescent reaction product is proportional to the ALDH1 activity in the cells and is measured using a flow cytometer. The assay can be used to isolate a subpopulation of cells that display stem cell properties from normal human breast tissue or breast carcinomas [5].

The mammosphere assay is a modification of the neurosphere assay, which was developed to quantify neural stem cells [10]. Dontu and colleagues developed an in vitro culture system that allowed for propagation of human mammary epithelial cells (HMECs) under non-adherent culture conditions [6, 11]. Cells capable of surviving and proliferating in such conditions formed discrete clusters of cells termed "mammospheres." Such spheroids were enriched in progenitor cells capable of differentiating along multiple lineages, including luminal, myoepithelial, and alveolar cells. Secondary mammospheres generated from harvesting, digesting primary mammospheres into single cells, and replating can be used to demonstrate self-renewal capacity [12].

Several other assays have been used to identify or isolate cells with CSC properties, such as side population cells, which are characterized by the exclusion of Hoechst dye, and CD44$^+$CD24$^-$ cells, which mark BCSCs in luminal-type breast cancers [13–15]. However, in basal-type cancers and cell lines, the majority of cells are CD44$^+$CD24$^-$, such that this assay is not a reliable marker for BCSCs.

Hypoxia has been shown to increase the percentage of BCSCs, both in vitro and in vivo through the activity of HIF-1, which activates the transcription of multiple genes that promote the BCSC phenotype [16–22]. Hypoxia may also induce the BCSC phenotype via HIF-1-independent mechanisms [23, 24].

2   Materials

Prepare all solutions using ultrapure water (prepared by purifying deionized water to attain a sensitivity of 18 MΩ cm at 25 °C) and analytical grade reagents. Prepare and store all reagents at room temperature (unless indicated otherwise). Diligently follow all waste disposal regulations when disposing waste materials. We do not add sodium azide to reagents.

2.1 Media   All breast cancer cells are cultured in their recommended media as published in [15].

1. DMEM (1×).

2. RPMI-1640 (1×).

3. DMEM/F12 (50:50).

4. Penicillin/Streptomycin.

5. Heat-inactivated BenchMark fetal bovine serum (FBS).

6. Hank's balanced salt solution (HBSS) containing 2% FBS.

7. 0.05% Trypsin-EDTA (1×).

8. Prepare 1 L of 1× PBS as follows: Start with 800 mL of distilled water and then add the following: 8 g of NaCl, 0.2 g of KCl, 1.44 g of Na_2HPO_4, and 0.24 g of KH_2PO_4. Adjust the pH to 7.4 with HCl. Finally, add distilled water to a total volume of 1 L. Autoclave and cool to room temperature before use.

9. Add appropriate amount of FBS (5–10% v/v, depending on the cell line) and penicillin/streptomycin (1% v/v) to the medium and store at 4 °C.

10. Prior to use, warm the medium and trypsin in a water bath maintained at 37 °C for 10 min.

2.2 ALDH Assay   1. ALDEFLUOR Kit (StemCell Technologies).

2. 5-mL polystyrene round-bottomed tube with cell strainer (35 μm) cap (Corning).

2.3 Mammosphere Assay   1. MammoCult™ Medium (StemCell Technologies).

2. Ultralow adherence dishes (6-well plates).

3. Heparin.

4. Hydrocortisone.

2.4 Hypoxia   1. Modular Incubator Chamber.

2. Gas cylinders containing mixture of 5% CO_2, 1% O_2, and balance N_2.

3 Methods

Carry out all procedures at room temperature unless otherwise specified.

3.1 ALDEFLUOR Kit Preparation

1. Allow kit reagents to come to room temperature (15–25 °C) before use for the first time.

2. Add 25 µL of DMSO to the ALDEFLUOR reagent bottle, pipette up and down, and let it stand for 1 min at room temperature (*see* **Note 1**).

3. Add 25 µL of 2 N HCl (supplied) to the ALDEFLUOR reagent vial and pipette up and down several times (*see* **Note 2**).

4. Incubate this mixture at room temperature for 15 min.

5. Add 360 µL of ALDEFLUOR Assay Buffer (supplied) to the vial and mix well making the total volume of ALDEFLUOR reagent 410 µL.

6. Aliquot the ALDEFLUOR reagent into separate microcentrifuge tubes (depending on your experimental plans) and store at −20 °C. Avoid repeated freezing and thawing of aliquots (*see* **Note 3**).

3.2 Preparation of Complete Mammosphere Medium

1. Prepare complete MammoCult Medium (Human) by adding 50 mL of thawed MammoCult Proliferation Supplements to 450 mL of MammoCult Basal Medium.

2. Add the following to obtain complete medium:

   (a) Heparin to 4 µg/mL (0.0004%).

   (b) Hydrocortisone to 0.48 µg/mL (add 100–500 mL of medium).

3.3 Exposure of Cells to Hypoxia

1. Plate appropriate number of cells so that they are 85–90% confluent at time of harvesting. Determine the number of cells before so that after induction of hypoxia for 72 h, the cells are not overconfluent (*see* **Note 4**).

2. The next day aspirate culture medium from the dish and replenish with fresh medium appropriate to each cell line.

3. Induce hypoxia in the cells by placing them in the modular incubator. All components of the modular incubator are non-toxic and can be alcohol-sterilized.

4. In one of the two outlets of the modular incubator, connect with the nozzle from cylinder with air containing 5% CO_2, 1% O_2, and balance N_2. Keep the other outlet free so that the air goes out from the other nozzle. The air pressure in the cylinder should not exceed 2 p.s.i. Continue flushing the air in the modular incubator containing 20% O_2 for 3 min.

5. First close the outlet of the modular chamber which was letting the air out. Only after you have secured this outlet, close the inlet quickly (*see* **Note 5**).

6. Seal the Chamber and place at 37 °C in your laboratory incubator or warm room.

7. For checking the BCSC phenotype, incubate the cells under hypoxia for 72 h. Afterward, take out the modular incubator. First release either one of the two valves. When the pressure is released, release the stainless steel ring clamp and open the incubator (*see* **Note 6**).

3.4 Cell Sample Preparation

1. Aspirate culture medium from the dish and wash the cells with 1× PBS at room temperature (*see* **Note 7**).

2. Add 1× 0.05% trypsin to cover the entire flask. Gently rock the flask to get complete coverage of the cell layer (*see* **Note 8**).

3. Place the flask in the tissue culture incubator for 2 min.

4. Observe the cells under the microscope for detachment. If cells are less than 90% detached, increase the incubation time a few more minutes, checking for dissociation every 30 s. You may also tap the vessel to expedite cell detachment.

5. Add a volume of medium containing FBS that is at least 5 times the volume of trypsin to stop the trypsin from acting.

6. Homogenize the mixture by pipetting up and down several times.

7. Determine the total number of cells and percent viability using a hemocytometer, cell counter and trypan blue exclusion, or the Countess automated cell counter (Invitrogen).

3.5 ALDEFLUOR Assay

1. Aliquot from 2.5×10^5 to 1×10^6 cells into pre-labeled microcentrifuge tubes. (*see* **Notes 9** and **10**).

2. Label one "test" and one "control" tube for each sample to be tested.

3. Label an additional tube as the unstained as negative control.

4. Centrifuge the cells $250 \times g$ at room temperature for 5 min.

5. Aspirate the supernatant and wash the cells with 1 mL of PBS.

6. Centrifuge the cells $300 \times g$ at room temperature for 5 min and aspirate the supernatant, ensuring that the cell pellet is not disturbed. Carefully aspirate and discard the supernatant with a pipette.

7. Dissolve the pellet in 500 μL of ALDEFLUOR Assay Buffer.

8. For 2.5×10^5 cells, add 1 μL of the activated ALDEFLUOR reagent to each tube (*see* **Note 11**).

9. For the control tubes, add 2 μL of ALDEFLUOR DEAB reagent (*see* **Note 12**).

10. Place the cells in a tissue culture incubator at 37 °C for 45 min (*see* **Note 13**).

11. Following incubation, centrifuge all tubes for 5 min at $250 \times g$ and remove the supernatant. Resuspend the cell pellets in 0.5 mL of ALDEFLUOR™ Assay Buffer.

12. Pass the cells through the cell strainer into the 5-mL polystyrene round-bottomed tube (*see* **Note 13**).

13. Analyze the cells using a flow cytometer.

3.6 Flow Cytometer Setup

1. Create a forward scatter (FSC) vs. side scatter (SSC) dot plot.

2. In setup mode, place your unstained tube on the cytometer.

3. Adjust the FSC and SSC voltage, such that the majority of the nucleated cell population is at the center of the FSC vs. SSC plot.

4. Gate on all nucleated cells, excluding debris (R1).

5. Create a FITC vs. SSC dot plot, gated on R1.

6. Adjust the FITC photomultiplier tube voltage such that there is significant shift along the FITC axis between the unstained and DEAB stained sample.

7. Align the rightmost edge of the stained DEAB control population with the second log decade on the FITC axis.

8. Keep the FSC, SSC, and FITC voltages constant for the remaining of the experiment.

9. Change the cytometer from setup to active mode, and collect at least 50,000 events in R1 for each DEAB and sample tube, without changing the instrument settings (Fig. 1).

3.7 Primary Mammosphere Assay

1. Resuspend a predetermined number of cells in 2 mL of complete MammoCult Medium (in triplicate) in each well of a 6-well ultralow adherent plate. The seeding density for cell lines is typically in the range of 5000–20,000 cells per well.

2. Incubate cultures in a 5% CO_2, humidified incubator at 37 °C for 7 days without replenishing the medium (*see* **Note 14**).

3. Photograph the mammospheres under an Olympus TH4-100 microscope with 4× apochromat objective lens (Fig. 2).

4. Determine the mammosphere number and volume using ImageJ software.

5. Count the mammospheres with area > 500 pixels in images of three fields per well in triplicate wells and determine the mean number of mammospheres per field (*see* **Note 15**).

3.8 Secondary Mammosphere Assay

1. Harvest mammospheres after 7 days in culture. Collect the entire culture from three wells into a 15-mL conical tube and centrifuge at $87 \times g$ for 5 min.

2. Aspirate as much supernatant as possible with a pipette without disturbing the pellet.

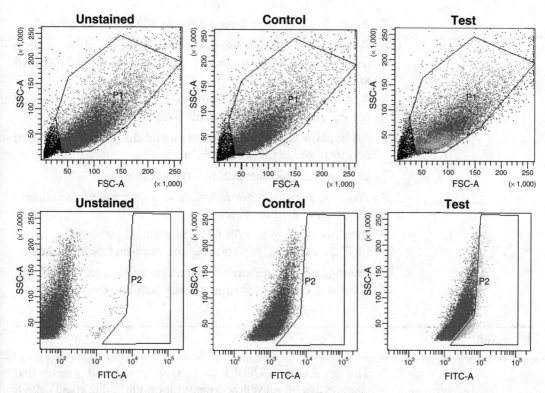

Fig. 1 Typical FACS plot of ALDEFLUOR Assay. In the top panel shows the plot of Forward Scatter Cells (FSC) vs. Side Scatter Cells (SSC). Only the single cell population is gated P1 for further analysis. In the bottom panel, the gated cells are plotted as SSC vs. FITC, since excitation wavelength of the activated ALDEFLUOR reagent is 488 nm which corresponds to FITC. Note how in the middle panel (control) the gate P2 is put at the rightmost edge of the population. Any cells having more fluorescence would be scored positive. The ALDEFLUOR positive cells are colored green

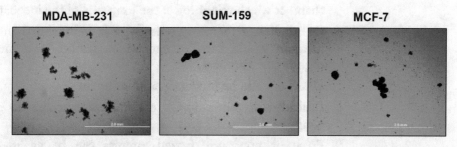

Fig. 2 Micrographs demonstrating the characterization of mammospheres derived from breast cancer cell lines MDA-231, SUM-159 and MCF-7, (4×) 7 days after plating. 5000 MDA231; SUM159 and 10000 MCF7 cells were plated. All scale bars are 2 mm

3. Add 0.5 mL of pre-warmed Trypsin-EDTA for each tube.

4. Break up the mammospheres into individual cells by passage in and out of a 25-gauge syringe three times (*see* **Notes 16** and **17**).

5. Add 5 mL of cold HBSS containing 2% FBS and centrifuge the cell suspension at 87 × *g* for 5 min.

6. Aspirate the supernatant and resuspend the pellet in 0.5 mL of complete MammoCult Medium.

7. Perform the cell count using Trypan Blue.

8. Plate the same number of live cells as was plated to generate the primary mammospheres.

9. Incubate cultures in a 5% CO_2, humidified incubator at 37 °C for 7 days without replenishing the medium (*see* **Note 14**).

10. Photograph and quantify the number of secondary mammospheres as described for the primary mammospheres.

4   Notes

1. The dry ALDEFLUOR reagent is an orange-red powder that changes to a bright yellow-green color upon addition of DMSO.

2. 2 N HCl must be added after addition of DMSO.

3. When frozen aliquots (−20 °C) of the activated ALDEFLUOR reagent are thawed, a small precipitate may be observed. Before use, mix the thawed reagent to resuspend the precipitate. This precipitate does not affect assay performance.

4. Confluent cultures will have increased cell death.

5. If you are not fast, then the pressure might build up in the chamber which might lower the longevity of the incubator.

6. One indicator for the proper hypoxia induction is that the medium would turn orange which denotes the medium slightly turns acidic.

7. HBSS can also be used for washing the cells.

8. Instead of trypsinization, use a sterile cell scraper to gently scrape cells from the dish.

9. Fresh or previously frozen samples can be analyzed for ALDEFLUOR assay. However, ALDH activity will only be detected in viable cells. The assay can also be performed on blood samples and hematopoietic cells (e.g., peripheral blood, apheresis product, and bone marrow or cord blood).

10. If using blood samples where the red blood cell to leukocyte ratio (RBC/WBC) of the specimen is >2:1, lyse the erythrocytes with ammonium chloride solution (StemCell Technologies).

11. For human breast cancer cell lines MDA-MB-231, MCF-7, SUM-159, and SUM-149, this volume of ALDEFLUOR reagent works well. For other cell lines, the optimal volume of ALDEFLUOR reagent used for the experiment should be determined.

12. DEAB (diethylaminobenzaldehyde) is a specific inhibitor of ALDH. It is used to control for background fluorescence.

13. Optimal incubation times may vary between different cell types and should be determined. Do not exceed an incubation time of 60 min.

14. After dissolving the cells in ALDEFLUOR buffer, take measurement as soon as possible. Keep the cells chilled (2–8 °C or on ice) until measurement to slow down the product efflux.

15. The number of days in culture must be kept to a minimum and optimized for each cell type. Plates must not be moved during the culture period.

16. Primary and secondary mammospheres can also be quantitated using the size of the mammospheres, such as greater than 50-μm diameter.

17. More sensitive cells may simply require gentle pipetting rather than syringe passage to obtain a single cell suspension. Make sure that the cells are not overexposed to trypsin.

Acknowledgments

Cancer research in the authors' laboratory is supported by grants from the American Cancer Society, Armstrong Family Foundation, Department of Defense Breast Cancer Research Program, and the Cindy Rosencrans Fund for Triple Negative Breast Cancer. G.L.S. is an American Cancer Society Research Professor and the C. Michael Armstrong Professor at the Johns Hopkins University School of Medicine.

References

1. Al-Hajj M, Wicha MS, Benito-Hernandez A et al (2003) Prospective identification of tumorigenic breast cancer cells. Proc Natl Acad Sci U S A 100(7):3983–3988

2. Oskarsson T, Batlle E, Massague J (2014) Metastatic stem cells: sources, niches, and vital pathways. Cell Stem Cell 14(3):306–321

3. Vaupel P, Hockel M, Mayer A (2007) Detection and characterization of tumor hypoxia using pO₂ histography. Antioxid Redox Signal 9(8):1221–1235

4. Semenza GL (2015) The hypoxic tumor microenvironment: a driving force for breast cancer progression. Biochim Biophys Acta 1863(3):382–391

5. Ginestier C, Hur MH, Charafe-Jauffret E et al (2007) ALDH1 is a marker of normal and malignant human mammary stem cells and a predictor of poor clinical outcome. Cell Stem Cell 1(5):555–567

6. Dontu G, Abdallah WM, Foley JM et al (2003) In vitro propagation and transcriptional profil-

ing of human mammary stem/progenitor cells. Genes Dev 17(10):1253–1270

7. Jones RJ, Barber JP, Vala MS et al (1995) Assessment of aldehyde dehydrogenase in viable cells. Blood 85(10):2742–2746

8. Chute JP, Muramoto GG, Whitesides J et al (2006) Inhibition of aldehyde dehydrogenase and retinoid signaling induces the expansion of human hematopoietic stem cells. Proc Natl Acad Sci U S A 103:11707–11712

9. Ghiaur G, Yegnasubramanian S, Perkins B et al (2013) Regulation of human hematopoietic stem cell self-renewal by the microenvironment's control of retinoic acid signaling. Proc Natl Acad Sci U S A 110(40):16121–16126

10. Pastrana E, Silva-Vargas V, Doetsch F (2011) Eyes wide open: a critical review of sphere-formation as an assay for stem cells. Cell Stem Cell 8(5):486–498

11. Dontu G, Al-Hajj M, Abdallah WM et al (2003) Stem cells in normal breast development and breast cancer. Cell Prolif 36(Suppl 1):59–72

12. Shaw FL, Harrison H, Spence K et al (2012) A detailed mammosphere assay protocol for the quantification of breast stem cell activity. J Mammary Gland Biol Neoplasia 17(2):111–117

13. Patrawala L, Calhoun T, Schneider-Broussard R et al (2005) Side population is enriched in tumorigenic, stem-like cancer cells, whereas ABCG2+ and ABCG2− cancer cells are similarly tumorigenic. Cancer Res 65:6207–6219

14. Christgen M, Ballmaier M, Bruchhardt H et al (2007) Identification of a distinct side population of cancer cells in the Cal-51 human breast carcinoma cell line. Mol Cell Biochem 306:201–212

15. Fillmore CM, Kuperwasser C (2008) Human breast cancer cell lines contain stem-like cells that self-renew, give rise to phenotypically diverse progeny and survive chemotherapy. Breast Cancer Res 10:R25

16. Conley SJ, Gheordunescu E, Kakarala P et al (2012) Antiangiogenic agents increase breast cancer stem cells via the generation of tumor hypoxia. Proc Natl Acad Sci U S A 109980:2784–2789

17. Xiang L, Gilkes DM, Hu H et al (2014) Hypoxia-inducible factor 1 mediates TAZ expression and nuclear localization to induce the breast cancer stem cell phenotype. Oncotarget 5(24):12509–12527

18. Zhang H, Lu H, Xiang L et al (2015) HIF-1 regulates CD47 expression in breast cancer cells to promote evasion of phagocytosis and maintenance of cancer stem cells. Proc Natl Acad Sci U S A 112(45):E6215–E6223

19. Brooks DL, Schwab LP, Krutilina R et al (2016) ITGA6 is directly regulated by hypoxia-inducible factors and enriches for cancer stem cell activity and invasion in metastatic breast cancer models. Mol Cancer 15:26

20. Samanta D, Park Y, Andrabi SA et al (2016) PHGDH expression is required for mitochondrial redox homeostasis, breast cancer stem cell maintenance, and lung metastasis. Cancer Res 76(15):4430–4442

21. Zhang C, Samanta D, Lu H et al (2016) Hypoxia induces the breast cancer stem cell phenotype by HIF-dependent and ALKBH5-mediated m^{6}A-demethylation of NANOG mRNA. Proc Natl Acad Sci U S A 113(14):E2047–E2056

22. Zhang C, Zhi WI, Lu H et al (2016) Hypoxia-inducible factors regulate pluripotency expression by ZNF217- and ALKBH5-mediated modulation of RNA methylation in breast cancer cells. Oncotarget 7(40):64527–64542. 10.18632/oncotarget.11743

23. Van den Beucken T, Koch E, Chu K et al (2014) Hypoxia promotes stem cell phenotypes and poor prognosis through epigenetic regulation of DICER. Nat Commun 5:5203

24. Iriondo O, Rabano M, Domenici G et al (2015) Distinct breast cancer stem/progenitor cell populations require either HIF-1α or loss of PHD3 to expand under hypoxic conditions. Oncotarget 6(31):31721–31739

Chapter 22

Fluorescence-Activated Cell Sorting of Murine Mammary Cancer Stem-Like Cell Subpopulations with HIF Activity

Danielle L. Brooks and Tiffany N. Seagroves

Abstract

Fluorescence-activated cell sorting (FACS) is a common method to identify and to isolate subpopulations within a complex mixture of cells based on their light scatter and fluorescent staining profiles. FACS is widely used to enrich for normal tissue and tumor cells that have stem cell potential. Whereas FACS protocols using conventional breast cancer cell lines are relatively routine, additional technical challenges are encountered when sorting for cell populations from freshly digested solid tumors, particularly for use in downstream cancer stem cell (CSC) assays. First, it is more difficult to isolate live, single cells from whole tumors, and second, single tumor cells prepared from enzymatically digested tumors are typically more sensitive to cell death following the physical stresses of digestion, pipetting, and sorting. Herein methods are described that have been optimized to harvest and to FACS profile viable tumor epithelial cells digested from late-stage mammary tumors originating in the mouse mammary tumor virus (MMTV)-polyomavirus middle T antigen (PyMT) transgenic mouse. Protocols were designed to enrich for single, viable, MMTV-PyMT tumor cell populations sorted by FACS and to facilitate the collection of sorted cell subpopulations suitable for head-to-head comparison of CSC activity by tumorsphere assays in vitro or limiting dilution transplantation in vivo.

Key words Fluorescence-activated cell sorting (FACS), Antibody, Staining, Single cell, Tumor, Polyomavirus middle T antigen (PyMT), Hypoxia-inducible factor (HIF), Tumor epithelial cells, Tumor-initiating cells (TICs)

1 Introduction

A key question in solid tumor biology is how does the tumor microenvironment promote progression and metastasis? An increasing body of evidence indicates that hypoxic oxygen tensions, in general, and the Hypoxia-Inducible Factor (HIF) transcription factors, specifically, enhance cancer stem cell (CSC) activity in solid tumors, including breast cancers [1–3]. Breast cells are highly heterogeneous and are comprised of cells spanning the spectrum of the differentiation cascade—from stem cells to trans-amplifying cells, to lineage-specific precursor cells, and to fully differentiated cells [4, 5]. In general, the same markers used to profile

L. Eric Huang (ed.), *Hypoxia: Methods and Protocols,* Methods in Molecular Biology, vol. 1742,
https://doi.org/10.1007/978-1-4939-7665-2_22, © Springer Science+Business Media, LLC 2018

stem cell and lineage-committed cells in the normal breast have also been applied to breast cancers to enrich for cells with CSC-like phenotypes. In general, relative to "bulk" tumor cells, cells with CSC-like activity represent a relatively rare fraction of tumor cells that have enhanced potential to self-renew, to form multiple cell lineages, and to reinitiate tumors when transplanted into recipient mice. Like hypoxic cells, CSCs are more refractory to radiation and to chemotherapy than bulk tumor cells, suggesting direct links between the hypoxic response and CSC biology [6–9].

FACS was first used to enrich for breast CSCs in human patient xenografts profiled for CD44 and CD24, which revealed that the CD44^{high}/CD24$^{-/neg}$/hematopoietic lineage (Lin^{neg}) population contained tumor-initiating cell (TIC) activity [10]. As additional cell surface markers, or combinations of markers, were reported to refine enrichment of breast stem cells and breast epithelium lineage progenitors, it became clear that the markers used to detect CSC-like cells in human breast cancers often differed in murine tumors [11]. For example, sorting for CD44^{High}/CD24$^{-/low}$ tumor cells did not enrich for TIC activity in the Balb/C p53 knockout (KO) mammary tumor model; rather, TIC activity was highest in the CD29^{High}/CD24^{High} double-positive population [12]. In particular, unlike human breast cancers, in murine models, the CD24$^{+/high}$ cells, rather than CD24$^{-/low}$ cells, typically exhibit the greatest TIC potential, including MMTV-PyMT tumors that show expansion of CD24^{high}/CD49f$^+$ luminal progenitor cells [12–15]. Replicating published FACS protocols can be arduous, particularly since policies for the depth of detail required in the materials and methods section varies among journals. It is important to acknowledge that there is variability in procedures and in FACS equipment used from lab to lab, which is further complicated by the process of setting gates, which may be a subjective process. To address these concerns, recommendations for developing standardized FACS protocols for profiling solid tumors have been developed [16].

The MMTV-PyMT transgenic mouse is a widely used, well-characterized model that quickly and reproducibly generates aggressive, metastatic breast cancer [17], yet, characterization of breast CSC activity in MMTV-PyMT tumors lagged behind other transgenic models of breast cancer, approximately 5 years after Al-Hajj et al. [2, 14, 15]. We had previously shown that Cre/loxP-mediated conditional deletion of *Hif1α* in PyMT tumors inhibited lung metastasis [18]. However, since the expression of the MMTV-Cre transgene is mosaic [19], it could not be determined if the few lung metastases observed in animals bearing *Hif1α*-deleted mammary tumors were derived from non-recombined tumor cells [18]. Therefore, we refined our approach to delete *Hif1α* activity by ex vivo transduction of late-stage MMTV-PyMT mammary tumor cells established in culture to generate HIF-1 wild-type (WT) and knockout (KO) lines exposed to adenovirus-ß-galactosidase or adenovirus-Cre recombinase, respectively [2]. The HIF-1

WT and KO MMTV-PyMT tumor cells were then used to characterize the role of HIF-1 in tumorsphere formation in vitro and in TIC activity in vivo by limiting dilution transplantation. Procedures were optimized for the digestion, staining and flow sorting of MMTV-PyMT transgenic tumors, and HIF-1 WT and KO tumors derived from transplantation of these PyMT cell lines, in order to reproducibly measure differences in stem-like activities between HIF-1 WT and KO cells in downstream assays that require input of single, viable cells.

2   Materials

Prepare all solutions using ultrapure water (distilled water, dH_2O) and cell culture grade reagents. Sterilize all solutions through a 0.22 μm filter unit prior to storage at 4 °C. We do not add sodium azide to any reagents, as it may interfere with antibody staining.

2.1   Tumor Mincing

1. 50 mL conical tubes.

2. 15 cm tissue culture dish.

3. Surgical scissors.

4. Three #10 scalpels taped together.

5. Three to five razor blades taped together.

6. Sterile 250 mL flask.

7. Sterile spatula.

8. Autoclaved aluminum foil.

9. Filter cap strainers: 70 and 40 μm sizes (these fit into standard 50 mL centrifuge tubes).

10. Trypan blue solution, 0.4% in 1× PBS (prepared solutions are commercially available).

2.2   FACS Sample Preparation, Staining, and Collection Tubes

1. 2 mL round-bottom microcentrifuge tubes for antibody staining for control samples.

2. 5 mL polystyrene round-bottom FACS tubes with a cell strainer cap (for processing samples).

3. 5 mL polypropylene round-bottom tubes with cap (for staining sort samples and collecting sorted cells.

4. 5 mL polystyrene round-bottom tubes (without cap, for post-sort analysis).

2.3   FACS Instrumentation

1. Access to a FACS instrument: the cell sorter should be capable of detecting fluorophores in the UV/violet to near-infrared range, and performing two-way or four-way sorts collected into standard FACS tubes. For the methods we describe, all conditions were optimized using a BD Biosciences FACSAria II cell sorter (*see* **Note 1**).

2. When sorting tumor cells prepared from digested mammary tumors from MMTV-PyMT transgenic mice, optimal survival post-sorting was achieved using a nozzle pore size of 100 μm with a flow rate of 2500 events/s (*see* **Note 2**).

3. Use FACSFlow sheath buffer if using an Aria machine. Otherwise, a standard phosphate-buffered saline (PBS) solution containing EDTA/EGTA, as well as antibacterial agents (penicillin/streptomycin), should be used.

2.4 Buffers and Reagents for Tumor Digestion, Red Blood Cell Lysis, and Sample Processing

1. **Tumor digestion buffer**: DMEM/F12 cell culture medium containing gentamycin sulfate to final concentration of 100 μg/mL, antibiotic-antimycotic (AA) solution (Sigma-Aldrich, A5955), or penicillin/streptomycin (P/S) to 1× final concentration. Prepare 500 mL of stock buffer without collagenase, hyaluronidase (Sigma), and heat-inactivated fetal bovine serum (FBS-HI). Store the sterile digestion buffer at 4 °C for up to 6 months. The volume of buffer used to digest tumors is based on tumor wet weight prior to mincing (10 mL/g). After minced tumor tissue is added to the digestion buffer, collagenase III (from *Clostridium histolyticum*, Worthington Biochemical) and hyalurondiase (from bovine testes, Sigma-Aldrich, H3384) enzymes will be added to final concentrations of 300 U/mg and 100 U/mg, respectively (*see* **Note 3**), and the digestion buffer will then be supplemented with heat-inactivated fetal bovine serum (FBS-HI) to a final concentration of 2.5%.

2. **Wash buffer** (Hank's balanced salt solution, HBSS+): 500 mL HBSS containing Ca^{2+} and Mg^{2+}, 10 mM HEPES, and 2% FBS-HI. Store sterile wash buffer at 4 °C for up to 6 months.

3. **Red blood cell lysis buffer**: Prepare a 10× (0.8%) stock solution of ammonium chloride (NH_4Cl) by dissolving 8.02 g of NH_4Cl into 100 mL of dH_2O. Store sterile lysis buffer at 4°C for up to 6 months. Immediately prior to use, prepare a working solution by diluting the 0.8% NH_4Cl stock with HBSS+ at a 4:1 ratio (e.g., 30 mL of 0.8% NH_4Cl to 10 mL HBSS+) in order to lyse red blood cells present in the tumor epithelium fraction of digested tumor tissue.

4. **Trypsin**: 0.25% trypsin/EDTA

5. **Dispase/DNase I solution**: DMEM/F12 cell culture medium, dispase II (neutral-protease, grade II, Sigma-Aldrich, cat # 4942078001) to final concentration of 5 mg/mL, DNase I (grade II, Sigma-Aldrich cat# 10104159001) to final concentration of 1 mg/mL.

6. **Flow buffer**: HBSS (without Ca^{2+}, Mg^{2+}), 2 mM EDTA, 2 mM EGTA, 25 mM HEPES, 0.5% BSA (*see* **Note 4**).

2.5 Cancer Stem Cell (CSC) Cell Surface Marker Antibodies, Cell Viability Stain, and Compensation Beads

1. Rat/hamster anti-mouse lineage panel, biotin conjugated (BD Biosciences cat # 559971; *see* **Note 5**), 1:100 dilution.

2. Rat anti-mouse CD31-biotin (BD Biosciences cat. # 553371; *see* **Note 6**), 1:100 dilution.

3. Rat anti-mouse CD24-PE (BD Biosciences cat. # 553262) or rat anti-mouse CD24-FITC (BD Biosciences cat. # 553261), 1:100 dilution.

4. Hamster anti-mouse CD29-FITC (BD Biosciences cat. # 555005), 1:100 dilution.

5. Rat anti-mouse/human CD49f-FITC (BD Biosciences, cat. # 555735), 1:100 dilution.

6. Rat anti-mouse CD133/2-PE (eBiosciences, cat. # 12-1331-80), 1:100 dilution.

7. Strepavidin-APC (SA-APC, BD Biosciences, cat. # 554067), 1:100 dilution.

8. SYTOX™ Blue Dead Cell Stain (Thermo Fisher Scientific).

9. Chicken anti-rat IgG-biotin conjugate (Invitrogen).

10. BD Biosciences CompBeads compensation particles:

    (a) Anti-mouse IgG (cat# 552843).

    (b) Anti-rat IgG and anti-hamster IgG (cat# 552845).

3 Methods

Carry out all procedures at room temperature unless otherwise specified.

3.1 Tumor Harvest and Digestion

1. Add 20–30 mL of cold tumor digestion buffer (without FBS-HI and enzymes) to a 50 mL conical tube and place tube on ice.

2. Weigh the conical tube with the digestion buffer and record the weight ("a"). Replace tube on ice.

3. Remove tumors carefully from the mouse. Make sure to discard any obvious fat, muscle or necrotic areas before placing tumor material into the pre-weighed tube containing chilled digestion buffer, store the tube on ice (*see* **Note 7**).

4. Weigh the conical tube containing the digestion buffer and tumor pieces ("b"). Calculate the final wet tumor weight by subtracting "a" from "b."

5. Within the biosafety cabinet, remove tumor pieces from the conical tube and place into the bottom of a P15 culture dish. Avoid transferring the digestion buffer to the dish, and aspirate remaining buffer on the dish.

6. Mechanically digest tumors by chopping tissue in a rapid motion with 3 scalpels (#10) taped together for ~7 min. Tissue should be a rough white to pinkish paste at this point with some small fragments still visible. Next, chop tissue with the set of three to five razor blades taped together for a total of 5 min. Spread the mixture as you chop to make sure to evenly chop the material. At the end of the chopping, the tumor tissue should be a fine paste.

7. Using a sterile spatula, transfer tissue paste to either a sterile 250 mL flask or to a sterile 50 mL conical tube (depending on amount of starting material, *see* **Note 8**). Cover flask with sterile aluminum foil and tape closed.

8. Add 10 mL digestion buffer/g tissue (as calculated above). To the digestion buffer, add 2.5% FBS-HI and collagenase and hyaluronidase (*see* **Note 9**).

9. Place the flask or tube in a pre-warmed shaker style incubator. Shake at 37 °C for 1 h at 125 rpm (*see* **Note 10**).

10. If necessary, transfer the digested material to a 50 mL tube within a biosafety cabinet. Centrifuge at $350 \times g$ at RT for 5 min.

11. Discard the supernatant, which will contain lipid droplets and stromal material.

12. Immediately add 10 mL of 0.8% NH_4Cl/HBSS+ (4:1 ratio) to the tumor cell pellet and incubate in the biosafety cabinet at RT for 5 min. After the lysis period, fill the tube to 50 mL with cold HBSS+, shake gently, and centrifuge at $350 \times g$ for 5 min.

13. If red blood cells are still present in the pellet, then aspirate the supernatant and resuspend the pellet again in 0.8% NH_4Cl/HBSS+ (4:1) and incubate for 5 min. Next, fill the tube to a volume to 50 mL, shake gently, and spin it at $450 \, g$ for 5–10 s ("quick spin"; *see* **Note 11**).

14. After red blood cells have been lysed, add 500 µL of DNase stock to 10 mL of DMEM/F12 and resuspend the tumor cell pellet to digest DNA released from lysed cells. Incubate for 5 min at room temperature (*see* **Note 12**).

15. Transfer material to a 15 mL conical tube. Shake gently and then centrifuge at $350 \times g$ for 5 min at RT and aspirate supernatant.

16. Add 10 mL of 0.25%trypsin/EDTA to the pellet and incubate in a 37°C water bath for 3–5 min. Gently agitate by turning the tube upside down every minute (*see* **Note 13**).

17. Transfer mixture to a 50 mL tube and add 40 mL of DMEM/F12 containing 10% FBS-HI to inactivate the trypsin. Shake gently and centrifuge at $350 \times g$ at RT for 5 min.

18. Aspirate the supernatant. Wash the pellet by adding 40 mL of cold HBSS+. Invert the tube a few times to mix and centrifuge immediately at $350 \times g$ for 5 min.

19. Aspirate supernatant and resuspend the pellet in 5 mL of dispase/DNase solution. Incubate in 37°C water bath for 10 min. Gently agitate by turning the tube upside down every 2 min.

20. Add 40 mL cold HBSS+ and centrifuge at 350 × *g* for 5 min to pellet digested tumor cells.

21. Resuspend the pellet in 10 mL cold HBSS+ (*see* **Note 13**). Carefully filter suspension through a 70 μM cell strainer that fits into standard 50 mL centrifuge tubes (*see* **Note 14**). Discard used filters.

22. Keep the flow through and re-filter using a 40 μM cell strainer into a clean 50 mL centrifuge tube. Repeat filtration through 40 μM strainer once more into a clean 50 mL centrifuge tube (*see* **Note 15**).

23. To check for dead and single cells, pipet 10 μL of trypan blue into a microcentrifuge tube and add 10 μL of the filtered cell suspension. Wait for 1 min.

24. Load hemacytometer with 10 μL of the trypan blue/cell suspension mix and calculate cell density per mL per hemacytometer instructions, accounting for twofold dilution of the sample by the trypan blue solution.

25. Calculate yield and estimate the number of cells available for staining for the sort sample, the mock sort sample (*see* **Note 16**), and all controls. Use the example experiment setup sheet created in Excel as a guide to plan your sort experiment (Fig. 1).

26. Based on cell yield, aliquot the appropriate amount of cells needed for controls into a fresh tube (*see* **Note 17**). Aliquot the number of cells needed for the sort and mock sort sample into a separate tube. Pellet cells at 200 *g* for 5 min.

27. Resuspend the control cells to final concentration of one million per 200 μL using flow buffer. Resuspend the sort sample to a concentration of 10×10^6 per 1 mL flow buffer.

3.2 Staining Live, Single Tumor Cells with Antibodies for FACS

1. Label sets of 2 mL tubes for each control sample according to experiment template sheet (refer to example experiment to sort for CD24/CD49f in Fig. 1). Remember that the sort sample must be processed in the biosafety cabinet to maintain sterility.

2. Aliquot 200 μL of control cell suspension to each tube. Immediately place the unstained control tube on ice until ready for analysis. Additionally, place the tube to be used as the heat-killed control at 70°C for 1 h. At the end of 1 h, place heat-killed cells on ice until ready for analysis.

3. To remaining tubes, containing cells for isotype and full minus one (FMO) compensation controls (*see* **Note 18**), add antibodies according to the experimental setup template at a final dilution of 1:100 per antibody. Tap to mix and incubate on ice in the dark for 1 h.

				Antibodies and Dilutions Set-Up				
Tube	Cell Number	Sample	Staining Volume	Primary Antibody	Primary Ab Volume	Secondary Antibody	Secondary Ab Volume	Final Volume
1	1 x 10⁶ PyMT+	Unstained cells	-	none	none	-	-	300 µL
2	1 x 10⁶ PyMT+	HEAT-KILLED	-	none	none 1h 70°C, ice	-	-	300 µL
3	1 x 10⁶ PyMT+	Isotype	200 µL	FITC-rat IgG PE-rat IgG SA-APC	2 µL 2 µL 2 µL	- - -	- - -	300 µL
4	1 x 10⁶ PyMT+	FMO for CD49f FITC	200 µL	CD24-PE Lin+CD31 Biotin	2 µL 2 µL	- SA-APC	- 2 µL	300 µL
5	1 x 10⁶ PyMT+	FMO for CD24 PE	200 µL	CD49f-FITC Lin+CD31 Biotin	2 µL 2 µL	- SA-APC	- 2 µL	300 µL
6	1 x 10⁶ PyMT+	FMO for APC	200 µL	CD49f-FITC CD24-PE	2 µL 2 µL	- -	- -	300 µL
7	Rat/Hamster Beads	FITC Compensation	200 µL	CD49f-FITC	2 µL	-	-	300 µL
8	Rat/Hamster Beads	PE Compensation	200 µL	CD24-PE	2 µL	-	-	300 µL
9	Rat/Hamster Beads	APC Compensation	200 µL	Anti-Rat IgG Biotin	2 µL	SA-APC	2 ml	300 µL
10	50 x 10⁶ PyMT+	SORT sample	5 mL	CD49f-FITC CD24-PE Lin-Biotin+CD31-Biotin	50 mL 50 mL 50 mL	- - SA-APC	- - 50 ml	2 mL
11	20 x 10⁶ PyMT+	MOCK SORT sample	2 mL	CD49f-FITC CD24-PE Lin-Biotin+CD31-Biotin	20 mL 20 mL 20 mL	- - SA-APC	- - 20 ml	2 mL

Fig. 1 Sorting setup template for primary and secondary antibody staining of controls and sort samples. An example Excel worksheet for planning staining of single PyMT cells isolated from freshly digested tumors is shown in order to sort for CD24-PE and CD49f-FITC. First, one million cells are added to a 2.0 mL microcentrifuge tube per control in a volume of 200 µL. For the sort and mock sort samples, ten million cells per mL of staining volume are added to 15 mL conical tubes. If using compensation beads, add 2 drops (~200 µL) of beads per tube. All antibodies are added at a dilution of 1:100 as shown. Primary antibody is incubated for 1 h on ice. After washing, samples to be stained with secondary antibody are resuspended in the appropriate volume according to the worksheet, and the secondary antibody is added at a dilution of 1:100. Secondary antibody samples are incubated on ice for 45 min. Samples that do not require being stained in a secondary reagent are resuspended in the final volume as shown and then stored on ice. After secondary staining, samples are washed and then resuspended in the final volume as indicated in the worksheet

4. Keep sort sample in a sterile 15 mL conical tube (*see* **Note 19**).

5. In a biosafety cabinet, add 1 µL antibody per 1×10^6 cells of each antibody used for sorting (e.g., 50 million cells would be incubated with 50 µL of each antibody).

6. Incubate all sort samples containing primary antibodies on ice for 1 h.

7. While samples are incubating in primary antibody, fill ~6 polypropylene FACS tubes with caps with 1 mL of 100% heat-inactivated FBS each. These tubes will be used to collect cells during sorting and then recapped before returning sorted samples back to the laboratory.

8. Rotate tubes with FBS for ~1 h to coat tubes, which will prevent sorted cells from sticking to sides of tubes.

9. Remove FBS just prior to traveling to the FACS core, and store the pre-coated tubes on ice.

10. For control and sort samples: To wash after incubation with primary antibodies, fill each microcentrifuge tube, or 15 mL conical tube, containing the sort sample with flow buffer. Spin microcentrifuge tubes at $850 \times g$ for 5 min in benchtop centrifuge and spin the sort sample at $200 \times g \times 5$ min in the clinical centrifuge.

11. Aspirate buffer from cell pellets, which will be very loosely attached, being careful not to disturb the pellet.

12. For any samples incubated with reagents that are directly conjugated to a fluorophore and do not require a secondary antibody staining step, resuspend the pellet in 300 μL buffer and keep these tubes in the dark on ice until use in cytometry.

13. For control samples requiring secondary staining (e.g., cells stained with the Lin panel and CD31), resuspend cell pellet in 200 μL flow buffer. Then, add the SA-APC secondary antibody and incubate on ice in dark for 45 min.

14. For sort sample, resuspend pellet in 5 mL of flow buffer and add SA-APC secondary antibody and incubate on ice in dark for 45 min.

15. Wash all tubes exposed to the secondary antibody following the procedures in **steps 10** and **11**.

16. Resuspend control samples in 300 μL flow buffer and resuspend sort sample in 2 mL (*see* **Note 20**) of flow buffer.

17. Filter each sample into clean FACS tubes by pipetting slowly through the FACS tube filter cap using a P1000 pipettor. This step will remove cells that have aggregated during the staining procedures.

18. Confirm overall cell viability after the staining procedures by removing a small aliquot from the unstained control and stain 1:1 with trypan blue using the hemacytometer. Cells should be greater than 90% viable prior to flow sorting.

3.3 Flow Sorting Procedures for Controls and Setting Gates for Sorted Populations

1. For control samples: As soon as samples are transported to the flow sorter, add 0.5 μL of SYTOX™ Blue directly to each *control* sample (or 0.5 μL per 300 μL total volume/sample). It is important that these samples are not incubated more than 30 min with SYTOX™ Blue to prevent overestimation of cell death!

2. Set compensation for sorter lasers using unstained cells, followed by isotype and FMO controls (*see* **Note 18**). Correctly setting compensation is essential to minimize spectral overlap among fluorophores excited by the same laser, such that only signal from FITC is only detected by the FITC detector after

excitation by the 488 nm laser, and only signal from PE is detected by the PE detector after excitation by the 488 nm laser, etc. It is important to realize that different cell types, and even different genotypes from the same parent population of cells, may have different fluorescence profiles that require setting up compensation for each sample or genotype; this issue is magnified when attempting to sort for rarer subpopulations of tumor cells (*see* **Note 21** and Fig. 2, panel f).

3. Set Gates (refer to Fig. 2 for an example gating strategy sequence and example cytometry profiles for MMTV-PyMT tumor cells stained with CD133-PE and CD24-FITC):

   – GATE 1 → Unstained cells profile (Fig. 2, panel a).

   FSC-Area vs. SSC-Area, use binomial axes. Use freeform tool to gate cells thought to represent most singlets and viable cells, avoiding dead cells and debris (left graph). Confirm placement of gate by using heat-killed cells (right graph). All dead cells should lie outside of your chosen gate area.

   – GATE 2 → Live/dead cell confirmation (Fig. 2, panel b).

   To select for live cells, gate on SYTOX™ Blue (SB) negative cells using the unstained cells plus SYTOX™ Blue plotted with the histogram tool. SYTOX™ Blue is detected using the violet laser.

   – GATE 3 → Set first singlets gate (Fig. 2, panel c).

   Using the unstained sample, set axes to FSC-H vs. FSC-W. Using the freeform tool again, draw around the tightest part of the cluster. Avoid including cells that are more scattered from the center.

   – GATE 4 → Set second singlets gate (Fig. 2, panel d).

   Still using the same unstained sample, set axes to SSC-H vs. SSC-W. Follow the same technique as above.

   – GATE 5 → Gate against lineage/CD31-positive cells (Fig. 2, panel e).

   Switch sample to the sort sample (*see* **Note 22**) and apply previous gate settings. Select for APC (lineage and CD31)-negative cells using the histogram tool. Base gate placement on APC expression from compensation controls.

   – GATE 6 → Using gate placement from FMO controls, set quartile gate on sort sample (Fig. 2, panel f). Use the box tool gate to select the sort population gates

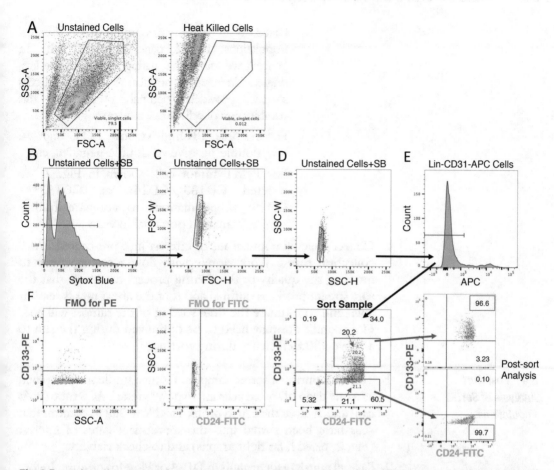

Fig. 2 Example gating strategy for collecting sorted cell subpopulations from MMTV-PyMT digested tumor cells stained with CD133-PE and CD24-FITC. Beginning with unstained cells (**a**, left panel), use binomial axes set to FSC-Area vs. SSC-Area and select the population most representative of single, viable cells. Placement of gates can be confirmed using heat-killed cells to ensure dead cells and debris are not included in the live cells gate (**a**, right panel). Using unstained cells spiked with SYTOX™ Blue (SB), using the histogram tool, select for cells that stain negative for SB (**b**). From the live cell population, select the first singlet gate using binomial axes set to FSC-H vs. FSC-W. Using the freeform tool, draw a gate around the tightest part of the cluster (**c**). Repeat this process for the second singlet gate with axes set to SSC-H vs. SSC-W (**d**). For the sort sample, apply these gates to select for the live, single-cell population. Then, gate for cells negative for APC (Lin panel- and CD31-negative cells, **e**). Using the gate placement as determined from the FMO control for PE and FITC expression (**f**, left two panels), set quartile gates on the sort sample. Next, adjust the quartile gates for the sort populations of interest. In this example, we are gating for CD133-PE-positive/CD24-FITC-positive (CD133+/CD24+) and CD133-PE-negative/CD24-FITC-positive (CD133neg/CD24+) populations, such that all sorted cells are CD24+. This strategy was chosen because Sleeman et al. previously found that, in the normal mouse mammary gland, all CD133+ cells are CD24High/+ [20], and previous studies using the MMTV-PyMT model found that almost all (~97%) tumor cells are CD24+ [14, 15]. CD133-PE gates were selected at the extreme ends of the CD24+ population, such that ~20% of the parent population is selected to be CD133-PE+ (20.2%) or CD133-PE negative (21.1%, **f**, third panel). A two-way sort was performed to collect CD133-PE+/CD24-FITC+ and CD133-PEneg/CD24-FITC+ cells. An aliquot of the sorted cells were then analyzed immediately post-sorting to ensure purity (**f**, right panel). Using the gating strategy outlined above, we found that CD133 levels are downregulated in response to HIF-1 knock-out (KO) in the MMTV-PyMT model and that CD133+/CD24+ PyMT cells have higher tumorsphere formation efficiency compared to CD133neg/CD24+ cells [2]. Of note, inclusion of CD133 along with CD44 and CD49f further enriches for tumor-initiating cells in human triple-negative breast cancer xenografts [21]

based on thedesired enrichment of specific sub-populations (Fig. 2, panel f, third graph). In general, we found that gating at 20% of the live, singlet, Lin^{neg} parent population (5% of total cells sorted) provided optimal purity for this combination of cell surface markers.

For example, PE high vs. low, FITC high vs. low, double-positive vs. double-negative, etc.

For PyMT tumor cells shown in Fig. 2, we selected CD133$^+$/CD24$^+$ vs. CD133^{neg}/CD24$^+$ subpopulations to compare tumor-sphere formation potential in vitro.

4. Collect selected populations of interest into pre-coated FACS collection tubes. Work closely with flow sorter operator to monitor the quality of the sorting process to ensure that the sort stream profile is optimal and that the abort rate is reasonable, and to estimate the time sorting of the sample will end. Flow conditions may need to be optimized during the run to achieve >80% efficiency during sorting.

3.4 Post-sort Analysis of Sorted Populations

1. To check purity of sorted samples: Using a sterile transfer pipet, add one drop of sorted cells into a polystyrene FACS tube (w/o cap) and analyze the FACS profile (*see* **Note 22**) to confirm that cells have been sorted into the appropriate gates of interest (Fig. 2, panel f, far right graphs) and to check viability.

2. Save all raw data for analysis in DIVA or FlowJo software. Before leaving instrument, print sort reports and note the number of events estimated to be contained in each sorted population.

3. Before use of sorted cells in downstream assays, confirm cell viability and cell yield by trypan blue staining of an aliquot of sorted cells and viewing on a hemacytometer (*see* **Note 23**).

4 Notes

1. The sorter used to develop these methods is the BD Biosciences FACS Aria II equipped with four lasers: a 355 nm solid-state UV laser, a 405 nm violet diode laser, a 488 nm blue diode laser, and a 638 nm red diode laser. The sorter has two- and four-way sort capability into standard FACS tubes, microplates, or microcentrifuge tubes.

2. The nozzle size and flow rate were optimized for MMTV-PyMT murine mammary tumor cells isolated from female transgenic stocks or from the transplantation of HIF-1 WT and KO tumor cell lines into the mammary fat pad to regenerate tumors as in [2]. Harvest of tumors smaller than 1.5 cm in diameter generated the highest proportion of live cells prior to staining

and sorting. For FACS, we observed that use of a smaller nozzle pore size (70 μM) resulted in poor cell viability post-sorting, whereas use of a larger nozzle (130 μM), and/or increasing the sample flow rate, resulted in poor post-sort purity (non-distinct populations were observed in post-sort samples).

3. The final amount of enzyme needed depends on the specific activity of the enzyme stock (which varies by lot number). The concentration listed in the protocol refers to the final working concentration.

4. Inclusion of EDTA and EGTA will prevent the mammary epithelial cells from clumping together during staining or analysis, which may clog the sorter nozzle.

5. The lineage panel (Lin) must be specific to the species of the tumor cells. The mouse lineage panel available from BD Biosciences contains a mixture of antibodies against mouse antigens to the major hematopoietic lineages: CD3, CD11b, CD45, Ly-6G and Ly-6C, and TER-119. To isolate tumor epithelial cells, gate cells into the lineage-negative (Lin^{neg}) population.

6. CD31-biotin conjugate is added to the Lin panel in order to gate against endothelial cells.

7. It may be useful to cut the tumor into small fragments with surgical scissors before adding the tumor material to the digestion buffer for weighing. It is important to minimize the time after the tumor is harvested and beginning the digestion process. Process tumor tissue immediately after harvest. In accordance with our local approved animal protocols, we isolate tumor material from anesthetized animals in the laboratory space before euthanizing the animals by an approved secondary method in order to rapidly transition to the tissue mincing and digestion steps.

8. Volumes above 15 mL should be digested in a 250 mL flask to allow sufficient mixing and aeration.

9. It may be difficult to weigh small quantities of digestion enzymes. Use a microscale and transfer powder to a sterile microcentrifuge tube. We find it is more convenient to resuspend enzymes in 1 mL of digestion buffer and then add this volume back to the digestion mixture rather than adding small amounts of enzyme powder stocks directly to the main digestion volume.

10. Removing the flask or tube containing the digesting tumor material after 30 min and shaking vigorously by hand for a few seconds can help to dissolve any large clumps. If the mixture appears viscous, it may also be beneficial to add DNase I enzyme to 50 μg/mL during the final 30 min of digestion. The flask (or tube) should then be immediately returned to the shaking incubator.

11. This step should eliminate most of the red blood cells layer from the tumor cell pellet. If a layer of red blood cells still remains, repeat this step once more.

12. At this step, check the progress of the digestion of the tumor to a single-cell population by loading 10 µL of material onto a hemacytometer and viewing under a microscope.

13. At this step, an aliquot may be taken to view using the hemacytometer to confirm there are single cells.

14. The pellet may be resuspended in more volume if the tumor cell pellet is large. This will aid in straining the suspension through the strainers, which fit loosely on top of 50 mL conical centrifuge tubes. It is beneficial to angle the filter slightly so that the suspension does not become stuck between the filter and the sides of the centrifuge tube. If several grams of tumor material have been digested, it is possible that the filters may become clogged and more than one filter cap will need to be used to process all of the digested material.

15. At this point, the majority of cells should be single cells or doublets, and no large, multicellular organoids should be present.

16. One important control to include for downstream experiments using sorted cell populations (sphere assays, invasion assays, animal experiments, etc.) is to include a mock-sorted control. Sorted cells are subjected to several physical stresses, including manipulation by pipetting and shear stress from the sorting process that may impact viability relative to non-sorted cells. Mock-sorted cells are isolated and are handled using the same procedures as the sort sample and are processed through the same gating scheme as sorted cells. However, mock-sorted cells are not stained with any antibodies and are not gated into the high- or low-marker subpopulations. Instead all live, singlet, Lin^{neg} cells are collected in a mock sort.

17. When using the cell surface markers we have indicated in the methods for PyMT tumor cells, to obtain sufficient numbers of cells to use in downstream assays, it is ideal to stain 50×10^6 cells per sort sample. To optimize antibody staining, try to stain the sort sample in no more than 2 mL of total volume (1.5 mL is optimal for up to 50 million cells). For more abundant cell surface markers (such as CD29/CD24/CD49f), fewer cells may be used; however, at least 50 million cells should be stained when sorting for rarer subpopulations, such as CD133+/CD24+ cells. Next, account for the number of cells needed for all controls (unstained, heat-killed + SYTOX™ Blue, isotype antibody controls (if used instead of full minus one, FMO, controls), FMO controls for FITC, PE, APC fluorophores, etc.). Typically, one million cells should be set aside for the unstained control sample and for each of the other controls (*see* Fig. 1; one million cells in a 200 µL volume/control sample). If there are remaining cells from the tumor digestion other than those needed to set up the sorting experiment, it is recommend to freeze aliquots for molecular analysis (such as qPCR or western blot) or to cytospin cells onto slides for immunostaining experiments.

18. In addition to isotype only or FMO controls, it is also possible to include compensation beads that are incubated with each individual fluorophore-conjugated primary antibody in the sort profile. Compensation beads are uniform in size and produce distinct histograms with sharp peaks, facilitating setting of positive gates. In our protocol, a mixture of anti-rat and anti-hamster beads that are stained individually with each antibody were used (*see* Fig. 1); for the APC control, first chicken anti-rat IgG-biotin was incubated with beads, which were then washed and incubated with SA-APC (*see* Fig. 1). FMOs (full minus one) are stained with all antibodies minus the one of interest (for example, APC and PE, but not FITC; *see* Fig. 1). FMO controls will be negative for staining of the antibody not added, and allow gating based on the lack of fluorophore expression in the context of spectral overlap due to cells' autofluorescence (*see* Fig. 2, panel f). Since cells are used for the FMO controls, often there is not a clear negative population to be able to gate on the negative. Therefore, we found it helpful to use the compensation beads to define the positive gates. Including both methods may be beneficial in refining the setting of compensation during the sorting run or for refining gates post-sort in third-party analysis software, such as FlowJo.

19. The sort sample must be kept sterile in order to culture the cells in vitro or to transplant to rodent recipients in vivo.

20. The final volume to bring to the sorter for the sort sample depends on the number of cells stained. A volume of 2 mL is optimal for 50×10^6 cells; at this concentration, cells are dilute enough to reduce risk of clogging the sorter but are concentrated enough to prevent a prolonged sort time. Therefore, if using more or fewer cells, the final volume will need to be adjusted accordingly for optimal density during the sorting.

21. If planning to analyze or to sort tumors from different HIF genotypes (e.g., HIF wild-type, WT, vs. HIF knockout, KO), then each genotype will need its own set of compensation controls. We have observed that HIF-1 WT and KO PyMT cells generated in [2] exhibited different levels autofluorescence, such that gates set for WT cells were not always optimal for gating of the KO cells.

22. Request that the operator clean the sorter before adding any of the sort samples. This will avoid contamination from the previous samples run in the core. Additionally, the machine will need to be cleaned before each post-sort analysis to avoid contamination when determining post-sort purity.

23. Note that the number of events (~cell number) reported by the sorter software is rarely accurate and typically overestimates yield as compared to the cell yield estimated using a hemacytometer in the laboratory; cell yield may be up to 1.5- to 2.0-fold lower than expected based on reported event numbers.

Acknowledgments

This work was supported by the NIH (CA138488), the Department of Defense (IDEA award BC083846), and the UTHSC Gerwin Cancer Research fund to T.N.S. Luciana P. Schwab created the HIF-1 WT and KO PyMT tumor cell line models and developed initial flow sorting protocols. All experiments were conducted at the UTHSC Flow Cytometry and Flow Sorting (FCCS) core facility, which is supported by the UTHSC campus Office of Research. Expert technical assistance was provided by Drs. Tony Marion and Dan Rosson. The BD Biosciences Aria II sorter was purchased with the support of a NIH instrumentation award (S10 RR022465).

References

1. Keith B, Simon MC (2007) Hypoxia-inducible factors, stem cells, and cancer. Cell 129(3):465–472

2. Schwab LP, Peacock DL, Majumdar D, Ingels JF, Jensen LC, Smith KD, Cushing RC, Seagroves TN (2012) Hypoxia-inducible factor 1alpha promotes primary tumor growth and tumor-initiating cell activity in breast cancer. Breast Cancer Res 14(1):R6. https://doi.org/10.1186/bcr3087

3. Semenza GL (2016) Dynamic regulation of stem cell specification and maintenance by hypoxia-inducible factors. Mol Asp Med 47–48:15–23. https://doi.org/10.1016/j.mam.2015.09.004

4. Sreekumar A, Roarty K, Rosen JM (2015) The mammary stem cell hierarchy: a looking glass into heterogeneous breast cancer landscapes. Endocr Relat Cancer 22(6):T161–T176. https://doi.org/10.1530/ERC-15-0263

5. Visvader JE, Stingl J (2014) Mammary stem cells and the differentiation hierarchy: current status and perspectives. Genes Dev 28(11):1143–1158. https://doi.org/10.1101/gad.242511.114

6. Brown JM (2000) Exploiting the hypoxic cancer cell: mechanisms and therapeutic strategies. Mol Med Today 6(4):157–162

7. Semenza GL (2012) Hypoxia-inducible factors: mediators of cancer progression and targets for cancer therapy. Trends Pharmacol Sci 33(4):207–214. https://doi.org/10.1016/j.tips.2012.01.005

8. Tredan O, Galmarini CM, Patel K, Tannock IF (2007) Drug resistance and the solid tumor microenvironment. J Natl Cancer Inst 99(19):1441–1454

9. Wei W, Lewis MT (2015) Identifying and targeting tumor-initiating cells in the treatment of breast cancer. Endocr Relat Cancer 22(3):R135–R155. https://doi.org/10.1530/ERC-14-0447

10. Al-Hajj M, Wicha MS, Benito-Hernandez A, Morrison SJ, Clarke MF (2003) Prospective identification of tumorigenic breast cancer cells. Proc Natl Acad Sci U S A 100(7):3983–3988. https://doi.org/10.1073/pnas.0530291100

11. Visvader JE (2009) Keeping abreast of the mammary epithelial hierarchy and breast tumorigenesis. Genes Dev 23(22):2563–2577. https://doi.org/10.1101/gad.1849509

12. Zhang M, Behbod F, Atkinson RL, Landis MD, Kittrell F, Edwards D, Medina D, Tsimelzon A, Hilsenbeck S, Green JE, Michalowska AM, Rosen JM (2008) Identification of tumor-initiating cells in a p53-null mouse model of breast cancer. Cancer Res 68(12):4674–4682. https://doi.org/10.1158/0008-5472.CAN-07-6353

13. Bao L, Cardiff RD, Steinbach P, Messer KS, Ellies LG (2015) Multipotent luminal mammary cancer stem cells model tumor heterogeneity. Breast Cancer Res 17(1):137. https://doi.org/10.1186/s13058-015-0615-y

14. Kouros-Mehr H, Bechis SK, Slorach EM, Littlepage LE, Egeblad M, Ewald AJ, Pai SY, Ho IC, Werb Z (2008) GATA-3 links tumor differentiation and dissemination in a luminal breast cancer model. Cancer Cell 13(2):141–152

15. Ma J, Lanza DG, Guest I, Uk-Lim C, Glinskii A, Glinsky G, Sell S (2012) Characterization of mammary cancer stem cells in the MMTV-PyMT mouse model. Tumour Biol 33(6):1983–1996. https://doi.org/10.1007/s13277-012-0458-4

16. Alexander CM, Puchalski J, Klos KS, Badders N, Ailles L, Kim CF, Dirks P, Smalley MJ (2009) Separating stem cells by flow cytometry: reducing variability for solid tissues. Cell Stem Cell 5(6):579–583. https://doi.org/10.1016/j.stem.2009.11.008

17. Lin EY, Jones JG, Li P, Zhu L, Whitney KD, Muller WJ, Pollard JW (2003) Progression to malignancy in the polyoma middle T oncoprotein mouse breast cancer model provides a reliable model for human diseases. Am J Pathol 163(5):2113–2126. https://doi.org/10.1016/S0002-9440(10)63568-7

18. Liao D, Corle C, Seagroves TN, Johnson RS (2007) Hypoxia-inducible factor-1alpha is a key regulator of metastasis in a transgenic model of cancer initiation and progression. Cancer Res 67(2):563–572. https://doi.org/10.1158/0008-5472.CAN-06-2701

19. Seagroves TN, Hadsell D, McManaman J, Palmer C, Liao D, McNulty W, Welm B, Wagner KU, Neville M, Johnson RS (2003) HIF1alpha is a critical regulator of secretory differentiation and activation, but not vascular expansion, in the mouse mammary gland. Development 130(8):1713–1724

20. Sleeman KE, Kendrick H, Robertson D, Isacke CM, Ashworth A, Smalley MJ (2007) Dissociation of estrogen receptor expression and in vivo stem cell activity in the mammary gland. J Cell Biol 176(1):19–26

21. Meyer MJ, Fleming JM, Lin AF, Hussnain SA, Ginsburg E, Vonderhaar BK (2010) CD44posCD49fhiCD133/2hi defines xenograft-initiating cells in estrogen receptor-negative breast cancer. Cancer Res 70(11):4624–4633. https://doi.org/10.1158/0008-5472.CAN-09-3619

Chapter 23

Evaluation of Macrophage Polarization in Pancreatic Cancer Microenvironment Under Hypoxia

Kuldeep S. Attri, Kamiya Mehla, and Pankaj K. Singh

Abstract

Hypoxic microenvironment found in pancreatic ductal adenocarcinoma and other solid tumors is central to physiological and metabolic alterations of immune cells that significantly impact tumor growth dynamics. Hypoxic adaptations in the immune cells are primarily mediated by the stabilization of hypoxia-inducible factor-1 alpha (HIF-1α), which regulates cellular metabolism by modulating glycolysis and other interconnected metabolic pathways. HIF-1α plays distinct roles in M1 and M2 macrophage polarization, which, in turn, regulates tumor cell immune escape and growth. In this chapter, we describe a real-time PCR-based assay to monitor the transcript levels of *Arg1* and *Nos2* to assess the status of tumor-induced macrophage polarization under hypoxic conditions. This method can be effectively utilized to delineate the genes critical for M1/M2 polarization in the hypoxic tumor microenvironment and would provide opportunities to develop immunomodulating therapies to regulate the tumor growth, progression, and metastatic dissemination.

Key words Pancreatic cancer, Macrophages, Metabolism, Hypoxia, HIF-1α

1 Introduction

Pancreatic tumors manifest significant desmoplasia and hypoperfusion, which make the tumor microenvironment very hypoxic [1]. The pancreatic cancer cells adapt to the limited oxygen delivery by stabilizing hypoxia-inducible factor-1 alpha (HIF-1α), a master regulator of tumor cell physiology and metabolism [2, 3]. The tumor microenvironment is created and modulated by the interactions between the tumor cells and the non-transformed cells such as the immune cells, activated fibroblasts, and adipocytes [4]. Macrophages infiltrating the tumor, commonly known as tumor-associated macrophages (TAMs), perform key homeostatic functions to accelerate tumor growth and progression [5]. In addition to its role in regulating tumor progression, hypoxia also regulates macrophage polarization that favors tumor development [6]. Thus, the macrophages play an important role in modulating

L. Eric Huang (ed.), *Hypoxia: Methods and Protocols*, Methods in Molecular Biology, vol. 1742,
https://doi.org/10.1007/978-1-4939-7665-2_23, © Springer Science+Business Media, LLC 2018

pancreatic cancer development under hypoxia. Hence, understanding the changes in macrophage dynamics presents promising therapeutic opportunities to reprogram the tumor microenvironment and to target hypoxia-adapted cancer cells.

Macrophages can be polarized into classically activated (M1) macrophages, which are involved in evoking the anti-pathogenic immune response, or into activated (M2) macrophages, which are involved mainly in wound healing and regeneration [7]. Besides differences in invasive, migratory, and immunomodulatory properties, the M1 and M2 macrophages also differ metabolically [8]. While M1 macrophages are fueled by glycolysis for ATP production and have augmented expression of iNOS, M2 macrophages rely on OXPHOS and have increased levels of arginase-1. In the tumor microenvironment, M1 macrophages possess anti-tumorigenic properties, and the M2 macrophages exhibit pro-tumorigenic features [9]. HIFs have been implicated in macrophage polarization, wherein HIF-1a is primarily associated with M1 macrophage polarization, whereas HIF-2a regulates M2 macrophage polarization [10]. Identification of the genes responsible for macrophage polarization under hypoxia would provide opportunities to modulate the tumor microenvironment to decelerate pancreatic cancer progression.

Macrophages apart from being polarized into M1 or M2 populations by cytokines can also be polarized by other bioactive agents like metabolites present in tumor cell-conditioned medium [6]. Investigation of genes responsible for tumor cell-conditioned medium-induced macrophage polarization would have a significant impact on tumor biology. Macrophage polarization can be studied by a variety of methods including flow cytometry, analysis of subtype-specific macrophage cytokine secretion, real-time PCR for evaluating *Arg1* and *Nos2* transcript levels, western blotting for polarization markers, and kit-based enzymatic activity assays [11, 12]. Real-time PCR-based analysis of *Arg1* and *Nos2*, which are classical markers for M2 and M1 macrophage polarization, respectively, offers a fast, reliable, reproducible, and efficient assay to monitor polarization. This chapter illustrates a detailed methodology for evaluating macrophage polarization upon tumor cell-conditioned medium treatment under normoxic and hypoxic conditions by a real-time PCR-based method. Here, we describe in detail an in vitro macrophage polarization assay in a series of steps, starting from tumor cell-conditioned medium preparation, culturing of primary bone marrow-derived macrophages, setting up the hypoxia experiment in macrophages treated with tumor cell-conditioned medium, real-time PCR-based analysis of M1/M2 macrophage polarization genes, and analysis of the generated data. The method described here can also be extended to study the biology of cytokine-polarized M1/M2 macrophages to decipher tumor-immune crosstalk.

2   Materials

2.1   Cell Culture Reagents

1. DMEM/HIGH with L-glutamine, without sodium pyruvate, powdered medium (Hyclone). Reconstitute as per manufacturer's instructions with the addition of L-sodium bicarbonate (3.7 g/L) and sodium pyruvate (0.110 g/L).

2. Fetal bovine serum (FBS).

3. Phosphate-buffered saline.

4. Trypsin (0.25%).

5. Penicillin-streptomycin solution: use 1× penicillin-streptomycin in all the cell culture media (100×) (penicillin 10,000 units/mL/streptomycin 10,000 μg/mL).

6. Histopaque, Sigma-Aldrich.

2.2   RNA Isolation Reagents

1. RNaseZap, Ambion.

2. Nuclease-free water.

3. TRIzol, Invitrogen.

4. Chloroform.

5. Isopropanol.

6. Ethanol absolute.

2.3   cDNA Synthesis and Real-Time PCR Reagents

1. Verso cDNA synthesis kit, Applied Biosystems.

2. FastStart Universal SYBR Green Master (ROX) dye, Roche.

3. MicroAmp Optical Adhesive Film, Applied Biosystems.

4. Primers are custom synthesized from Eurofins Genomics at a concentration of 100 μM. The primers are diluted at a dilution of 1:100. The sequence of primers is as follows:

Gene	Forward primer (5′ − 3′)	Reverse primer (5′ − 3′)
Actb	GGCTGTATTCCCCTCCATCG	CCAGTTGGTAACAATGCCATGT
Arg1	CCACAGTCTGGCAGTTGGAAG	GGTTGTCAGGGGAGTGTTGATG
Nos2	GTTCTCAGCCCAACAATACAAGA	GTGGACGGGTCGATGTCAC

2.4   Equipment

1. 6-well plate.

2. Cell scrapers.

3. O_2/CO_2 incubator, equipped with gas-tight split inner doors for each rack.

4. 10 cm tissue culture dish.

5. T-75 cell culture flask.

6. 15 mL polypropylene tube.

7. 50 mL polypropylene tube.

8. 384-well plate, Applied Biosystems.

9. 1.5 mL microcentrifuge tubes.

10. 200 μL PCR tubes.

11. EZFlow 0.22 μm syringe filter.

12. Pipettes.

13. Multichannel pipettes, Gilson PIPETMAN Neo.

14. Pipette tips—0.1–20 μL, 200 μL, and 1000 μL.

15. 10 mL syringe.

16. 21 gauge needle.

17. Disposable scalpel.

18. Mini vortex mixer.

19. Refrigerated centrifuge.

20. Cytation 3 Imaging Reader by BioTek (for RNA quantification).

21. Thermal cycler.

22. QuantStudio 5 Real-Time PCR, Applied Biosystems.

2.5 Software

1. GEN5 3.0 Microplate Reader and Imager Software.

2. QuantStudio™ design and analysis software V1.3.1.

3. GraphPad Prism 5.0B.

3 Methods

3.1 Preparation of
L929 Cell-Conditioned
Medium

1. Seed 1×10^6 L929 cells in a 100 mm cell culture dish in 25 mL DMEM complete medium containing 10% FBS.

2. Incubate cells at 37 °C in an incubator containing 5% CO_2 and 95% humidity for 7 days without changing the medium. Maintain the oxygen consumption at 20%.

3. After 7 days, collect the spent medium in 50 mL polypropylene tube (*see* **Note 1**).

4. Centrifuge the medium in 50 mL polypropylene tube at $300 \times g$ for 5 min, and filter using 0.22 μm syringe filter in the biosafety cabinet.

5. Use 20% L929-conditioned medium as a supplement with DMEM medium for culturing primary bone marrow-derived macrophages.

6. Use the filtered conditioned medium immediately, or it can be snap-frozen and stored in aliquots at −80 °C (*see* **Note 2**).

3.2 Pancreatic Cancer Cell-Conditioned Medium Preparation

1. Seed 2×10^6 pancreatic cancer cells (S2-013 or any other pancreatic cancer cell line) in a 100 mm cell culture dish containing 10 mL DMEM complete medium with 10% FBS.

2. Incubate cells at 37 °C in an incubator containing 5% CO_2, 20% O_2, and 95% humidity until the cells reach 70% confluence (*see* **Note 3**).

3. Aspirate and discard the supernatant when cells are 70% confluent. Wash the cells three times with 1× PBS and discard PBS.

4. Wash the cells three times with 3 mL of DMEM serum-free medium.

5. Add 10 mL of serum-free medium in the culture dish, and incubate cells at 37 °C in a 5% CO_2 incubator for the next 48 h or desired time duration (*see* **Note 4**).

6. After 48 h, collect the spent medium, now called as the conditioned medium in 15 mL polypropylene tube.

7. Centrifuge the medium in 50 mL polypropylene tube at $300 \times g$ for 5 min, and filter using 0.22-μm syringe filter in the biosafety cabinet.

8. Treat bone marrow-derived macrophages with this conditioned medium, or store the conditioned medium as 10 mL aliquots in 15 mL polypropylene tubes at −80 °C after snap freezing.

3.3 Isolation and Culturing of Bone Marrow-Derived Macrophages

1. Euthanize 6–8-week-old C57BL/6 or BALB/c mice as per the institutional IACUC guidelines.

2. Dissect the mice, and aseptically isolate the femur and tibia bones from the animal.

3. Remove the adherent soft tissue from the bone with the help of the scalpel, and cut the epiphysis of the bones.

4. Flush out the contents of bone marrow cavity with serum-free DMEM medium from one end of the bone using a sterile 21-gauge needle to take out the cells in a 100 mm cell culture dish (*see* **Note 5**).

5. Pass the collected cells through a syringe with 21-gauge needle to make single-cell suspension, and collect the cells in a 15 mL polypropylene tube.

6. Wash the cells with DMEM medium containing 10% FBS by centrifuging the cells at $500 \times g$ for 5 min at room temperature.

7. Resuspend the pellet thus obtained in 6 mL of DMEM medium containing 10% FBS by pipetting the contents up and down.

8. Layer the 6 mL of resuspended cells on a Ficoll-Hypaque gradient in a ratio of 1:4 (*see* **Note 6**).

9. Centrifuge the layered cells at $850 \times g$ for 20 min at room temperature.

10. Macrophage precursor cells form a buffy coat at the interface of the two layers. Therefore, carefully take out the buffy coat cells by inserting the pipette into the tube, and gently take out the middle layer contents only. Add 5 mL of DMEM medium containing 10% FBS, and mix the contents well with the pipette.

11. Centrifuge the cells in 15 mL polypropylene tube at $500 \times g$ for 7 min at room temperature.

12. Resuspend the pellet thus obtained in 5 mL of DMEM medium containing 10% FBS and 20% L929-conditioned medium.

13. Mix 20 μL of the cell suspension with 20 μL of trypan blue dye. Load 10 μL of this mixture to the hemocytometer, and count the cells.

14. Seed 4×10^6 bone marrow-derived macrophages in a 100-mm cell culture dish containing 10 mL DMEM medium containing 10% FBS and 20% L929-conditioned medium. Incubate the cells at 37 °C in an incubator containing 5% CO_2, 20% O_2, and 95% humidity.

15. After 24 h, add 10 mL of DMEM medium containing 10% FBS and 20% L929-conditioned medium from the side of the dish, and incubate cells in 5% CO_2 incubator at 37 °C for 3 days.

16. After 3 days, replace with fresh 10 mL of DMEM medium containing 10% FBS and 20% L929-conditioned medium, and incubate cells at 37 °C in 5% CO_2 incubator for next 3 days (*see* **Note 7**).

17. Detach the macrophages using 1 mL of trypsin, and harvest with the help of cell scrapper. Count the macrophages using hemocytometer, and reseed 1×10^6 macrophages in each well of the 6-well plate in 2 mL of DMEM medium containing 10% FBS and 20% L929-conditioned medium. Incubate the cells at 37 °C in 5% CO_2 incubator till the cells attach, and attain desired morphology.

3.4 Conditioned Medium Treatment and Hypoxia Exposure

1. Aspirate the medium from cells seeded in 6-well plates, and wash the cells with 1× PBS.

2. Wash the cells again with 3 mL of serum-free DMEM medium twice.

3. Prepare tumor cell-conditioned medium for feeding in macrophages by adding 10% L929-conditioned medium to the serum-free DMEM medium (*see* **Note 8**).

4. Add 2 mL of tumor cell-conditioned medium in the treatment well of the 6-well plate, while in control well, add 2 mL of

serum-free DMEM medium containing 10% L929-conditioned medium. The cells were incubated in 5% CO_2 incubator having 20% O_2 at 37 °C for 40 h.

5. Similarly, for hypoxia treatment, add 2 mL of tumor cell-conditioned medium containing 10% L929-conditioned medium to the treatment cells, and add 2 mL of serum-free DMEM medium containing 10% L929-conditioned medium to the control cells. Incubate the cells at 37 °C in hypoxia incubator having 5% CO_2 and 1% O_2 for 40 h.

6. At the indicated time points, remove the cells from the normoxia and hypoxia incubators, and immediately aspirate the medium.

7. Wash the cells with 2 mL of 1× PBS quickly, and proceed for RNA isolation (*see* **Note 9**).

3.5 RNA Isolation and Quality Assessment

1. Add 1 mL of TRIzol to each well of the 6-well plate inside the chemical hood. Keep the 6-well plate at room temperature for 5 min (*see* **Note 10**).

2. Pipette the TRIzol fraction up and down several times till the cells are lysed completely and a free-flowing solution is made (*see* **Note 11**).

3. Transfer the TRIzol fraction in 1.5 mL microcentrifuge tubes, and add 200 μL of chloroform to the TRIzol fraction.

4. Vortex the microcentrifuge tubes vigorously for 15 s, and incubate at room temperature for 5 min with intermittent mixing.

5. Centrifuge the microcentrifuge tubes at 15,000 × g for 10 min at 4 °C.

6. After centrifugation, three layers are formed in the tube—upper aqueous layer, lower organic layer, and the middle layer containing genomic DNA. Carefully transfer the upper aqueous layer containing RNA in a new 1.5 mL microcentrifuge tube (*see* **Note 12**).

7. Add 500 μL chloroform to the RNA fraction, and incubate at room temperature for 20 min.

8. After 20 min, centrifuge the microcentrifuge tubes at 15,000 × g for 10 min at 4 °C.

9. Decant supernatant from the microcentrifuge tube by quickly inverting the contents out of the tube.

10. Place the tubes on ice, and add 1 mL 70% ethanol for precipitation at −20 °C overnight.

11. Centrifuge the reaction tubes at 15,000 × g for 10 min at 4 °C.

12. Decant the supernatant from the microcentrifuge tube by quickly inverting the contents out of the tube.

13. Centrifuge the reaction tubes again at $15,000 \times g$ for 5 min at 4 °C. Carefully take out the supernatant with 200 µL pipette without disturbing the pellet.

14. Air-dry the pellets at 37 °C for 20 min (*see* **Note 13**).

15. Resuspend the pellet in 35 µL nuclease-free water. Keep the reaction tube at room temperature for 5 min. Tap mix the contents, briefly mix, and keep the RNA on ice until quantification. The RNA should be stored at −80 °C for long-term storage.

16. Measure the absorbance at 230, 260, and 280 nm using the Cytation 3 multimode reader. The good quality RNA should have both the ratios—A260/280 and A260/230 close to 2 (*see* **Note 14**).

3.6   cDNA Synthesis

1. Before starting the cDNA synthesis, ensure that all the plasticware and labware used are RNase- and DNase-free.

2. To synthesize the cDNA, reverse transcribe 1 µg of RNA isolated by the TRIzol method using Verso cDNA synthesis kit (*see* **Note 15**).

3. Prepare the 20 µL reaction mix in a 200 µL PCR tube by adding the components provided in Table 1 below.

4. A typical reverse transcription program comprises of two steps—cDNA synthesis, followed by heat inactivation of the enzyme. Carry out the synthesis of cDNA at 42 °C for 30 min, followed by heat inactivation at 95 °C for 2 min in the thermal cycler.

5. Dilute the cDNA 20 times using nuclease-free water. Use the diluted cDNA for performing quantitative real-time PCR to analyze gene expression changes in control medium- and tumor cell-conditioned medium-treated macrophages cultured in normoxic and hypoxic conditions (*see* **Note 16**).

Table 1
Components

Components	Volume (µL)
5× cDNA synthesis buffer	4
dNTP Mix	2
RNA primer (random hexamer)	0.75
RNA primer (oligo dT)	0.25
RT enhancer	1
Verso enzyme mix	1
Template RNA	Volume having 1 µg RNA
Nuclease-free water	Makeup to 20

3.7 Gene Expression Analysis by Real-Time PCR

1. Thaw the solutions before starting the experiment. It is advisable to spin the vials in a microcentrifuge tube before opening to get the maximum recovery of contents. Mix the contents of the vial carefully by pipetting up and down on the ice.

2. Design the real-time PCR primers using NCBI Primer-BLAST software (https://www.ncbi.nlm.nih.gov/tools/primer-blast/), or pick validated primers from the PrimerBank database (https://pga.mgh.harvard.edu/primerbank/) (*see* **Note 17**).

3. First, add 3 μL of the 1:20 diluted synthesized cDNA to each well of the 384-well plate using a multichannel pipette (*see* **Note 18**).

4. Prepare the PCR mix for one 10 μL reaction by adding 5 μL FastStart Universal SYBR Green Master mix, 1 μL forward primer, and 1 μL reverse primer. Mix the PCR reaction components carefully by pipetting up and down. Do not vortex.

5. For a 384-well plate PCR, where more than one reaction of the PCR mix needs to be prepared, multiply the amount required for one reaction with a total number of reactions to be run. Dispense 7 μL master mix into different cDNA samples in the 384-well plate for the same primer reaction (*see* **Note 19**).

6. Seal the multi-well plate using optically clear self-adhesive film.

7. Spin the 384-well plate at 850 × *g* for 2 min to bring down the contents and to ensure the absence of any air bubbles.

8. Place the plate in the QuantStudio 5 Real-Time PCR Instrument (*see* **Note 20**).

9. Perform the quantitative real-time PCR using QuantStudio™ design and analysis software V1.3.1. Use comparative Ct method with standard ddCt SYBR with melt continuous template. Start the real-time PCR reaction (*see* **Note 21**).

10. At the end of the reaction, check the melting curve for each reaction. A single product peak indicates primer-specific amplification. Export the Ct value data in a .xml file.

3.8 Data Analysis and Representation

1. Calculate the average Ct values of *Actb* expression in primary bone marrow-derived control macrophages cultured under normoxic and hypoxic conditions separately from .xml file. Similarly, calculate the average Ct values for tumor cell-conditioned medium-treated macrophages cultured under normoxic and hypoxic conditions

2. Subtract the average Ct value of housekeeping gene *Actb* from the Ct value of target genes *Arg1* and *Nos2* separately to obtain ΔCt values. The ΔCt values are calculated separately for all the four samples—control and tumor cell-conditioned medium-treated macrophages under normoxic and hypoxic conditions.

3. Calculate the $\Delta\Delta Ct$ values from ΔCt values by subtracting the ΔCt value of control macrophages cultured under normoxia from the ΔCt value of all the other three samples separately (*see* **Note 22**).

4. Apply the formula $2^{(-\Delta\Delta Ct)}$ to obtain relative fold changes. These values are normalized relative fold changes with respect to control macrophages cultured under normoxia.

5. Use the fold values obtained in **step 4** for the graphical data representation using GraphPad Prism. The data can be plotted as either mean ± standard deviation or mean ± standard error of the mean.

6. The increased expression of *Arg1* indicates increased M2 macrophage polarization, whereas increased *Nos2* expression indicates increased M1 macrophage polarization (*see* **Note 23**).

4   Notes

1. Spent media should be collected carefully from the side of the dish so that cells do not detach and come in spent media. The cells in the spent media can clog the filter.

2. Do not store more than 40 mL conditional medium in a 50 mL polypropylene tube, as it can leak out from the tube during thawing due to expansion.

3. Since different cancer cells have different growth kinetics, the seeding densities may vary depending upon the cell line used. Therefore, it is advisable to standardize the initial seeding density.

4. The time duration of serum-free medium incubation for preparing the conditioned medium may need to be optimized depending upon the choice of a cell line. The cells should not be cultured beyond their confluence as they can start floating in conditioned media. This can lead to variation in reproducibility of experiments.

5. Flush the bones with serum-free DMEM medium from both the ends until all the content comes out and the bones become translucent.

6. Add the cell suspension dropwise to the 2 mL of a Histopaque solution in the 15 mL polypropylene tube, and make sure that there is no mixing up of the two layers.

7. While taking out the old spent media, tilt the plate, and aspirate slowly from the side of the dish without disturbing the attached cells.

8. The tumor cell-conditioned medium can also be supplemented with 1, 2, 5, or 10% FBS to provide additional nutrients as

conditioned medium prepared after a very long incubation can deplete all nutrients from the medium.

9. The cells should be washed as fast as possible and should be processed immediately after taking out from the incubator.

10. Wipe all the surfaces with RNaseZap before starting the work.

11. Incubation of TRIzol fraction at room temperature and proper pipetting are critical steps in the protocol to ensure that the cells have lysed properly and RNA is released out of the cell.

12. While pipetting the upper layer, utmost care should be taken not to disturb the middle layer as the mixing of middle layer contents can contaminate RNA fraction with genomic DNA. Leave around 50 μL contents from the upper layer so that middle layer is not disturbed.

13. The drying time can vary depending upon the residual ethanol in the tube. Air-dry the pellets till the odor of ethanol vanishes. Care should also be taken not to overdry the pellets.

14. If the RNA quality is poor, then the DNase treatment followed by column purification using Qiagen RNeasy mini kit should be done to get rid of contaminants and improve the quality of RNA before proceeding to further downstream experiments.

15. Carry out all the procedures on ice so as to ensure that the RNA does not degrade.

16. A (−)RT reaction, comprising all the components of cDNA synthesis except RNA, can be put to check for genomic DNA contamination. If the Ct value is not detected by real-time PCR, it indicates pure RNA free of any genomic DNA contamination.

17. The optimal design of the PCR primers is critical for quantitative real-time PCR as it determines the length and melting temperature of the PCR amplicon that can affect the PCR efficiency and yield.

18. The optimal amount of cDNA template in the PCR reaction is determined by running real-time PCR with undiluted, 1:10, 1:100, and 1:1000 diluted cDNA template.

19. Always prepare 10% additional PCR reaction mix to account for pipetting errors.

20. Check the alignment and direction of the plate carefully.

21. For optimal results, make sure that the instrument is calibrated correctly.

22. The ΔΔCt values are calculated separately for all the target genes.

23. Additional M1/M2 macrophage marker genes like *Fizz1*, *Ym1*, *Tnfa*, *Icam1*, etc. can also be probed from the same cDNA.

References

1. Whatcott CJ, Diep CH, Jiang P, Watanabe A, LoBello J, Sima C, Hostetter G, Shepard HM, Von Hoff DD, Han H (2015) Desmoplasia in primary tumors and metastatic lesions of pancreatic cancer. Clin Cancer Res 21:3561–3568

2. Chaika NV, Gebregiworgis T, Lewallen ME, Purohit V, Radhakrishnan P, Liu X, Zhang B, Mehla K, Brown RB, Caffrey T, Yu F, Johnson KR, Powers R, Hollingsworth MA, Singh PK (2012) MUC1 mucin stabilizes and activates hypoxia-inducible factor 1 alpha to regulate metabolism in pancreatic cancer. Proc Natl Acad Sci U S A 109:13787–13792

3. Surendra K. Shukla, Vinee Purohit, Kamiya Mehla, Venugopal Gunda, Nina V. Chaika, Enza Vernucci, Ryan J. King, Jaime Abrego, Gennifer D. Goode, Aneesha Dasgupta, Alysha L. Illies, Teklab Gebregiworgis, Bingbing Dai, Jithesh J. Augustine, Divya Murthy, Kuldeep S. Attri, Oksana Mashadova, Paul M. Grandgenett, Robert Powers, Quan P. Ly, Audrey J. Lazenby, Jean L. Grem, Fang Yu, José M. Matés, John M. Asara, Jung-whan Kim, Jordan H. Hankins, Colin Weekes, Michael A. Hollingsworth, Natalie J. Serkova, Aaron R. Sasson, Jason B. Fleming, Jennifer M. Oliveto, Costas A. Lyssiotis, Lewis C. Cantley, Lyudmyla Berim, Pankaj K. Singh, (2017) MUC1 and HIF-1alpha Signaling Crosstalk Induces Anabolic Glucose Metabolism to Impart Gemcitabine Resistance to Pancreatic Cancer. Cancer Cell 32(1):71-87.e7

4. Joyce JA, Pollard JW (2009) Microenvironmental regulation of metastasis. Nat Rev Cancer 9:239–252

5. Mantovani A, Sozzani S, Locati M, Allavena P, Sica A (2002) Macrophage polarization: tumor-associated macrophages as a paradigm for polarized M2 mononuclear phagocytes. Trends Immunol 23:549–555

6. Colegio OR, Chu NQ, Szabo AL, Chu T, Rhebergen AM, Jairam V, Cyrus N, Brokowski CE, Eisenbarth SC, Phillips GM, Cline GW, Phillips AJ, Medzhitov R (2014) Functional polarization of tumour-associated macrophages by tumour-derived lactic acid. Nature 513:559–563

7. Galvan-Pena S, O'Neill LA (2014) Metabolic reprograming in macrophage polarization. Front Immunol 5:420

8. Geeraerts X, Bolli E, Fendt SM, Van Ginderachter JA (2017) Macrophage metabolism as therapeutic target for cancer, atherosclerosis, and obesity. Front Immunol 8:289

9. Qian BZ, Pollard JW (2010) Macrophage diversity enhances tumor progression and metastasis. Cell 141:39–51

10. Fang HY, Hughes R, Murdoch C, Coffelt SB, Biswas SK, Harris AL, Johnson RS, Imityaz HZ, Simon MC, Fredlund E, Greten FR, Rius J, Lewis CE (2009) Hypoxia-inducible factors 1 and 2 are important transcriptional effectors in primary macrophages experiencing hypoxia. Blood 114:844–859

11. Jablonski KA, Amici SA, Webb LM, Ruiz-Rosado Jde D, Popovich PG, Partida-Sanchez S, Guerau-de-Arellano M (2015) Novel markers to delineate murine M1 and M2 macrophages. PLoS One 10:e0145342

12. Zajac E, Schweighofer B, Kupriyanova TA, Juncker-Jensen A, Minder P, Quigley JP, Deryugina EI (2013) Angiogenic capacity of M1- and M2-polarized macrophages is determined by the levels of TIMP-1 complexed with their secreted proMMP-9. Blood 122:4054–4067

Chapter 24

Detection of Hypoxia and HIF in Paraffin-Embedded Tumor Tissues

Fuming Li, Kyoung Eun Lee, and M. Celeste Simon

Abstract

Hypoxia (insufficient O_2 availability) is involved in various biological processes, such as tumorigenesis and inflammation. Hypoxia results in stabilization of hypoxia-inducible factors (HIFs) including HIF1α and HIF2α. Here we describe a protocol to detect mouse and human tissue hypoxia by using Hypoxyprobe and immunohistochemical staining for HIF1α and HIF2α.

Key words Hypoxia, HIF1α, HIF2α, Hypoxyprobe, Immunohistochemistry

1 Introduction

Molecular oxygen (O_2) is an essential substrate for mitochondrial ATP production and universal electron receptor for numerous biochemical pathways. Therefore, maintenance of O_2 homeostasis is critical for the survival of most prokaryotic and eukaryotic species [1, 2]. O_2 deprivation (hypoxia) triggers complex adaptive responses to match O_2 supply with metabolic and bioenergetic demands. When faced with hypoxia, mammalian cells engage multiple evolutionarily conserved molecular responses, including those coordinated by the hypoxia-inducible factors (HIF) transcriptional regulators. Under normoxic conditions, HIFs are rapidly degraded through an ubiquitin-proteasome pathway, whereas hypoxia results in the stabilization and nuclear translocation of HIF [1, 2]. HIFs, including HIF1α and HIF2α, transactivate a wide range of genes involved in the regulation of angiogenesis, metabolism, cell survival, and inflammation [3]. There are numerous approaches to detect hypoxia, for example, through directly detecting HIF proteins [4, 5], or HIF targets at the mRNA and/or protein levels [6]. Alternatively, Hypoxyprobe can be

---

Fuming Li and Kyoung Eun Lee contributed equally to this work.

L. Eric Huang (ed.), *Hypoxia: Methods and Protocols*, Methods in Molecular Biology, vol. 1742,
https://doi.org/10.1007/978-1-4939-7665-2_24, © Springer Science+Business Media, LLC 2018

used to detect in vivo tissue hypoxia [5, 6]. Here, we describe the protocol to detect hypoxia in mouse and human paraffin-embedded tissues by immunohistochemical staining for Hypoxyprobe, HIF1α, and HIF2α (*see* **Note 1**).

2 Materials

2.1 Antibody

Mouse anti-human HIF1α (HA111, #NB100-296, 1:250; Novus), mouse anti-human HIF2α (#NB100-132, 1:5000; Novus), rabbit anti-mouse HIF1α (#ab2185, 1: 5000; Abcam), rabbit anti-mouse HIF2α (#ab199, 1: 2000; Abcam), biotinylated goat anti-mouse IgG antibody (#BA9200, 1:500; Vector Laboratories), biotinylated goat anti-rabbit IgG antibody (#BA1000, 1:500; Vector Laboratories).

2.2 Reagents

Prepare all solutions using ultrapure water and analytical grade reagents. Unless indicated otherwise, prepare and store all reagents at room temperature. Diligently follow waste disposal regulations when disposing waste materials.

1. Hypoxyprobe Kit: 200 mg pimonidazole HCl plus 1.0 mL of 4.3.11.3 mouse MAb (#HP1-200 Kit; Hypoxyprobe, Inc.).

2. 4% paraformaldehyde (PFA) in 1× PBS.

3. Paraffin.

4. 100% ethanol, 95% ethanol, 70% ethanol.

5. Methanol.

6. Xylenes (#247642-4L-CB; Sigma).

7. Normal goat serum (#S-1000; Vector Laboratories).

8. Antigen unmasking solution (#H3300; Vector Laboratories).

9. Hydrogen peroxide (H_2O_2) solution (30% w/w).

10. VECTASTAIN ABC HRP Kit (Peroxidase, Standard, #PK-4000; Vector Laboratories).

11. DAB Peroxidase (HRP) Substrate Kit (with Nickel), 3,3'-diaminobenzidine (#SK-4100; Vector Laboratories).

12. Hematoxylin (#SH26-500D; Fisher Scientific). 1:5 dilutions with Millipore H_2O.

13. Fisher Chemical™ Permount™ Mounting Medium (#SP15-500; Fisher Scientific).

2.3 Buffers

1. 1× PBS.

2. 1× TT buffer: 25 mM NaCl, 200 mM Tris–HCl, pH 7.6, 0.05% Tween 20 in Millipore H_2O.

3. Blocking buffer: 1× TT buffer containing 4% BSA and 2% goat serum.

2.4 Others	1. Tissue-Tek manual slide staining set.
	2. Coverslips.
	3. Microtome.
	4. PAP pen.

3 Methods

3.1 Hypoxyprobe Injection (See Note 2)	1. Prepare Hypoxyprobe solutions (100 mg/mL) in 0.9% saline (*see* **Note 3**).
	2. Inject Hypoxyprobe solution intraperitoneally into mice at 60 mg/kg.
	3. 1.5 to 2 h after injection, harvest tissue for fixation and processing (*see* below).

3.2 Tissue Fixation and Processing	1. Dissect tissue of interest from animal, and place immediately into cold PBS.
	2. Fix tissue in 4% PFA overnight at 4 °C with gentle shaking.
	3. Transfer tissue into a plastic cassette, and trim the tissue with a new razor blade if necessary.
	4. Wash tissue in PBS two times with 30 min each time.
	5. Tissue dehydration in 70% ethanol for 1 h. Shake if possible.
	6. Tissue dehydration in 95% ethanol two times with 1 h each. Shake if possible.
	7. Tissue dehydration in 100% ethanol for 1 h (*see* **Note 4**).
	8. Tissue dehydration in new 100% ethanol three to four times with 1 h each and gentle shaking.
	9. Transfer tissue to xylenes and soak two times with 1 h each.
	10. Remove xylenes as much as possible, and place tissue in paraffin solution three times with 1 h each (*see* **Note 5**).
	11. Embed tissue with fresh paraffin solution. Orient if necessary. Be sure to place the cutting face downward (bottom of the base mold).
	12. Cool the block for 20 min. Remove the molds, and clean the blocks.
	13. Paraffin-embedded tissues are sectioned at 5 μm using a microtome.

1.1 Immunohisto-chemistry (Carry Out All Procedures at Room Temperature Unless Otherwise Specified)	1. Prewarm the slides in a dry oven at 55 °C for 20 min.
	2. Deparaffinize the slides in xylenes two times with 20 min each.

3. Rehydrate the slides: 100% ethanol for 2 min, two times; 95% ethanol for 2 min, two times; 70% ethanol for 2 min; tap water for 5 min.

4. Antigen unmasking: We use commercial antigen unmasking solution and follow the manufacturer's instructions.

5. Block endogenous peroxidase activity: incubate the slides in 0.6% H_2O_2 for 15 min (*see* **Note 6**).

6. Wash slides in TT buffer for three times with 5 min each.

7. Remove extra buffer from the slides, and draw a barrier around the section with a PAP pen (*see* **Note 7**).

8. Lay slides flat in a humidified chamber, add blocking buffer to cover the sections, and incubate for 1 h.

9. Tip off the blocking buffer, and incubate the slides with primary antibody diluted in blocking buffer overnight at 4 °C.

10. Wash the slides with TT buffer for three times with 5 min each (*see* **Note 8**).

11. Incubate the slides with secondary antibody diluted in blocking buffer for 1 h.

12. While the secondary antibody is incubating, prepare the ABC-HRP reagent following the kit instructions (*see* **Note 9**).

13. Wash the slides with TT buffer three times with 5 min each.

14. Incubate the slides with ABC-HRP for 30 min.

15. Wash the slides with TT buffer for three times with 5 min each.

16. Prepare the DAB substrate following the kit instructions.

17. Wash the slides once with tap water.

18. Lay slides flat on a paper towel or bench paper, and apply DAB substrate solution to cover the sample area. Let the reaction run 0.5–10 min away from direct light. Monitor the staining under microscope. DAB substrate incubation time should be determined empirically for optimal staining.

19. Rinse the slides in tap water three times with 5 min each.

20. Counterstain the slides with hematoxylin solution for 0.5–1 min.

21. Rinse the slides in tap water for three times with 5 min each.

22. Dehydrate the slides: 70% ethanol for 1 min, 95% ethanol for 1 min, 100% ethanol two times with 1 min each, and xylene two times with 5 min each.

23. Mount the slides.

24. See the following: Fig. 1 for representative IHC staining of Hypoxyprobe (a), human HIF1α (b), mouse HIF1α (c), and HIF2α (d). Please refer to Reference 4 for human HIF2α IHC staining.

A Hypoxyprobe B human HIF1α

C mouse HIF1α D mouse HIF2α

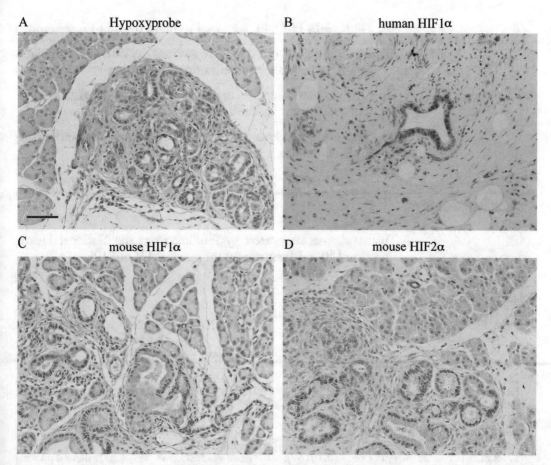

Fig. 1 (**a**) Immunohistochemical staining for Hypoxyprobe in pancreata from 2-month-old p48-Cre; Lox-Stop-Lox-Kras[G12D] mice with areas of pancreatic intraepithelial neoplasia (PanIN). (**b**) Immunohistochemical staining for human HIF1α from human pancreatic tissue with pancreatic ductal adenocarcinoma. (**c**) Immunohistochemical staining for mouse HIF1α from 2-month-old p48-Cre; Lox-Stop-Lox-Kras[G12D] mice with areas of PanINs. (**d**) Immunohistochemical staining of mouse HIF2α from 2-month-old p48-Cre; Lox-Stop-Lox-Kras[G12D] mice with areas of PanINs. Scale bar: 50 μm

4 Notes

1. HIF1α and HIF2α proteins can be accumulated to at least low levels under certain pathological conditions, making Hypoxyprobe the more reliable approach to detect hypoxia.

2. All animal protocols must be reviewed and approved by the institutional animal care and use committee.

3. We do a further tenfold dilution from stock before each injection.

4. The tissue can be stored in 100% ethanol at −20 °C before further processing.

5. Make sure the paraffin temperature doesn't exceed 58 °C.

6. Dilute 30% H_2O_2 in methanol to make 0.6% H_2O_2 solution.

7. Be careful not to let the tissue sections dry out. Keep the slides in TT buffer before and after drawing barrier.

8. Do not cross contaminate the samples, i.e., wash negative control or different antibodies separately.

9. The mixed ABC reagents must sit for 30–40 min before adding to the slides.

Acknowledgments

This work was supported by funding from the National Heart, Lung, and Blood Institute (grant number HL66310).

References

1. Nakazawa MS, Keith B, Simon MC (2016) Oxygen availability and metabolic adaptations. Nat Rev Cancer 16(10):663–673. https://doi.org/10.1038/nrc.2016.84

2. Lin N, Simon MC (2016) Hypoxia-inducible factors: key regulators of myeloid cells during inflammation. J Clin Invest 126(10):3661–3671. https://doi.org/10.1172/JCI84426

3. Lee KE, Simon MC (2015) SnapShot: hypoxia-inducible factors. Cell 163(5):1288–1288.e1281. https://doi.org/10.1016/j.cell.2015.11.011

4. Gordan JD, Lal P, Dondeti VR, Letrero R, Parekh KN, Oquendo CE, Greenberg RA, Flaherty KT, Rathmell WK, Keith B, Simon MC, Nathanson KL (2008) HIF-alpha effects on c-Myc distinguish two subtypes of sporadic VHL-deficient clear cell renal carcinoma. Cancer Cell 14(6):435–446. https://doi.org/10.1016/j.ccr.2008.10.016

5. Lee KE, Spata M, Bayne LJ, Buza EL, Durham AC, Allman D, Vonderheide RH, Simon MC (2016) Hif1a deletion reveals pro-neoplastic function of b cells in pancreatic neoplasia. Cancer Discov 6(3):256–269. https://doi.org/10.1158/2159-8290.CD-15-0822

6. Eisinger-Mathason TS, Zhang M, Qiu Q, Skuli N, Nakazawa MS, Karakasheva T, Mucaj V, Shay JE, Stangenberg L, Sadri N, Pure E, Yoon SS, Kirsch DG, Simon MC (2013) Hypoxia-dependent modification of collagen networks promotes sarcoma metastasis. Cancer Discov 3(10):1190–1205. https://doi.org/10.1158/2159-8290.CD-13-0118

Chapter 25

Analysis of Hypoxia and the Hypoxic Response in Tumor Xenografts

Nuray Böğürcü, Sascha Seidel, Boyan K. Garvalov, and Till Acker

Abstract

Solid tumors are often characterized by insufficient oxygen supply (hypoxia), as a result of inadequate vascularization, which cannot keep up with the rapid growth rate of the tumor. Tumor hypoxia is a negative prognostic and predictive factor and is associated with a more aggressive phenotype in various tumor entities. Activation of the hypoxic response in tumors, which is centered around the hypoxia-inducible transcription factors (HIFs), has been causally linked to neovascularization, increased radio- and chemoresistance, altered cell metabolism, genomic instability, increased metastatic potential, and tumor stem cell characteristics. Thus, the hypoxic tumor microenvironment represents a main driving force for tumor progression and a potential target for therapeutic interventions. Here, we describe several methods for the analysis of tumor hypoxia and the hypoxic response in vivo in tumor xenograft models. These methods can be applied to various tumor models, including brain tumor xenotransplants, and allow simultaneously determining the extent and distribution of hypoxia within the tumor, analyzing HIF levels by immunohistochemistry and immunoblot, and quantifying the expression of HIF target genes in tumor tissue. The combination of these approaches provides an important tool to assess the role of the hypoxic tumor microenvironment in vivo.

Key words Hypoxia, Hypoxyprobe, Hypoxic response, Hypoxia-inducible factor, VEGF, Glioblastoma, Brain tumor, Xenograft, Transplantation model, Tumor resection, Immunohistochemistry, Immunoblot, Real-time PCR

1 Introduction

Hypoxia is a common feature of rapidly growing tumors with high metabolic activity. As the tumor expands beyond the diffusion range of oxygen from the supplying vasculature, local foci with insufficient oxygen availability emerge within the tumor [1, 2]. This shortage of oxygen leads to widespread adaptive cellular responses, such as the activation of pro-survival, proliferative, metastatic, but also pro-apoptotic pathways [3]. One of the imme-

---

Nuray Böğürcü and Sascha Seidel contributed equally to this work.
Boyan K. Garvalov and Till Acker jointly supervised this work.

L. Eric Huang (ed.), *Hypoxia: Methods and Protocols*, Methods in Molecular Biology, vol. 1742,
https://doi.org/10.1007/978-1-4939-7665-2_25, © Springer Science+Business Media, LLC 2018

diate cellular responses to overcome the lack of oxygen is the secretion of pro-angiogenic factors such as vascular endothelial growth factor (VEGF), resulting in the recruitment of new blood vessels to reconstitute the oxygen supply of the tumor tissue [4]. Nevertheless, the newly formed vasculature in rapidly proliferating tumors is often poorly organized and leaky due to abnormal pericyte coverage [5] and therefore heterogeneously perfused [6, 7], overall leading to sustained reduction of the oxygen supply of tumor cells. Reduced oxygenation has been linked to numerous aspects of tumor development and pathophysiology. Moreover, hypoxia correlates with the survival probability of patients [8] and is associated with resistance to traditional radiotherapies and chemotherapies [1, 7, 9–11].

At the molecular level, hypoxia induces a number of responses that modulate cell proliferation, survival, migration, and inflammation. The main response of cellular adaption to hypoxia in tumors is mediated by the heterodimeric hypoxia-inducible transcription factors HIF1 and HIF2, consisting of an O_2-regulated α subunit (HIF1α or HIF2α) and an oxygen-insensitive β subunit [12]. The expression of HIF1α or HIF2α has been linked to a poor patient outcome in different tumor entities [13–15]. The hypoxia-inducible factors regulate a variety of target genes involved in diverse biological processes including tumor cell metabolism, pH regulation, proliferation, apoptosis, metastasis, and invasion [16–21]. Moreover, hypoxia-inducible factors are also the main drivers of tumor angiogenesis by activating pro-angiogenic factors such as VEGF, VEGF receptors, and angiopoietin 2, which stimulate blood vessel formation to overcome oxygen deprivation of tumor cells [22, 23]. Furthermore, HIF1α and HIF2α have been shown to increase the self-renewal capacity of cancer cells, thereby promoting aggressiveness and the stem cell phenotype in different tumor entities [24–28]. These adaptive responses are generally linked to hallmarks of cancer [29], underlining the pivotal role of tumor hypoxia and the hypoxic response for tumor development and progression. Typically, the hypoxic response of tumor cells is studied in cell culture under predefined oxygen concentrations. A deeper and proper understanding of the pathophysiological effects of hypoxia on tumor development, however, requires the assessment of the extent and spatial distribution of hypoxia, as well as the expression of HIFs and their target genes in vivo within the tumor tissue. Here, we describe several methods for the visualization of tumor hypoxia and the analysis of HIF1α or HIF2α expression and the hypoxic response in tumor xenograft models, allowing the detailed examination of hypoxic mechanisms in intact tumor tissues.

2 Materials

2.1 Tumor Resection

1. 0.9% saline.
2. 4% paraformaldehyde (PFA) solution (*see* **Note 1**).
3. Hypoxyprobe Plus Kit (FITC-Mab) (NPI Inc., Hypoxyprobe-1, HP2-kit).
4. Sterile syringes (1 mL, 25G × 0.625 in.).
5. Blunt end surgical scissors (FST).
6. Iris spatula (FST).
7. Curved forceps (FST).
8. Single edge blade (Single Edge, Stainless Steel Uncoated GEM Blades).
9. 1.5 mL reaction tubes.
10. Liquid nitrogen.
11. Disposable scalpels.
12. Sterile 10 cm plastic petri dishes.

2.2 Histology

1. Phosphate-buffered saline (PBS): 140 mM NaCl, 2.7 mM KCl, 10 mM $Na_2HPO_4 \times 2H_2O$, 1.8 mM KH_2PO_4, pH 7.4.
2. Paraffin embedding machine.
3. Vacuum infiltration processor.
4. Paraffin.
5. Microtome.
6. Microscope slides.
7. Coverslips (24 mm × 50 mm).
8. Staining cuvettes.
9. Xylene.
10. 100% ethanol (EtOH).
11. 96% EtOH.
12. 70% EtOH.
13. 50% EtOH.
14. Tris-buffered saline (TBS, 10×): 1.5 M NaCl, 0.1 M Tris-HCl, pH 7.4.
15. TBS (1×) containing 0.1% Tween-20 (TBST).
16. PBS containing 0.1% Tween-20 (PBST).
17. ABC reagent from the VECTASTAIN Elite ABC HRP Kit (Vector Laboratories).

18. Citrate buffer: 10 mM sodium citrate, pH 6.0.

19. Deionized H_2O (dH_2O).

20. 0.6% hydrogen peroxide (H_2O_2) in dH_2O.

21. PAP Pen.

22. Normal goat serum (NGS).

23. Liquid DAB+ Substrate Chromogen System (Dako).

24. Hematoxylin.

25. Tap water.

26. Cytoseal XYL mounting medium (Thermo Scientific).

27. BenchMark XT Automated IHC/ISH slide staining system (Ventana/Roche).

28. HIF1α antibody (1:200) (Cayman Chemical, 10006421).

29. FITC-Mab1 antibody (1:100) (part of the Hypoxyprobe Plus Kit; *see* Subheading 2.1).

30. HRP-conjugated anti-FITC (1:200) (part of the Hypoxyprobe Plus Kit; *see* Subheading 2.1).

31. M.O.M. Mouse IgG blocking reagent (part of the Vector M.O.M. Immunodetection Kit, Vector Laboratories).

32. Staining cuvettes.

33. Humidified staining tray (chamber).

34. Steamer (Braun).

35. Orbital shaker.

2.3 Protein Isolation and Immunoblotting

1. Laemmli sample buffer: 10 mM Tris, 2% SDS, 2 mM EGTA, and 20 mM NaF in dH_2O.

2. ULTRA-TURRAX disperser (IKA).

3. Sonifier (Bandelin).

4. Centrifuge for 1.5–2 mL tubes.

5. Heating block.

6. HIF1α antibody (Cayman Chemical, 10006421, 1:3000).

7. HIF2α antibody (Novus, NB100-122, 1:1000).

8. Alpha tubulin antibody (Dianova, DLN-09992, 1:8000).

2.4 RNA Isolation

1. RNeasy Mini Kit including RLT lysis buffer (Qiagen).

2. RNase-Free DNase Set (Qiagen).

3. Precellys 24 Homogenizer (Bertin Instruments).

4. Ceramic beads (VWR, CK14 Soft Tissue Homogenizing Kit, beads in bulk format, 325 g of inert 1.4 mm ceramic beads).

5. Screw cap microtubes.

6. Human-specific VEGF-A primers.

   VEGF-A forward: 5′-AGCCTTGCCTTGCTGCTCTA-3′

   VEGF-A reverse: 5′-GTGCTGGCCTTGGTGAGG-3′.

7. Human-specific β-actin primers

   β-actin forward: 5′-AGAAAATCTGGCACCACACC-3′

   β-actin reverse: 5′-AGAGGCGTACAGGGATAGCA-3′.

3 Methods

3.1 Perfusion and Tumor Resection

In the following section, we describe a method for the resection of brain tumor tissue from an orthotopic intracranial transplantation mouse model and the subsequent sample preparation for the analysis of tumor hypoxia by histology and of hypoxia-induced gene expression by immunoblotting and real-time PCR (*see* **Note 2**). The described procedures can also be applied to subcutaneous and mammary fat pad transplantation tumor models by adapting the tumor resection step accordingly (*see* **Note 3**). For the isolation of tumor tissue, it is important to remove as much of the surrounding nonneoplastic mouse tissue as possible.

1. Perform an intracranial xenograft experiment as previously described [30].

2. At the endpoint of your experiment, inject the tumor-bearing mice with 60 μg/g Hypoxyprobe in 0.9% saline intraperitoneally 90 min prior to cardiac perfusion (*see* **Note 4**).

3. For anesthesia, inject the mice intraperitoneally with a mixture of 100 mg/kg ketamine and 10 mg/kg xylazine.

4. Before the perfusion, make sure that the mouse is deeply anesthetized by checking for absence of the withdrawal reflex of the toe.

5. Perfuse the mouse intracardially using a 23G butterfly needle with cold 0.9% saline.

6. Decapitate the perfused animal. In order to remove the brain, perform a midline incision at the skin close to the hindbrain. Flip the skin aside to expose the skull.

7. Carefully make a small incision at the base of the skull with blunt end surgical scissors, and cut along the sagittal suture without damaging the brain (Fig. 1a) (*see* **Note 5**).

8. Tilt one half of the skull bone to the side, and break it off using a curved forceps. Repeat this step with the other side of the skull bone. If the frontal bone remains, make a careful cut through the most anterior part of the skull between the eyes, and remove the bone (Fig. 1b).

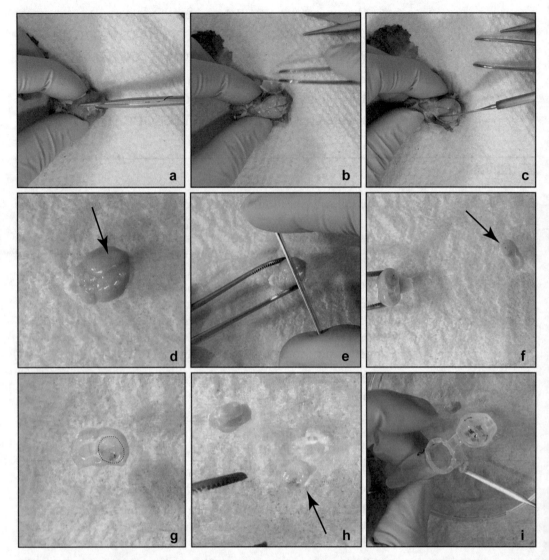

Fig. 1 Illustration of tumor resection and sample preservation. (**a–c**) Excision of the brain includes cutting and removing the cranium, as well as removal of the meninges and cutting of the cranial nerves. (**d**) After brain excision, the injection site where the tumor cells were transplanted can be located as a small, slightly darker spot, as indicated with the arrow. (**e–f**) Bisecting the brain at the injection site divides the tumor at its largest diameter. The frontal part, containing around half of the tumor, which should be used for further histological analysis, is indicated with an arrow (**f**). (**g–i**) In the posterior part, the tumor tissue, encircled with a dotted line (**g**), can be easily identified by its structure and texture. After removing the healthy brain tissue, the tumor, indicated with an arrow (**h**), is divided into two parts, transferred to two separate 1.5 mL reaction tubes, and snap-frozen for subsequent protein and RNA isolation, respectively

9. Use an iris spatula to remove the meninges around the brain, turn the skull upside down, and carefully slide the iris spatula under the posterior part of the brain. Disrupt the optic nerves and other cranial nerves. Carefully take out the brain, and place it into a sterile petri dish (Fig. 1c, d).

10. Locate the injection site where the tumor cells were transplanted (Fig. 1d, indicated with an arrow).

11. Fix the brain with the help of forceps, and use a single edge blade to cut the brain along the frontal axis at the injection site (Fig. 1e) (*see* **Note 6**).

    For subcutaneous and mammary fat pad tumor models, resect the tumor, and divide it with the blade approximately in the middle.

12. Transfer the frontal part containing one half of the tumor (Fig. 1f, indicated with an arrow) into 5 mL of 4% PFA solution in a 15 mL Falcon tube for overnight fixation and subsequent analysis by histology (Fig. 1f).

13. Locate the tumor in the posterior part of the brain (Fig. 1g, indicated by a dotted line) (*see* **Note 7**).

14. Remove as much of the surrounding healthy brain tissue as possible with the help of sterile forceps and a scalpel (Fig. 1h) (*see* **Note 8**).

15. After removing the healthy brain tissue, divide the isolated tumor tissue into two pieces using a scalpel.

16. For subsequent protein and RNA isolation, transfer the tumor samples into 1.5 mL reaction tubes, and snap freeze them immediately in liquid nitrogen (Fig. 1i).

3.2 Detection of Tumor Hypoxia by Histology

This section describes how to detect tumor hypoxia by immunohistochemical Hypoxyprobe and HIF1α staining (Fig. 2).

3.2.1 Sample Preparation for Histology

1. Fix the frontal part containing the tumor in 4% PFA at 4 °C overnight (*see* **step 12** in Subheading 3.1).

2. On the following day, transfer the brain to PBS, and store the tumors at 4 °C until paraffin embedding (*see* **Note 9**).

3. Perform tissue dehydration and paraffin infiltration. By using the Sakura Tissue-Tek VIP 5 Jr. vacuum infiltration processor, the following protocol was applied:

   - 1.5 h postfixation in 4% PFA at 35 °C
   - 30 min dH$_2$O
   - 45 min 50% EtOH
   - 45 min 70% EtOH
   - 1 h 96% EtOH
   - 1 h 96% EtOH
   - 1 h 100% EtOH

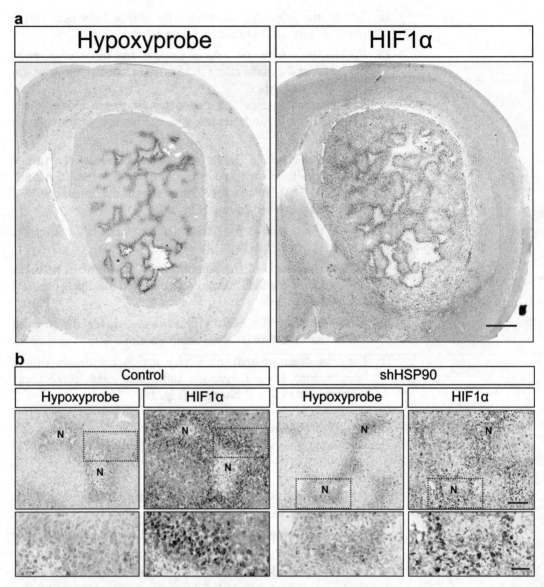

Fig. 2 Analysis of tumor hypoxia and the hypoxic response in orthotopic glioblastoma xenotransplants by immunohistochemistry. (**a**) Hypoxic areas and the hypoxic response around necrotic regions within the tumor can clearly be identified by Hypoxyprobe and HIF1α staining, respectively. Note the similar patterns of hypoxic areas (Hypoxyprobe) and HIF1α expression in consecutive sections of tumor containing brains. Scale bar 500 μm. (**b**) To examine the role of HSP90 for the hypoxic response, tumors in immunodeficient mice that were orthotopically transplanted with control and HSP90 knockdown (shHSP90) G55 glioblastoma cells, were stained for Hypoxyprobe and HIF1α [31]. G55 control and shHSP90 tumors displayed similar levels of tumor hypoxia, as revealed by Hypoxyprobe staining, while HIF1α expression was markedly reduced in HSP90 knockdown tumors, indicating that HSP90 is an important regulator of the hypoxic response. **N**, necrosis. Scale bars 100 μm (top, lower magnification) and 40 μm (bottom, higher magnification)

- 1 h 100% EtOH
- 1 h xylene
- 1 h xylene
- Four times 30 min paraffin

  For subsequent embedding, the Sakura Tissue-Tek TEC tissue embedding console was used.

4. Cut thin sections (4–6 µm) of the paraffin embedded tissue using a microtome and transfer them on glass slides (*see* **Note 10**).

3.2.2 Hypoxyprobe Detection

All the steps for dewaxing, rehydration, antigen retrieval, and peroxidase blocking are done in staining cuvettes. Washing steps are performed in staining cuvettes on an orbital shaker at moderate speed. Blocking of unspecific antibody binding, incubation with the FITC-Mab1 and anti-FITC HRP antibodies, as well as the DAB chromogen reaction are performed in a humidified staining chamber.

1. Place the paraffin sections that were prepared in the previous step into the staining cuvette.

2. For dewaxing, incubate the paraffin sections two times for 10 min in xylene.

3. For rehydration of the tissue, incubate the sections in a series of descending alcohol concentration.
   - Two times with 100% EtOH for 5 min.
   - Two times with 96% EtOH for 5 min.
   - Two times with 70% EtOH for 5 min.

4. Rehydrate the sections by washing in dH_2O for 5 min and in 1× TBS buffer for additional 5 min.

5. Preheat the citrate buffer (pH 6.0) in a steamer for 10 min.

6. For antigen retrieval, boil the sections in preheated citrate buffer for 10 min in a steamer.

7. Cool down the sections in the cuvette for 20 min at room temperature.

8. Wash the sections two times with 1× TBS buffer for 5 min.

9. To block endogenous peroxidase activity, incubate the sections in 0.6% H_2O_2 for 30 min.

10. Wash the sections three times with 1× TBS buffer for 5 min.

11. Encircle the tissue with a PAP Pen to minimize the amount of antibody used. For a typical brain section encircled with a PAP Pen, 200 µL blocking/antibody solution is used.

12. Incubate the sections with 20% normal goat serum (NGS) in TBST buffer to block unspecific antibody binding.

13. Incubate the section with FITC-Mab1 antibody (1:100 dilution in 10% NGS in TBST) for 1 h at room temperature.

14. Wash the sections three times with TBST buffer for 5 min.

15. Incubate the section with anti-FITC HRP antibody (1:200 dilution in 10% NGS in TBST) for 2 h at room temperature.

16. Wash the sections three times with TBST buffer for 5 min.

17. Prepare the diaminobenzidine (DAB+) substrate according to the manufacturer's recommendations.

18. To perform the chromogen reaction, incubate the sections with DAB+ for up to 5 min, depending on the desired chromogen intensity.

19. To stop the chromogen reaction, transfer the sections to TBS immediately.

20. Perform nuclear counter staining with hematoxylin for 2 min.

21. Rinse the sections with tap water two times for bluing.

22. Rinse the sections with dH_2O once.

23. Dehydrate the sections in a series of ascending alcohol concentration.

    • Two times with 70% EtOH for 5 min
    • Two times with 96% EtOH for 5 min
    • Two times with 100% EtOH for 5 min

24. Incubate the sections in xylene two times for 5 min.

25. Mount the sections permanently with coverslips by using Cytoseal XYL mounting medium.

3.2.3 HIF1α Staining

All the steps for the HIF1α staining were done on an automated staining system (Ventana BenchMark XT IHC/ISH Staining module), using the ultraView universal DAB v3 kit according to the manufacturer's instructions, with the following modifications (*see* **Note 11**):

1. Paraffin sections.

2. Deparaffinization.

3. Incubation with Cell Conditioner 2 (CC2, pH 6, Roche) for 36 min at 94 °C, for antigen retrieval.

4. Incubation with M.O.M. Mouse IgG blocking reagent (part of the Vector M.O.M. Immunodetection Kit, Vector Laboratories) (four drops in 2.5 mL PBS) for 32 min.

5. Incubation with the primary HIF1α antibody (1:200 dilution) for 1 h at 37 °C.

6. The staining reaction was performed by using the ultraView universal DAB v3 kit for 8 min at 37 °C.

7. Counter staining with hematoxylin for 8 min.

8. Bluing reagent for 8 min.

We observed the best staining results for HIF1α using the above-mentioned procedure with the automated staining system. Alternatively, the staining could be performed manually using the following protocol:

1. For dewaxing, incubate the paraffin sections two times in xylene for 10 min each.

2. For rehydration of the tissue, incubate the sections in a series of descending alcohol concentration.
   - Two times with 100% EtOH for 5 min.
   - Two times with 96% EtOH for 5 min.
   - Two times with 70% EtOH for 5 min.
   - Rehydrate the sections by washing in dH$_2$O for 5 min and in 1× PBS buffer for 5 additional min.

3. Preheat the citrate buffer (pH 6.0) in a steamer for 10 min.

4. For antigen retrieval, boil the sections in preheated citrate buffer for 10 min in a steamer.

5. Cool down the sections in the cuvette for 20 min at room temperature.

6. Wash the sections two times with 1× PBS buffer for 5 min.

7. To block endogenous peroxidase activity, incubate the sections in 0.6% H$_2$O$_2$ for 30 min.

8. Wash the sections three times with 1× PBS buffer for 5 min.

9. Encircle the tissue with a PAP Pen to minimize the amount of antibody use. For a typical brain section encircled with a PAP Pen, 200 μL blocking/antibody solution is used.

10. Incubate the sections with 20% normal goat serum (NGS) in PBST buffer to block unspecific antibody binding.

11. Incubate the section with HIF1α antibody (1:200 dilution in 10% NGS in PBST) for 2 h at room temperature.

12. Wash the sections three times with PBST buffer for 5 min.

13. Incubate the section with biotinylated anti-rabbit antibody from the ABC HRP kit (1:200 dilution in 10% NGS in PBST) for 1 h at room temperature.

14. Wash the sections three times with PBST buffer for 5 min.

15. Incubate with ABC reagent according to the manufacturer's instructions for 45 min at room temperature.

16. Wash the sections three times with PBST buffer for 5 min.

17. Prepare the diaminobenzidine (DAB+) substrate according to the manufacturer's recommendations.

18. To perform the chromogen reaction, incubate the sections with DAB+ for up to 5 min, depending on the desired chromogen intensity.

19. To stop the chromogen reaction, transfer the sections to PBS immediately.

20. Perform nuclear counter staining with hematoxylin (1:4 dilution in dH_2O) for 2 min.

21. Rinse the sections with tap water two times for bluing.

22. Rinse the sections with dH_2O once.

23. Dehydrate the sections in a series of ascending alcohol concentration.
    - Two times with 70% EtOH for 5 min.
    - Two times with 96% EtOH for 5 min.
    - Two times with 100% EtOH for 5 min.

24. Incubate the sections in xylene two times for 5 min.

25. Mount the sections permanently with coverslips by using Cytoseal XYL.

3.3 Protein Isolation from Tumor Samples to Assess HIF1α and HIF2α Levels

In this section, we describe how to isolate total protein from tumor samples for the detection of proteins of interest by immunoblotting. After protein isolation, a standard method for immunoblotting can be applied (Fig. 3).

1. Weigh the cryopreserved tumor samples for protein isolation.

2. Add 20 times the amount of Laemmli buffer to the tumor samples (*see* **Note 12**).

3. Mechanically disrupt the tumor samples in Laemmli buffer using an ULTRA-TURRAX disperser for 1 min until the sample is homogenized (*see* **Note 13**).

4. Sonicate the tumor samples at 90% amplitude with 0.5 s pulse period durations for 1 min.

5. Centrifuge the samples at $17,000 \times g$ for 5 min at room temperature.

6. Collect the supernatant in a fresh reaction tube, and discard the pellet.

7. Heat the protein samples at 95 °C for 5 min in a heating block.

8. Measure the protein concentration.

9. Proceed with Western blotting.

10. Use HIF1α (1:3000 dilution) and HIF2α (1:1000 dilution) antibodies to detect protein expression (*see* **Note 14**).

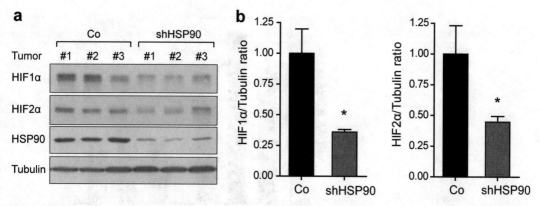

Fig. 3 Analysis of the hypoxic response in tumor samples by immunoblotting. Protein samples isolated from G55 control (Co) and HSP90 knockdown (shHSP90) tumors [31] were analyzed by immunoblotting ($n = 3$ separate tumors per group). A pronounced reduction in HIF1α and HIF2α expression levels was observed upon HSP90 silencing (**a**). Densitometric quantification of HIF1α and HIF2α levels normalized to Tubulin using ImageJ confirmed the significant reduction of HIF1α and HIF2α protein levels in these tumor samples (**b**). All values are mean + SEM. $^*P < 0.05$

3.4 RNA Isolation from Tumors to Detect Hypoxia-Dependent Gene Expression

In this section, we describe how to prepare the tumor samples for RNA isolation to detect the expression levels of HIF target genes, such as VEGF-A, as a measure of the hypoxic response within the tumor sample (Fig. 4). Subsequently, every laboratory may use their established RNA isolation kit and protocol. Our protocol is based on the RLT lysis buffer from the RNeasy kit (Qiagen). Depending on the supplier, buffers in the kit could be named differently and should be used according to the manufacturer's instructions.

1. Weigh the cryopreserved tumor samples for RNA isolation in the reaction tube (*see* **Note 15**).

2. Add 350 µL RLT buffer (lysis buffer).

3. Mechanically homogenize the tumor samples for 1 min in RLT buffer using a Precellys sample homogenizer with ceramic beads in microtubes (*see* **Note 16**).

4. Spin down the beads and collect the supernatant.

5. Proceed with RNA isolation following the manufacturer's recommendations for the corresponding isolation kit of your choice.

6. Transcribe 1 µg of RNA to cDNA.

7. Analyze HIF target gene expression by real-time PCR using human-specific primers (*see* **Note 17**).

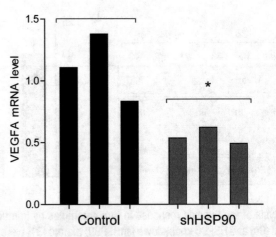

Fig. 4 Analysis of the hypoxic response by quantifying the expression level of the HIF1α target gene VEGF-A in G55 control and HSP90 knockdown (shHSP90) tumors [31]. VEGF-A mRNA levels from tumors were assessed by real-time PCR ($n = 3$ separate tumors per group) and are shown for each individual tumor. VEGF-A expression is significantly downregulated in HSP90 silenced tumors. *$P < 0.05$

---

4   Notes

1. Dissolve 80 g PFA in 1.5 L dH$_2$O and adjust the pH to 11 with 5 M NaOH and stir until the solution gets clear. Add 100 mL 20× PBS, adjust the pH to 7.4 with HCl, and fill up to 2 L with dH$_2$O. Aliquots can be stored at −20 °C. It is important to use only freshly thawed PFA for the perfusions.

2. Since HIFs get degraded very rapidly upon exposure to increased oxygen levels, it is important to isolate and process the tumor samples as quickly as possible until the fixation or lysis steps. We find it optimal to perform these experiments in pairs of two people. We recommend that one person keeps track of equal Hypoxyprobe incubation times and performs the intracardial perfusions, whereas the second person immediately isolates the brain and resects the tumor tissue for subsequent sample preparation.

3. The protocols described above can also be applied to tumor transplantation models other than intracranial, such as subcutaneous or mammary fat pad transplantation. In this case, an incision to the skin surrounding the tumor should be made by blunt end surgical scissors. Using a curved forceps, evert the skin including the tumor, and carefully separate the tumor tissue from the epidermis or mammary tissue with a scalpel.

Generally, subcutaneous and mammary fat pad xenotransplants are easier to fully separate from the surrounding host tissue compared to intracranial tumors.

4. It is important to keep Hypoxyprobe incubation times equal for all injected animals.

5. To prevent any damage to the brain while cutting the skull, the scissor should be located along the midline of the brain, and small cuts should be made.

6. It is important to keep the blade orthogonal to the surface. Avoid having several attempts to cut the brain. It is important to use single edge blade to protect the laboratory personnel performing the procedure.

7. At this step it is possible to take a photo of the tumor next to a ruler for calculating the tumor volume. Cutting the brain at the injection site enables the measurement of the largest tumor diameter. The tumor volume can be calculated using the section of the biggest tumor area by measuring the largest diameter (L) and largest perpendicular diameter (W), using the formula $L \times W^2/2$.

8. To prevent cross contamination between the samples, use new petri dishes, blades, and scalpels for every individual tumor, and clean the forceps and iris spatula carefully after every animal.

9. We strongly recommend embedding the brain in paraffin as soon as possible after fixation.

10. We observed better staining results using freshly cut paraffin sections.

11. The BenchMark XT Staining System (Ventana) has a predefined protocol for the ultraView universal DAB v3 kit, comprising over 90 steps. The indicated modifications refer to the steps that can be modified by the user; the remaining steps are fixed.

12. For every mg of tumor tissue, 20 μL Laemmli buffer should be added, e.g., for 50 mg of tumor tissue, 1000 μL Laemmli buffer should be used. Depending on the sample weight, the volume of buffer therefore varies for every tumor sample. We recommend to use a minimum of 10 mg and maximum of 75 mg of tumor sample to effectively homogenize and lyse the samples and to be able to perform all steps in 1.5 mL or 2 mL reaction tubes. If the tumor samples are heavier than 75 mg, they can be divided with a scalpel. In this case, it is important to work on ice and process the sample quickly, to prevent thawing and tissue degradation, especially for the analysis of HIF levels.

Note that the Laemmli buffer used for sample lysis does not contain bromophenol blue, to enable measurement of the protein concentration after lysis using colorimetric assays. Before running an SDS-PAGE gel for the Western blot, equal amounts of protein from all samples should be supplemented with 1/5 volume of 5× Laemmli buffer with bromophenol blue.

13. If you observe large unhomogenized tumor pieces in the buffer, we recommend to repeat the homogenization with the ULTRA-TURRAX. Keep in mind that really small tumor pieces will not be homogenized with the ULTRA-TURRAX, but this will not reduce the protein yield significantly. Handling the samples on ice will lead to precipitation of SDS in the Laemmli buffer and should therefore be avoided. Carefully clean the ULTRA-TURRAX after every tumor sample to avoid cross contamination.

14. Antibody dilutions may need to be optimized. Depending on the expression levels of HIF1α and HIF2α in the tumor cells, the amount of proteins loaded on the gel, and the efficiency of transfer during the Western blotting, different dilutions of the antibody may produce optimal results.

15. We recommend using between 5 mg and 30 mg of the tumor samples. Do not use samples larger than 30 mg, otherwise lysis efficiency will decline. The RNA yield will not increase linearly with increasing amounts of tumor sample.

16. Other mechanical methods such as using a mortar and pestle can also be used for lysing the tissue, but we strongly recommend using the Precellys tissue homogenizer. Using a mortar and pestle may lead to cross contamination between the samples, which can be minimized by cleaning the mortar and pestle properly to ensure that no RNA is left from the previous sample.

17. When analyzing human xenotransplants, it is important to use primers that are human-specific and do not amplify the mouse orthologue of the gene of interest. Since all tumor xenotransplants will contain stromal cells from the mouse host, selecting human-specific primers ensures that only changes in gene expression that occur in the tumor cells (but not in the stromal cells) will be quantified. The primers should be designed so that both the forward and the reverse primer sequence have at least one mismatch to the closest mouse orthologue, and it is recommended that at least one of the primers has ≥2 mismatches.

References

1. Evans SM, Judy KD, Dunphy I, Jenkins WT, Hwang WT, Nelson PT, Lustig RA, Jenkins K, Magarelli DP, Hahn SM, Collins RA, Grady MS, Koch CJ (2004) Hypoxia is important in the biology and aggression of human glial brain tumors. Clin Cancer Res 10:8177–8184

2. Rong Y, Durden DL, Van Meir EG, Brat DJ (2006) 'Pseudopalisading' necrosis in glioblastoma: a familiar morphologic feature that links vascular pathology, hypoxia, and angiogenesis. J Neuropathol Exp Neurol 65:529–539

3. Henze AT, Acker T (2010) Feedback regulators of hypoxia-inducible factors and their role in cancer biology. Cell Cycle 9:2749–2763

4. Liao D, Johnson RS (2007) Hypoxia: a key regulator of angiogenesis in cancer. Cancer Metastasis Rev 26:281–290

5. Morikawa S, Baluk P, Kaidoh T, Haskell A, Jain RK, McDonald DM (2002) Abnormalities in pericytes on blood vessels and endothelial sprouts in tumors. Am J Pathol 160:985–1000

6. Dewhirst MW, Kimura H, Rehmus SW, Braun RD, Papahadjopoulos D, Hong K, Secomb TW (1996) Microvascular studies on the origins of perfusion-limited hypoxia. Br J Cancer Suppl 27:S247–S251

7. Vaupel P, Thews O, Hoeckel M (2001) Treatment resistance of solid tumors: role of hypoxia and anemia. Med Oncol 18:243–259

8. Vaupel P, Mayer A (2007) Hypoxia in cancer: significance and impact on clinical outcome. Cancer Metastasis Rev 26:225–239

9. Chaplin DJ, Durand RE, Olive PL (1986) Acute hypoxia in tumors: implications for modifiers of radiation effects. Int J Radiat Oncol Biol Phys 12:1279–1282

10. Dewhirst MW, Cao Y, Moeller B (2008) Cycling hypoxia and free radicals regulate angiogenesis and radiotherapy response. Nat Rev Cancer 8:425–437

11. Vaupel P, Mayer A (2005) Hypoxia and anemia: effects on tumor biology and treatment resistance. Transfus Clin Biol 12:5–10

12. Semenza GL (2003) Targeting HIF-1 for cancer therapy. Nat Rev Cancer 3:721–732

13. Aebersold DM, Burri P, Beer KT, Laissue J, Djonov V, Greiner RH, Semenza GL (2001) Expression of hypoxia-inducible factor-1alpha: a novel predictive and prognostic parameter in the radiotherapy of oropharyngeal cancer. Cancer Res 61:2911–2916

14. Burri P, Djonov V, Aebersold DM, Lindel K, Studer U, Altermatt HJ, Mazzucchelli L, Greiner RH, Gruber G (2003) Significant correlation of hypoxia-inducible factor-1alpha with treatment outcome in cervical cancer treated with radical radiotherapy. Int J Radiat Oncol Biol Phys 56:494–501

15. Holmquist-Mengelbier L, Fredlund E, Lofstedt T, Noguera R, Navarro S, Nilsson H, Pietras A, Vallon-Christersson J, Borg A, Gradin K, Poellinger L, Pahlman S (2006) Recruitment of HIF-1alpha and HIF-2alpha to common target genes is differentially regulated in neuroblastoma: HIF-2alpha promotes an aggressive phenotype. Cancer Cell 10:413–423

16. Benita Y, Kikuchi H, Smith AD, Zhang MQ, Chung DC, Xavier RJ (2009) An integrative genomics approach identifies hypoxia inducible Factor-1 (HIF-1)-target genes that form the core response to hypoxia. Nucleic Acids Res 37:4587–4602

17. Dengler VL, Galbraith MD, Espinosa JM (2014) Transcriptional regulation by hypoxia inducible factors. Crit Rev Biochem Mol Biol 49:1–15

18. Denko NC (2008) Hypoxia, HIF1 and glucose metabolism in the solid tumour. Nat Rev Cancer 8:705–713

19. Iyer NV, Kotch LE, Agani F, Leung SW, Laughner E, Wenger RH, Gassmann M, Gearhart JD, Lawler AM, AY Y, Semenza GL (1998) Cellular and developmental control of O2 homeostasis by hypoxia-inducible factor 1 alpha. Genes Dev 12:149–162

20. Kim JW, Tchernyshyov I, Semenza GL, Dang CV (2006) HIF-1-mediated expression of pyruvate dehydrogenase kinase: a metabolic switch required for cellular adaptation to hypoxia. Cell Metab 3:177–185

21. Semenza GL (2012) Hypoxia-inducible factors: mediators of cancer progression and targets for cancer therapy. Trends Pharmacol Sci 33:207–214

22. Pugh CW, Ratcliffe PJ (2003) Regulation of angiogenesis by hypoxia: role of the HIF system. Nat Med 9:677–684

23. Takeda N, Maemura K, Imai Y, Harada T, Kawanami D, Nojiri T, Manabe I, Nagai R (2004) Endothelial PAS domain protein 1 gene promotes angiogenesis through the transactivation of both vascular endothelial growth factor and its receptor, Flt-1. Circ Res 95:146–153

24. Bar EE, Lin A, Mahairaki V, Matsui W, Eberhart CG (2010) Hypoxia increases the expression of stem-cell markers and promotes clonogenicity in glioblastoma neurospheres. Am J Pathol 177:1491–1502

25. Heddleston JM, Li Z, McLendon RE, Hjelmeland AB, Rich JN (2009) The hypoxic microenvironment maintains glioblastoma stem cells and promotes reprogramming towards a cancer stem cell phenotype. Cell Cycle 8:3274–3284

26. Li Z, Bao S, Wu Q, Wang H, Eyler C, Sathornsumetee S, Shi Q, Cao Y, Lathia J, McLendon RE, Hjelmeland AB, Rich JN (2009) Hypoxia-inducible factors regulate tumorigenic capacity of glioma stem cells. Cancer Cell 15:501–513

27. Seidel S, Garvalov BK, Wirta V, von Stechow L, Schanzer A, Meletis K, Wolter M, Sommerlad D, Henze AT, Nister M, Reifenberger G, Lundeberg J, Frisen J, Acker T (2010) A hypoxic niche regulates glioblastoma stem cells through hypoxia inducible factor 2 alpha. Brain 133:983–995

28. Semenza GL (2015) Regulation of the breast cancer stem cell phenotype by hypoxia-inducible factors. Clin Sci (Lond) 129:1037–1045

29. Hanahan D, Weinberg RA (2011) Hallmarks of cancer: the next generation. Cell 144:646–674

30. Sawamiphak S, Seidel S, Essmann CL, Wilkinson GA, Pitulescu ME, Acker T, Acker-Palmer A (2010) Ephrin-B2 regulates VEGFR2 function in developmental and tumour angiogenesis. Nature 465:487–491

31. Filatova A, Seidel S, Bogurcu N, Graf S, Garvalov BK, Acker T (2016) Acidosis acts through HSP90 in a PHD/VHL-independent manner to promote HIF function and stem cell maintenance in Glioma. Cancer Res 76:5845–5856

Chapter 26

Correlation of Glioma Proliferation and Hypoxia by Luciferase, Magnetic Resonance, and Positron Emission Tomography Imaging

Michael Karsy, David L. Gillespie, Kevin P. Horn, Lance D. Burrell, Jeffery T. Yap, and Randy L. Jensen

Abstract

Gliomas are the most common type of primary, malignant brain tumor and significantly impact patients, who have a median survival of ~1 year depending on mutational background. Novel imaging modalities such as luciferase bioluminescence, micro-magnetic resonance imaging (micro-MRI), micro-computerized tomography (micro-CT), and micro-positron emission tomography (micro-PET) have expanded the portfolio of tools available to study this disease. Hypoxia, a key oncogenic driver of glioma and mechanism of resistance, can be studied in vivo by the concomitant use of noninvasive MRI and PET imaging. We present a protocol involving stereotactic injection of syngenic F98 luciferase-expressing glioma cells generated by our laboratory into Fischer 344 rat brains and imaging using luciferase. In addition, 18-F-fludeoxyglucose, 18F–fluoromisonidazole, and 18F–fluorothymidine PET imaging are compared with quantified luciferase flux. These tools can potentially be used for assessing tumor growth characteristics, hypoxia, mutational effects, and treatment effects.

Key words Luciferase, Bioluminescence, Glioma, Animal magnetic resonance imaging, ^{18}F–fluoromisonidazole, ^{18}F–fluorothymidine

1 Introduction

Gliomas are the most common type of malignant brain tumor and range in severity based on the World Health Organization classification, mutational characteristics (e.g., isocitrate dehydrogenase, O6-methylguanine methyltransferase), and microenvironment changes [1–3]. Although several primary glioma models exist, the evaluation of cell mutation, imaging performance, and treatment options can be more rapidly assessed via a syngenic orthotopic glioma model [4–6]. Bioluminescence imaging with luciferase-capturing cameras can be used to evaluate luciferase-containing

Michael Karsy, David L. Gillespie and Kevin P. Horn contributed equally to this work.

L. Eric Huang (ed.), *Hypoxia: Methods and Protocols*, Methods in Molecular Biology, vol. 1742,
https://doi.org/10.1007/978-1-4939-7665-2_26, © Springer Science+Business Media, LLC 2018

gene expression plasmids that have been transfected into gliomas [7–9]. Several imaging platforms exist for the evaluation of animals, including micro-magnetic resonance imaging (micro-MRI), micro-computerized tomography (micro-CT), and micro-positron emission tomography (micro-PET). We present a protocol evaluating multimodal imaging of rodent gliomas, which can be used for monitoring tumor growth and evaluating treatment efficacy. The F98/Fischer 344 rat syngeneic implantation model has been well documented to reproduce some of the hallmarks of glioblastoma (GBM) in an immunocompetent animal, which is more clinically relevant than immunocompromised models [10, 11].

The complex tumor biology of GBM and its response to therapy can be readily evaluated with positron emission tomography (PET). Among the aspects of tumor biology that can be characterized and quantified with various radiotracers are glucose metabolism with ^{18}F–fluorodeoxyglucose (FDG), proliferation with ^{18}F–fluorothymidine (FLT) , and hypoxia with ^{18}F–fluoromisonidazole (FMISO) [12]. FDG is a radiolabeled analog of glucose that has been validated in the diagnosis, staging, and response assessment of patients with cancer; most malignant tissues have increased FDG uptake associated with an increased rate of glycolysis as well as increased glucose transport [13, 14]. FLT is a structural analog of the DNA constituent thymidine; it is not incorporated into DNA but instead is trapped intracellularly after phosphorylation by thymidine kinase and has increased uptake in proliferating cells [15]. FMISO is an azomycin-based hypoxic cell sensitizer that binds covalently to cellular molecules at rates that are inversely proportional to intracellular oxygen concentration [16].

2   Materials

2.1   Cell Culture and Luciferase Plasmid Transfection

1. F98 rat glioma cell line (ATCC).

2. Cell culture media: Dulbecco's Modified Eagle Medium, antibiotics (penicillin, streptomycin 10,000 U/mL), 5% fetal bovine serum.

3. Selective cell culture media: Dulbecco's Modified Eagle Medium, antibiotics (penicillin, streptomycin, 10,000 U/mL), 10% fetal bovine serum, selection antibiotic, such as hygromycin or G418.

4. Phosphate-buffered saline (PBS).

5. TrypLE (Invitrogen).

6. Trypan blue.

7. Lipofectamine 2000.

8. Hemocytometer or other cell counting device (e.g., Countess automated cell counter, Thermo Fisher Scientific).

9. Glass cloning cylinders.

10. Bright-Glo (Promega).

11. Dimethyl sulfoxide.

12. Hygromycin B antibiotic (50 mg/mL).

2.2 Stereotactic Cranial Injection

1. Fischer 344 rats (Envigo).

2. Matrigel (Corning).

3. Animal warming apparatus (recirculating water heating pad recommended).

4. Anesthesia (isoflurane) (*see* **Note 1**).

5. Medical grade oxygen and vaporizer.

6. Fur clippers with blade #50.

7. Scalpel (#15 blade, one per rat).

8. Retractor, Barraquer (Colibri) style.

9. Drill bit (0.6 mm) (Braintree Scientific).

10. Drill (Ideal, Braintree Scientific).

11. Hamilton syringe (10 μL with 26-gauge large-bore needle).

12. Needle driver, clamp, and scissors.

13. Stereotactic frame with ear bars, drill, and needle attachments.

14. 0.5-mL syringe (for buprenorphine).

15. Cauterizer, battery operated (Fine Science Tools).

16. Cotton-tipped swabs.

17. Bone wax.

18. Suture (6–0 silk, CP Medical).

19. Rimadyl tablets (2 mg) (*see* **Note 1**).

20. Buprenorphine (0.3 mg/mL), (*see* **Note 1**).

21. Sterile PBS.

22. 70% ethanol diluted in distilled water (70% ETOH).

2.3 Rodent Bioluminescence/ Luciferase Imaging

1. IVIS Spectrum In Vivo Imaging System (PerkinElmer) or equivalent.

2. Animal warming apparatus (recirculating water heating pad recommended).

3. Anesthesia (isoflurane).

4. D-Luciferin, potassium salt: 30 mg/mL diluted in PBS, filter sterilized.

5. 1-mL syringe (one per pair of animals).

6. 25-gauge needle.

7. Timer.

8. Living Image software (PerkinElmer).

2.4 Rodent Tail Vein Catheter Construction

1. One-half-inch 30-gauge needles (two per catheter).

2. "10 PE" polyethylene tubing.

3. Clear tape.

4. Forceps.

5. Scissors.

6. A syringe with sterile saline.

7. A magnifying lamp or low-power procedure microscope.

2.5 Rodent Tail Vein Injection

1. Animal warming apparatus (recirculating water heating pad recommended).

2. Anesthesia (isoflurane).

3. Magnifying lamp or low-power procedure microscope.

4. Tail vein catheter.

5. Solution to be injected.

6. 1-mL non-Luer-Lok syringe.

7. Clear tape.

8. Reusable sodium acetate "handwarmer" (or other safe heat source).

2.6 Rodent MRI

1. Animal warming apparatus (recirculating water heating pad recommended).

2. Anesthesia (isoflurane).

3. Intravenous contrast agent (Bracco Diagnostics Inc.).

4. nanoScan PET/MRI scanner (Mediso USA) (*see* **Note 2**).

2.7 Rodent PET Imaging

1. Animal warming apparatus (recirculating water heating pad recommended).

2. Anesthesia (sevoflurane [VetOne]).

3. nanoScan PET/MRI scanner (Mediso USA) (*see* **Note 2**).

2.8 Rodent FDG-PET Imaging

1. Animal warming apparatus (recirculating water heating pad recommended).

2. Anesthesia (sevoflurane).

3. FDG (*see* **Note 3**).

4. nanoScan PET/MRI scanner (Mediso USA) (*see* **Note 2**).

2.9 Rodent FLT-PET Imaging

1. Animal warming apparatus (recirculating water heating pad recommended).
2. Anesthesia (sevoflurane).
3. Thymidine phosphorylase (Sigma-Aldrich).
4. FLT (CQCI) (*see* **Note 3**).
5. Mediso nanoScan PET/MRI scanner (Mediso USA) (*see* **Note 2**).

2.10 Rodent FMISO-PET Imaging

1. Animal warming apparatus (recirculating water heating pad recommended).
2. Anesthesia (sevoflurane).
3. FMISO (CQCI) (*see* **Note 3**).
4. nanoScan PET/MRI scanner (Mediso USA) (*see* **Note 2**).

2.11 Rodent Sacrifice and Brain Tumor Harvesting

1. Animal warming apparatus (recirculating water heating pad recommended).
2. Anesthesia (isoflurane).
3. Adult rat brain matrix, 1-mm coronal section intervals (Braintree Scientific).
4. Scalpel (#15 blade, one per rat).
5. Straight razor blades.
6. Formaldehyde (4%).
7. Tissue cassettes.
8. Needle driver, clamp, and scissors.
9. Small straight or Kerrison rongeur.

3 Methods

3.1 Cell Culture and Luciferase Plasmid Transfection

1. A luciferase plasmid vector can be acquired from online repositories such as Addgene or constructed from prior plasmids (Table 1). The plasmid should contain an antibiotic resistance gene and replication origin for growth in *E. coli* (usually ampicillin) as well as a resistance gene for mammalian cells (e.g., hygromycin).
2. Amplify the plasmid in *E. coli*, create glycerol frozen stocks as needed, and extract the DNA using commercially available kits.
3. Grow F98 rat glioma cells in culture media until 70% confluence, which signifies the linear growth phase of cells.
4. Detach the cells from the plate by washing with sterile PBS and adding trypsin (2 mL per 100-mm² flask) for 3 min at room temperature. Rinse cells off the flask with 8 mL of medium.

Table 1
List of available luciferase-expressing plasmids with hygromycin selection

Name	Vector type	Resistance marker	Bacterial resistance	Source
pGL4.33[luc2P/SRE/Hygro]	Luciferase	Hygromycin	Ampicillin	Promega
pGL4.26[luc2/minP/Hygro]	Luciferase	Hygromycin	Ampicillin	Promega
pGL4.28[luc2CP/minP/Hygro]	Luciferase	Hygromycin	Ampicillin	Promega
pGL4.34[luc2P/SRF-RE/Hygro]	Luciferase	Hygromycin	Ampicillin	Promega
pNL3.2.NF-κB-RE[NlucP/NF-κB-RE/Hygro]	Luciferase	Hygromycin	Ampicillin	Promega
pGL4.29[luc2P/CRE/Hygro]	Luciferase	Hygromycin	Ampicillin	Promega
pNL2.1[Nluc/Hygro]	Luciferase	Hygromycin	Ampicillin	Promega
pAC-Luc	Insect expression, retroviral, luciferase	Hygromycin, blasticidin	Ampicillin	Addgene
pGL4.27[luc2P/minP/Hygro]	Luciferase	Hygromycin	Ampicillin	Promega
pGL4.30[luc2P/NFAT-RE/Hygro]	Luciferase	Hygromycin	Ampicillin	Promega
pNL2.2[NlucP/Hygro]	Luciferase	Hygromycin	Ampicillin	Promega
MSCV Luciferase PGK-hygro	Mammalian expression, retroviral, luciferase	Hygromycin	Ampicillin	Addgene
pGL4.31[luc2P/GAL4UAS/Hygro]	Luciferase	Hygromycin	Ampicillin	Promega
pGL4.32[luc2P/NF-κB-RE/Hygro]	Luciferase	Hygromycin	Ampicillin	Promega
pNL2.3[secNluc/Hygro]	Luciferase	Hygromycin	Ampicillin	Promega

Data from https://www.addgene.org/vector-database/, accessed April 26, 2017

5. Count the cells using a hemocytometer or other method, e.g., Countess automated cell counter. Add cells (10 µL) to 10 µL of trypan blue, and then add 10 µL of the mixture to a counting cuvette (*see* **Note 4**).

6. Place ~2–4 × 10⁴ cells/well in a 96-well plate, and allow to adhere overnight prior to transfection. Adjust the cell concentration depending on the size of the transfection vessel.

7. Transfect the cells using Lipofectamine 2000 and the DNA plasmid according to the manufacturer's instructions. Select cells using antibiotic selection with hygromycin (200 µg/mL) 2 days after transfection. Hygromycin selection and Lipofectamine 2000 transfection should be optimized for cells (*see* **Note 5**).

8. Select single-cell colonies by cell isolation with glass cloning ring wells or flow cytometer cell sorting, and expand the cell line in antibiotic selection.

9. Screened cells for luciferase activity using Bright-Glo or other luciferase substrate. Freeze the identified cells in 10% dimethyl sulfoxide containing media using a cell-freezing container.

3.2 Stereotactic Cranial Injection

1. All procedures involving animals must be approved by the local Institutional Animal Care and Use Committee (IACUC) office.

2. Grow cells to 70% confluence.

3. Before starting, thaw Matrigel on ice.

4. Add trypsin to cells and resuspend in media, count cells, and place 1.5 × 10⁴ cells per injection in a sterile centrifuge tube. Use three times the number of cells for rats to be injected.

5. Spin cells down at 400 × g and resuspend in 8 µL of Matrigel per injection and keep on ice (*see* **Note 6**).

6. Anesthetize rats using isoflurane in an induction box, and maintain animals under anesthesia using a vaporizer at 2.5–3% isoflurane, 2.0 L/m oxygen. Keep animals warm using a water-heated animal warmer (*see* **Note 7**).

7. Somewhere outside the surgical area to avoid contamination, clip the top of the head of the rat from the ears to the eyes with fur clippers.

8. Position the rat on the stereotaxic apparatus at the right height (the head must be straight in the same axis as the spinal column) (Fig. 1). Put upper teeth in the tooth bar, the bottom teeth underneath, and the nose in the anesthesia nose cone. Position the ear bars just below the external auditory canal. Take care with positioning: The rat must be in the center of the apparatus, with the head straight, and the ear bars must press on the bone (not in the external auditory meatus). If needed, adjust the height of the tooth bar.

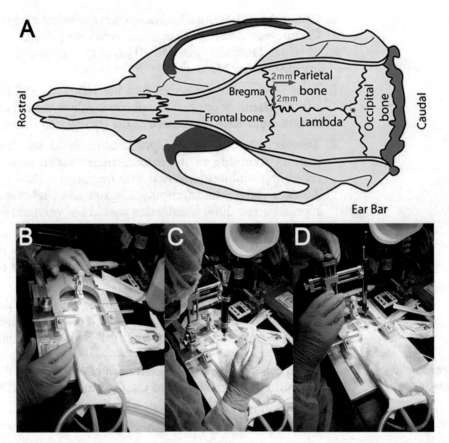

Fig. 1 Stereotactic injection of rodent brain. (**a**) Rodent sutures and cranial bones are shown with the intersection of the coronal and sagittal sutures marked as bregma. A burr hole is typically made 2 mm laterally and 2 mm posteriorly to the bregma. The depth is 3 mm deep to the skull. (**b**) Setup of the rodent in the stereotactic head frame is shown with a bite block for the teeth, a funnel to deliver anesthesia while on the frame, and ear prongs to maintain the head. A heated pad is placed under the animal. (**c**) After incision and dissection, a microdrill is positioned at the burr hole site. (**d**) Injection of the cells (8 μL total volume, 1.5–3 × 10^5 cells total) via the Hamilton syringe and needle are shown

9. Prepare the surgery tools on a sterile towel.

10. Clean the area of incision with 70% ETOH, from the ears to the eyes (avoid the eyes), and allow 2 min for it to be effective. While waiting, do a pinch test to verify whether the rat is under anesthesia. Put sterile gloves on.

11. Incise (not more than 2 cm) from the interpupillary area posteriorly to the mid-ear position. This incision is carried to the bone, and the galea is displaced laterally.

12. Insert a modified Barraquer (Colibri) retractor to maintain the exposure, and electrocauterize any skin or bone bleeding as needed.

13. Attach the micro drill to the frame, and line up the drill bit with the intersection of the coronal and sagittal sutures, i.e.,

the bregma. Move the drill 2 mm laterally and 2 mm posteriorly to the burr hole location (*see* **Note 8**).

14. Control the drilling depth manually by listening for the loss of cortical-sounding bone or observing the return of cerebrospinal fluid (CSF) (*see* **Note 9**).

15. Irrigate a Hamilton 10-μL delivery syringe with sterile PBS to remove air bubbles. Vortex cells for 20 s, and then draw up 10 μL of cells into the syringe for a planned delivery of 8 μL, accounting for 2 μL of dead space.

16. Attach the syringe to the stereotactic frame. Place the needle just above the hole previously drilled (the needle tip should be barely in the hole) (*see* **Note 10**).

17. Lower the needle 3 mm in depth (from the top of the skull) into the brain. Inject 10 μL (with actual delivery of 8 μL) over 1 min. Some CSF may come out during injection.

18. Leave the syringe in position for 1 min to let the Matrigel solidify. Retract the syringe carefully, and rinse immediately five times in ice-cold sterile PBS.

19. Close the drilled hole with bone wax.

20. After hemostasis is assured, close the skin with 3–4 simple interrupted knots using 4-0 to 6-0 silk suture.

21. Place surgical glue (cyanoacrylate) on the incision to protect the sutures, and inject the rat with buprenorphine (0.01–0.05 mg/kg, subcutaneously or intraperitoneally). Monitor the animals to ensure adequate recovery from anesthesia.

3.3 Rodent Bioluminescence/ Luciferase Imaging

1. Prior to imaging, make a stock solution of luciferin at a concentration of 30 mg/mL in 1× PBS in a biosafety cabinet, then filter sterilize, and aliquot it directly into 1.5-mL tubes for storage at −20 °C until utilized. Do not refreeze after thawing.

2. Thaw the required amount of luciferin in the biosafety cabinet. Draw up into 1-mL syringes.

3. Anesthetize the rats using isoflurane in an induction box, and maintain them under anesthesia using a vaporizer at 2.5–3% isoflurane, 2.0 L/m oxygen. Keep the rats warm using a water-heated animal warmer (*see* **Note 7**).

4. Prepare the IVIS Spectrum In Vivo Imaging System or equivalent. Set the system to initialize for approximately 5 min to allow adequate heating of the animal platform and cooling of the camera. Set the camera to a focal length of 2.5 cm for image capturing.

5. Inject the rats with 0.5 mL of 30 mg/mL (0.1 mg/kg) luciferin subcutaneously or intraperitoneally, and allow 15 min for drug distribution.

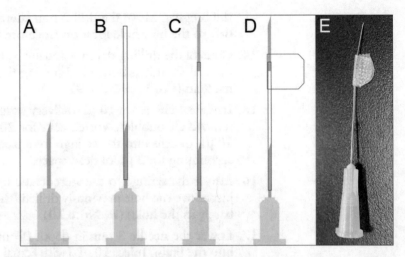

Fig. 2 Tail vein catheter construction. (**a**) Intact 30-gauge needle. (**b**) Intact needle is inserted into one end of the desired length of PE 10 tubing. (**c**) A second 30-gauge needle is removed from the hub, and the non-beveled end is inserted into the open end of the tubing. (**d**) A piece of tape is wrapped around the junction of the tubing and second needle. (**e**) Photograph of complete 30-gauge tail vein catheter

6. Place the rats under the imager, and align them, so the camera captures the entire head. Place the head in a neutral position so that both ears and eyes can be identified from the above camera.

7. Capture images over 1-5 min with medium binning. Analyze images with Living Image software. Luciferase imaging is performed 2–3 times a week to account for variability in tumor expression (*see* **Notes 11 and 12**).

3.4 Rodent Tail Vein Catheter Construction

1. Cut tubing to desired length. Approximately 3 cm is sufficient for temporary placement for injection; however, greater lengths may be used for injections of a contrast agent during imaging with MRI, etc.

2. While securing the needle by the hub, gently slide one end of the tubing onto the needle using forceps, being careful to avoid puncturing the tubing with the needle (Fig. 2).

3. Secure a second needle by the hub. Using forceps, grasp the needle near the hub, and gently rotate the forceps back and forth, bending the metal needle until it snaps loose.

4. Gently slide the "blunt" end (opposite the factory-beveled end) of the needle into the open end of the tubing using forceps, again carefully avoiding puncture of the tubing.

5. Wrap a small strip of tape around the end of the tubing at the second needle to form a small tab (that will act as a handle to hold the catheter during venipuncture).

6. Carefully attach the hub of the first needle to a syringe containing sterile saline and flush, examining the newly constructed catheter for any leaks or obstructions.

3.5   Rodent Tail Vein Injection

1. Place solution to be injected in a 1-mL non-Luer-Lok syringe. The concentration should be appropriate to allow for the desired dosage to be administered without exceeding the maximum recommended daily injection volume for the species and size of the animal.

2. Flush tail vein catheter with sterile saline and disconnect from the syringe.

3. Anesthetize the rats per laboratory protocol, and maintain the body temperature for the entire duration of anesthesia.

4. Place the animal on its side with the tail extending toward the magnifying lamp field of view.

5. Place the "handwarmer" (or other safe heat source) over the tail for 10–15 s to dilate the lateral tail vein.

6. Sterilize the skin with an alcohol wipe.

7. Localize the lateral tail vein visually.

8. Grasp the tape tab on the tail vein catheter with the dominant hand (between thumb and forefinger) and the tip of the tail with the other hand.

9. Using a flat trajectory, gently insert the needle through the skin directly over and parallel to the tail vein (Fig. 3). Once the needle tip is in the intravenous space, a small "flash" of blood should be visible in the catheter tubing when the tail is gently "milked" from the base toward the catheter.

10. Gently secure the catheter to tail with a piece of tape.

11. Carefully attach the non-Luer-Lok syringe to the hub of the catheter. The blood within the catheter tubing should clear as the solution is injected.

12. Slowly inject the desired volume of solution while being careful to not displace the catheter. There should be very little resistance to depressing the plunger of the syringe. Stop the injection if resistance is encountered, as there is likely either an obstruction of the catheter or the needle tip is no longer within the vein.

13. After the injection is complete, gently hold pressure on the tail with an alcohol wipe as the needle is withdrawn.

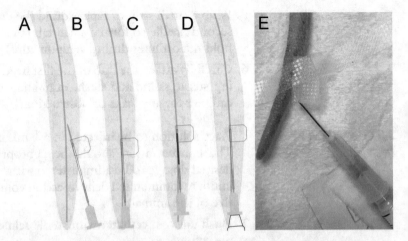

Fig. 3 Tail vein injection. (**a**) After the tail is briefly warmed, the lateral tail vein is identified through the skin. (**b**) The tail vein catheter is inserted through the skin (bevel facing up) and advanced into the vein. (**c**) Verification of intravenous placement with "flash" of venous blood into the catheter tubing after "milking" the tail (proximal to distal). (**d**) Syringe is attached, and the desired volume is injected, clearing most of the blood. (**e**) Photograph during successful tail vein cannulation and intravenous injection

14. If the injection is not successful, disconnect the syringe. Carefully remove the catheter while applying gentle pressure to the tail as above. Flush the catheter with sterile saline to clear any blood/clot as well as to test for any obstructions/leaks.

15. Initially, attempt to cannulate the tail vein as distally as possible. This allows for additional attempts more proximately should the vein be damaged or experience vasospasm during the earlier attempt(s) (*see* **Note 13**).

3.6 Rodent MRI Tips

1. Imaging parameters for the various MRI sequences (Figs. 4 and 5) are highly dependent on the equipment used: magnet field strength, spectrometer, transmitting and receiving coils, etc. Consult with the imaging center and/or collaborating laboratory to identify and optimize appropriate sequences.

2. In addition to the local imaging center experts and radiology departments, numerous MRI resources are available online for free.

3. When optimizing MRI sequences, pay particular attention to the duration of image acquisition. Depending upon the sequence type and the volume of tissue to be imaged, imaging times may range from less than a minute to more than an hour for a single sequence. Remember that animals must be kept under anesthesia and warm for the entire duration of imaging.

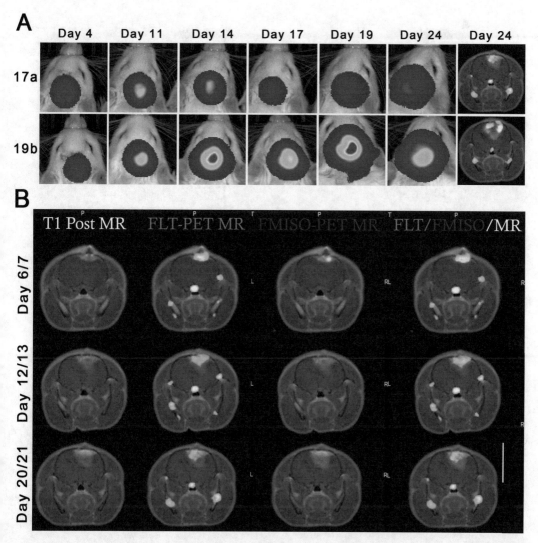

Fig. 4 Luciferase, MR, and PET imaging after tumor injection are demonstrated. (**a**) Luciferase imaging over time shows expansion of the cell line injected in the right frontal lobe. Variation in cell proliferation rates and total expression are seen. MRI/PET imaging at day 24 is shown for each animal (FMISO [red], FLT [green]). (**b**) Similarly, FMISO (red) and FLT (green) PET MRI are shown at three different time points. MRI involves a contrast-enhancing T1 sequence. FMISO expression tails FLT as expected since hypoxic areas follow tumor proliferation

4. Optimization of sequences can be time intensive, placing live animals at risk for anesthesia-related complications during prolonged imaging sessions. Raw chicken wings make useful, yet inexpensive, MRI phantoms, as they contain multiple tissues (skin/fat, bone, muscle, etc.).

5. Inhalational anesthesia is recommended for imaging as the concentration will need to be adjusted in real time to maintain an adequate level of anesthesia throughout imaging. Respiratory

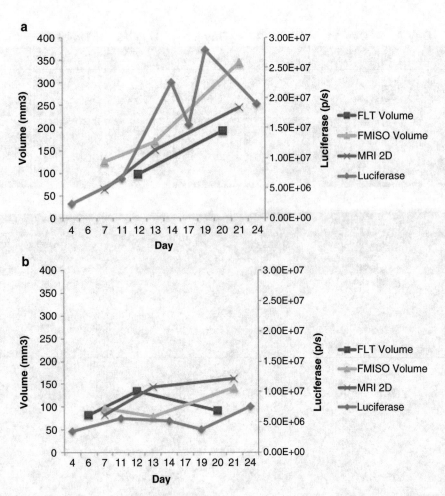

Fig. 5 Examples of quantified MR, PET, and luciferase imaging. (**a**, **b**) Quantification of two distinct animals (17a and 19b) from Fig. 4 is shown over time for FLT volume, FMISO volume, MRI volume, and luciferase expression

and/or cardiac monitoring equipment is highly recommended as the animal is often not directly visible while in the scanner. Animal body temperature should also be monitored and maintained throughout imaging and recovery.

6. Gadolinium-based contrast agents can be used to "label" tumors when using certain MRI sequences, such as a T1-weighted fast spin echo. Consult with the imaging center to determine the appropriate contrast solution concentration and volume of injection as well as the timing of post-contrast sequence acquisition (*see* **Notes 14–22**).

7. Although it is not always possible, the animal should not be manipulated or moved once placed on the scanner bed until imaging has been completed; any change in position will cause misregistration/colocalization errors between sequences.

3.7 Rodent PET Imaging Tips

1. Imaging parameters for PET imaging (Figs. 4 and 5) depend on the PET scanner and radiotracer used, among other variables. Consult with the imaging center and/or collaborating laboratory to identify and optimize appropriate uptake times, acquisition times, radiotracer doses, etc.

2. In addition to your local imaging center experts and radiology departments, numerous PET resources are available online for free.

3. PET radiotracers must be produced with a cyclotron facility, and some research PET radiotracers require special synthesis equipment. Consult the local imaging center and cyclotron facility to determine which PET radiotracers are available for use (*see* **Note 3**).

4. PET radiotracers are radioactive materials and are therefore highly regulated. Special handling precautions are necessary, and these agents should only be handled by properly trained individuals.

5. Some PET radiotracers require special preparation for the animals such as fasting, special diets, or premedication.

6. Inhalational anesthesia is recommended for imaging as the concentration will need to be adjusted in real time to maintain an adequate level of anesthesia throughout imaging. Sevoflurane is recommended for PET imaging, especially FDG-PET, as isoflurane alters cellular metabolism and FDG uptake [17, 18].

7. Respiratory and/or cardiac monitoring equipment is highly recommended because the animal is often not directly visible while in the scanner. Animal body temperature should also be monitored and maintained throughout imaging and recovery.

8. Unlike MRI, which allows for rapid image generation, PET data require extensive post-processing and reconstruction before images are available to view. The animal will likely be back in its cage and awake long before images are available. If there is uncertainty about the PET acquisition, consider repeating it, while the animal is still in the scanner.

3.8 Rodent FDG-PET Imaging

1. Animals will need to fast overnight prior to FDG-PET imaging.

2. Anesthetize the animal with ~3% sevoflurane in 2.0 L/m oxygen; adjust as necessary to maintain adequate anesthesia.

3. Inject FDG intravenously at a dose of 1.0 millicurie.

4. During the 60-min uptake time for FDG, any desired MRI and/or CT imaging may be performed (including the CT or MRI sequence to be used for attenuation correction).

5. Sixty minutes after injection of FLT, obtain a 20-min PET emission scan.

6. Perform desired PET data post-processing and reconstructions (work with the local imaging center to determine the appropriate parameters).

3.9 Rodent FLT-PET Imaging

1. Anesthetize the rat with ~3% sevoflurane in 2.0 L/m oxygen, and adjust as necessary to maintain adequate anesthesia.

2. Consider pretreatment with thymidine phosphorylase to clear endogenous (nonradioactive) thymidine for FLT-PET [19]. Inject thymidine phosphorylase intravenously at a body weight dose of 1.0 unit/g.

3. Allow the rat to awaken and recover for 45 min.

4. Reanesthetize the rat with sevoflurane.

5. Inject FLT intravenously at a dose of 1.0 millicurie.

6. During the 60-min uptake time for FLT, any desired MRI and/or CT imaging may be performed (including the CT or MRI sequence to be used for attenuation correction).

7. Sixty minutes after injection of FLT, obtain a 20-min PET emission scan.

8. Perform desired PET data post-processing and reconstructions (work with the local imaging center to determine the appropriate parameters).

3.10 Rodent FMISO-PET Imaging

1. Anesthetize the rat with ~3% sevoflurane in 2.0 L/m oxygen; adjust as necessary to maintain adequate anesthesia.

2. Inject FMISO intravenously at a dose of 1.0 millicurie.

3. There is a 120-min uptake time for FMISO. Because this is a long duration to keep an animal under anesthesia, the rat may be allowed to awaken and recover.

4. Reanesthetize the rat (if needed) and perform any desired MRI and/or CT imaging (including the CT or MRI sequence to be used for attenuation correction).

5. One hundred twenty minutes after injection of FMISO, obtain a 20-min PET emission scan.

6. Perform desired PET data post-processing and reconstructions (work with the local imaging center to determine the appropriate parameters).

3.11 Rodent Sacrifice and Brain Tumor Harvesting

1. Anesthetize the animals using isoflurane in an induction box with a vaporizer at 3–5% isoflurane, 2.0 L/m oxygen.

2. After adequate anesthesia as assessed by paw pinch, ear pinch, and blink reflex, dislocate the cervical spine. To do this, grasp the head firmly behind the base of the skull and the tail, and

swiftly twist in opposite directions. Dislocation is performed in one smooth motion. Alternately, an intracardiac injection of pentobarbital can be used, as approved by the local IACUC (*see* **Note 23**).

3. After ensuring the animal is dead, create a midline incision similarly to the prior surgical incision. Continue the incision deeply inferiorly to the skull to detach the neck muscles and ligaments from the skull.

4. Use rongeurs to remove the skull bone, starting at the foramen magnum and then biting along the lateral-most portions until the skull hinges open.

5. Detach the cerebellum sharply from the cerebrum, and move the cerebrum en bloc to the adult rodent brain slicer (*see* **Note 24**).

6. Use a razor blade to cut directly through the center of the tumor, and leave it in place. Use two more razor blades to cut 4- to 5-mm sections above and below the level of interest. Discard the rest of the brain.

7. Carefully remove the razor blades, and place the brain slices into cassettes for formalin (4%) fixation and paraffin embedding. If available, send the tissue to the clinical pathology laboratory to process the tissue and generate brain slices for histology and immunohistochemistry. The area of interest in a section is placed downwards into a tissue cassette.

4 Notes

1. These drugs require a US Drug Enforcement Agency license. If you do not have one, most animal facilities can order them for you.

2. The MRI and PET imaging described in this chapter was performed on a Mediso nanoScan PET/MRI scanner (Mediso USA) [20] using a 3D gradient recalled echo sequence for attenuation correction.

3. All PET radiotracers used in this chapter were produced by the cyclotron facility within the Center for Quantitative Cancer Imaging (CQCI) at the Huntsman Cancer Institute. There are alternative radiotracers to both FLT and FMISO as well as other radiotracers that assess different aspects of tumor biology [12].

4. Although Countess disposable cuvettes are recommended for single use, they can be rinsed with diluted soap in distilled water, rinsed in distilled water, dried using compressed air, and reused several times.

5. Different trials of cell transfection using Lipofectamine 2000 should be evaluated according to the manufacturer's instructions. Cell sensitivity to hygromycin should be evaluated using dilutional assays prior to cell transfection. Hygromycin selection at varying doses is performed with medium/antibiotic replacement every other day. Maximal cell apoptosis at the minimum concentration at 1 week should be identified.

6. Matrigel is viscous and forms bubbles easily, which will make it difficult to draw up cells later on. To avoid bubble formation while resuspending cells, do not eject all of the Matrigel out of the pipette tip, and do not draw up the full amount while pipetting up and down until cells are resuspended.

7. Ensuring adequate oxygen flow is important to maintain animal viability during procedures.

8. Our initial experiments involved placement of a burr hole 2 mm anteriorly and 2 mm laterally to the bregma, but several of these approaches showed insertion of cells anterior to the brain missing the brain entirely. It is important to also avoid insertion of cells into the ventricle. A rodent brain atlas can be used to evaluate cell placement.

9. Significant efflux of blood can indicate laceration of a cortical vein or the sinus if the drilling is off target. This can usually be controlled with placement of gauze, steady pressure, and time. Should the bleeding not stop, the burr hole can be closed with a piece of bone wax and a new position drilled.

10. To align the needle, place it centered over the hole, and use forward/back controls to check height such that the needle tip is barely restrained by the hole edge when moving the needle across it.

11. It is very important for images to be taken at exactly the same time, since the luciferase signal fluctuates over time. The rats can be moved to the IVIS stage and the next animals placed in the induction chamber to save time. Because the entire process takes at least 15 min (10 min for distribution, 5 min for imaging), it is faster for multiple animals to overlap the imaging of one set and the injection of the next by 5–8 min.

12. Total captured luminescence flux is captured for a specific region of interest automatically selected around the tumor. Total flux is compared among animals. Average flux or flux/time can also be calculated as alternative measures but have not shown a dramatic difference with total flux in our experience.

13. The animal can be flipped to its other side to access the opposite lateral tail vein if needed.

14. For contrast-enhanced MRI in this study, 300 μL of a 20% dilution (in sterile saline) of MultiHance® was injected intravenously via a tail vein catheter followed by a 300-μL sterile saline flush. Approximately 3 min after injection, a T1-weighted fast spin-echo sequence (TR 500 ms, TE 7.8 ms) was then acquired.

15. Contrast administration will typically be performed via tail vein injection, while the animal is positioned in the scanner. A long tail vein catheter (~60 cm) is needed because the tubing must extend from the animal to the syringe, which must be outside of the scanner and accessible to perform the injection.

16. The catheter will need to be flushed and preloaded with contrast solution prior to placement into the tail vein.

17. To prevent clotting of the intravenous needle while pre-contrast sequences are performed, draw up ~5 μL of heparin solution into the needle after the catheter has been flushed and loaded with contrast solution.

18. After the catheter is successfully placed in the tail vein, it will need to be secured to the tail (with tape, etc.). After the animal is positioned on the scanner bed, secure the catheter to the scanner bed at multiple points so that the tubing remains stationary with respect to the animal and scanner bed as they move into and out of the scanner.

19. Obtain all of the pre-contrast sequences prior to the injection of contrast; once contrast is administered, it cannot be undone.

20. If there is no contrast enhancement of a tumor on a T1-weighted post-contrast sequence (when enhancement is expected), acquire a T1-weighted sequence of the kidneys and bladder to determine if there was successful intravenous contrast administration (and subsequent renal excretion).

21. If there is no kidney/bladder enhancement, consider running a T1-weighted sequence of the tail as the catheter may have migrated out of the vein, resulting in a subcutaneous injection.

22. If intravenous administration of contrast was not successful, consider reattempting after removing and replacing the catheter.

23. For hypoxia-related studies, it is vital that the animal be sacrificed quickly, without inducing a hypoxic environment. Be aware the standard method of CO_2 inhalation causes hypoxia and is not appropriate.

24. Be aware there are several large cranial nerves attaching at the skull base. These can impede brain removal. Fine microscissors or tweezers can be inserted from posterior to anterior along the skull base, while the brain is slightly elevated to cut the nerves.

References

1. Ostrom QT, Gittleman H, Fulop J et al (2015) CBTRUS statistical report: primary brain and central nervous system tumors diagnosed in the United States in 2008-2012. Neuro-Oncology 17(Suppl 4):iv1–iv62

2. Karsy M, Guan J, Cohen AL et al (2017) New molecular considerations for glioma: IDH, ATRX, BRAF, TERT, H3 K27M. Curr Neurol Neurosci Rep 17(2):19

3. Karsy M, Neil JA, Guan J et al (2015) A practical review of prognostic correlations of molecular biomarkers in glioblastoma. Neurosurg Focus 38(3):E4

4. Gillespie DL, Aguirre MT, Ravichandran S et al (2015) RNA interference targeting hypoxia-inducible factor 1alpha via a novel multifunctional surfactant attenuates glioma growth in an intracranial mouse model. J Neurosurg 122(2):331–341

5. Ramachandran R, Junnuthula VR, Gowd GS et al (2017) Theranostic 3-dimensional nano brain-implant for prolonged and localized treatment of recurrent glioma. Sci Rep 7:43271

6. Xuesong D, Wei X, Heng L et al (2016) Evaluation of neovascularization patterns in an orthotopic rat glioma model with dynamic contrast-enhanced MRI. Acta Radiol. https://doi.org/10.1177/0284185116681038

7. Lehmann S, Stiehl DP, Honer M et al (2009) Longitudinal and multimodal in vivo imaging of tumor hypoxia and its downstream molecular events. Proc Natl Acad Sci U S A 106(33): 14004–14009

8. Lo Dico A, Valtorta S, Martelli C et al (2014) Validation of an engineered cell model for in vitro and in vivo HIF-1alpha evaluation by different imaging modalities. Mol Imaging Biol 16(2):210–223

9. Dhermain FG, Hau P, Lanfermann H et al (2010) Advanced MRI and PET imaging for assessment of treatment response in patients with gliomas. Lancet Neurol 9(9):906–920

10. Barth RF (1998) Rat brain tumor models in experimental neuro-oncology: the 9L, C6, T9, F98, RG2 (D74), RT-2 and CNS-1 gliomas. J Neuro-Oncol 36(1):91–102

11. Biasibetti E, Valazza A, Capucchio MT et al (2017) Comparison of allogeneic and syngeneic rat glioma models by using MRI and histopathologic evaluation. Comp Med 67(2):147–156

12. Bansal A, Wong TZ, Degrado TR (2011) PET imaging of gliomas. In: Chen C (ed) Advances in the biology, imaging and therapies for Glioblastoma. InTech, Rijeka, Croatia

13. Kelloff GJ, Hoffman JM, Johnson B et al (2005) Progress and promise of FDG-PET imaging for cancer patient management and oncologic drug development. Clin Cancer Res 11(8):2785–2808

14. Shankar LK, Hoffman JM, Bacharach S et al (2006) Consensus recommendations for the use of 18F-FDG PET as an indicator of therapeutic response in patients in National Cancer Institute trials. J Nucl Med 47(6):1059–1066

15. Tehrani OS, Shields AF (2013) PET imaging of proliferation with pyrimidines. J Nucl Med 54(6):903–912

16. Rasey JS, Grunbaum Z, Magee S et al (1987) Characterization of radiolabeled fluoromisonidazole as a probe for hypoxic cells. Radiat Res 111(2):292–304

17. Flores JE, McFarland LM, Vanderbilt A et al (2008) The effects of anesthetic agent and carrier gas on blood glucose and tissue uptake in mice undergoing dynamic FDG-PET imaging: sevoflurane and isoflurane compared in air and in oxygen. Mol Imaging Biol 10(4):192–200

18. Lee KH, Ko BH, Paik JY et al (2005) Effects of anesthetic agents and fasting duration on 18F-FDG biodistribution and insulin levels in tumor-bearing mice. J Nucl Med 46(9): 1531–1536

19. van Waarde A, Cobben DC, Suurmeijer AJ et al (2004) Selectivity of 18F-FLT and 18F-FDG for differentiating tumor from inflammation in a rodent model. J Nucl Med 45(4): 695–700

20. Nagy K, Toth M, Major P et al (2013) Performance evaluation of the small-animal nanoScan PET/MRI system. J Nucl Med 54(10):1825–1832

INDEX

L. Eric Huang (ed.), *Hypoxia: Methods and Protocols*, Methods in Molecular Biology, vol. 1742,
https://doi.org/10.1007/978-1-4939-7665-2, © Springer Science+Business Media, LLC 2018